神农架联合国教科文组织世界地质公园科普丛书

地质探秘神农架

李晓池 著

图书在版编目（CIP）数据

地质探秘神农架＝The Geologic Discovery of Shennongjia：汉英对照／李晓池著.——武汉：中国地质大学出版社，2017.5
（神农架联合国教科文组织世界地质公园科普丛书）
ISBN 978-7-5625-4029-8
Ⅰ.①地…
Ⅱ.①李…
Ⅲ.①神农架－地质－国家公园－普及读物－汉、英
Ⅳ.①S759.93-49

中国版本图书馆CIP数据核字（2017）第088829号

地质探秘神农架　李晓池　著

责任编辑：胡珞兰　责任校对：林泉
英文翻译：冯清高　英文校对：J.A.Grant-Mackie(NZ)
美术设计：赖亮鑫　美术编辑：广州亮丽品牌设计工作室

出版发行：中国地质大学出版社（武汉市洪山区鲁磨路388号）　邮政编码：430074
电　话：（027）67883511　传　真：67883580　E-mail:cbb@cug.edu.cn
经　销：全国新华书店　http://www.cugp.cug.edu.cn

开本：889毫米×1194毫米　1/20　字数：590千字　印张：20.4
版次：2017年5月第1版　印次：2017年5月第1次印刷
印刷：武汉中远印务有限公司　印数：1—2000册

ISBN 978-7-5625-4029-8　定价：128.00元

《地质探秘神农架》

编辑委员会

前 言

亲爱的读者，您面前的这本《地质探秘神农架》是中国神农架世界地质公园为您准备的一份丰厚的礼物。神农架一直给人一种梦幻神秘的感觉：炎帝神农的优美传说、中华民族的壮丽史诗《黑暗传》、郁郁葱葱的原始森林和构成“华中屋脊”的雄伟群山，能够媲美北美洲著名大峡谷的阴峪河谷、荒芜而壮观的老君山、神奇的野人故事……无一不具有强烈的吸引力，等待着人们去探索。

作为联合国教科文组织“人和生物圈计划”世界生物圈保护区网络和世界地质公园网络的成员，以及世界自然遗产地的中国神农架世界地质公园，肩负着保护地球自然资源和地质遗迹、促进地质学科研和科学普及、推动当地经济发展的重任。地球环境和地质遗迹是不可再生的宝贵资源。但是，随着社会生产力的发展和人们对大自然无止境的索取，地球环境的破坏、自然资源的消耗已经到了无以复加的危险程度。解决问题的关键在于教育民众，必须使人们懂得：为了我们自己，更为了我们后代的生存和幸福，我们每一个人都应该行动起来共同保卫我们的地球。

《地质探秘神农架》是一本通俗易懂的科普读物。它面对的不仅仅是青少年和儿童，对于所有前来神农架参观游览、旅游休闲的人们来讲，它也是一本富含科学知识、具有导游性质、趣味盎然的故事书。本书从科学的视角，较全面地介绍了神农架世界地质公园主要景点和地质遗迹的特点，以及这些地质遗迹形成的过程和它们所包含的科学原理，将科学知识的普及和趣味性的阅读结合起来，激发人们热爱大自然、了解大自然的情趣，教育和引导人们尊重大自然、自觉履行保护地球环境的应尽职责。

本书通过知识老人和小明乘坐神奇的飞毯游览神农架世界地质公园的故事，把读者带进童话般美丽的神农架，以一问一答的方式启发读者进行探索的兴趣。书中提出的问题涵盖了较宽的科学知识领域，既有适合于中小学生理解的简单问题，也有启发成年人进行思索和探讨的趣味性问题，此外，还提出了不少无法从阅读本书中直接发现结论的疑问。这是为了“抛砖引玉”，引导读者充分运用现代资讯手段，带着问题进行更加广泛的阅读或网上查询，以拓展读者的知识面。在自然界，地质与地球环境，就如同“皮”和“毛”的关系不可分割；特定的地质条件，必然造就特定的自然环境。大自然的一切，皆以地质为根本，很难界定地质与地理环境的区别。因此，虽然本书以“地质探秘神农架”冠名，但书中涉及到的科学内容已经超出了纯地质学的范畴。

亲爱的读者朋友，希望您喜爱这本书，并把它作为您了解神农架、打开神农架神秘大门的一把钥匙。神农架欢迎您！

李晓池

2016 年 12 月 22 日

01

清晨，知识老人带着小明和他的小狗汪汪乘坐飞毯，从江城武汉开始了小明盼望已久的神农架之旅。在朦朦胧胧的曙光中，城市的高楼逐渐变得清晰起来。龟山上的电视塔高耸入云，对面的黄鹤楼金碧辉煌；街上的汽车像小甲虫似地来往奔忙，人们开始了一天繁忙的生活和工作……

地质探秘神农架

爷爷，下面就是我们的武汉市吧？

对呀。你看，武汉市多美呀！它是湖北省最大、人口最多的城市，也是湖北省的省会。

省会？什么是省会？

省政府所在的城市就是这个省的省会。例如湖南省的省会是长沙，江西省的省会是南昌，湖北省的省会就是武汉。

我们是在朝哪个方向飞呀，爷爷？

早上的太阳现在正在我们的身后，那我们现在面朝哪个方向呢？

我们后面是东升的太阳，那我们现在是向西飞。

是的。神农架在武汉市的西北边，靠近重庆市。

爷爷，神农架远吗？

不远。神农架在湖北省的西北部。离武汉市只有480km。

哇！爷爷您快看，咱们左边的这条河好宽呀！

这就是长江呀。它在我们的左边，也就是南边。你看长江朝哪个方向流呀？

它向我们身后流去，也就是说，它向东边流去，对吗？

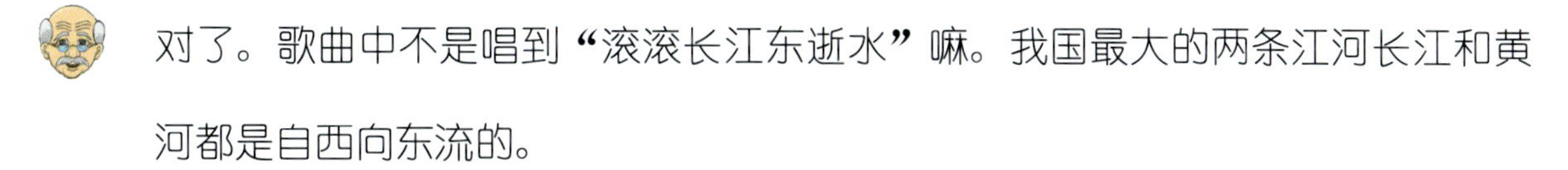

对了。歌曲中不是唱到“滚滚长江东逝水”嘛。我国最大的两条江河长江和黄河都是自西向东流的。

爷爷您快看，咱们右边的远处还有一条河呢！

那条河叫“汉江”。它是长江最大的一条支流。长江和汉江之间，以及它们汇合处周边的这片平原就是江汉平原，是湖北省的“粮仓”。你能告诉我为什么长江和汉江以及黄河都是向东边流吗？

让我想想…… 哦，我知道了，咱们中国的地势是西边高、东边低，所以大江大河都由西向东流。水往低处流嘛。

你回答得很正确！看到远处的那片高山了吗？那就是鄂西山区，神农架就在那里。

鄂西！“鄂西”是哪里？

地质探秘神农架

鄂西就是指湖北省西部。“鄂”，是湖北省的简称。咱们中国每一个省份都可以用一个字来简称。例如江西省的简称是“赣”、湖南省叫“湘”、河南省叫“豫”。

哦，我懂了。咱们中国的文字真神奇，既简练又准确。哇！爷爷，您看，鄂西的山好高呀！云彩只到它的半腰。

可不是吗。神农架有 6 座 3000m 以上的高峰，最高的一座山峰叫作“神农顶”，海拔高度是 3106.2m。

“海拔高度”是什么？

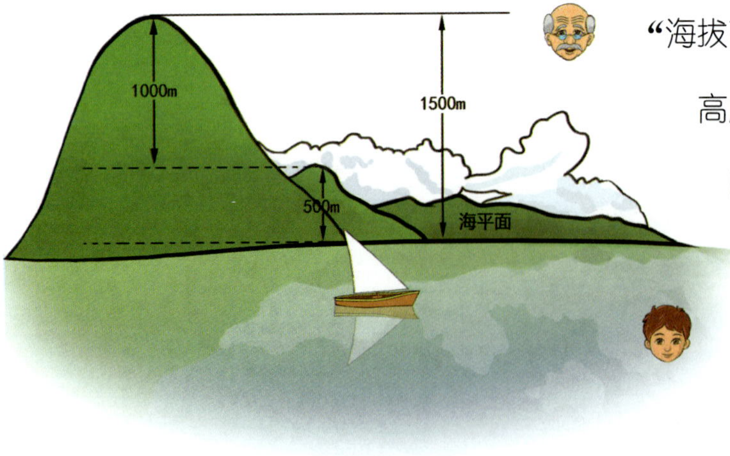

“海拔高度”是指从海平面开始计算的高度。神农架在这里拔地而起，如同擎天柱一般，人们把它叫作“华中屋脊”。

“屋脊”？！就像全球最高的“世界屋脊”喜马拉雅山那样，神农架是华中地区最高的地方了？

可不是嘛。

我们一下子就从江汉平原飞到了“华中屋脊”，变化太大了。

是啊。中国是一个地形复杂多样的国家，我们有高山、丘陵、平原，还有低于海平面的低地。

是呀。青藏高原世界闻名。

中国总的地势是西边高、东边低，向海洋倾斜。全国地形可以划分为 3 级阶梯。

3 级阶梯？怎么划分的呢？

第一级阶梯是青藏高原，平均海拔在 4500m 以上。

那第二级阶梯呢？

第二级阶梯上分布着大型的盆地和高原，平均海拔在 1000 ～ 2000m 之间，包括塔里木盆地、准噶尔盆地、内蒙古草原、黄土高原、四川盆地和云贵高原 6 个地形区。

第三级阶梯就是沿海的平原了吧？

是的。第三级阶梯上分布着广阔的平原，还有一些丘陵和低山，海拔多在 500m 以下。主要有东北平原、华北平原和长江中下游平原 3 个地形区。

哇，这 3 级阶梯就像是一栋三层楼房那样。我们武汉人住在“一楼”，第一阶梯嘛！

你很善于想象。虽然武汉人住在“一楼”，但是我们的神农架却上到了“二楼”，而且就在楼梯口呢。

是吗？什么意思呢？

神农架位于湖北省的西北部，正好处于第二阶梯与第三阶梯交会处。它的地势是从秦巴山脉东端起，自西南向东北延伸。地貌上具有山川交错、脊岭连绵、高低悬殊的特征。

看来我们湖北省得天独厚，占据了两层楼！

是的。我们马上就要“上楼”了。

好哦，“上楼”喽！

我们先到“神农顶”去看看。整个神农架林区太大了，面积超过 3000km^2 呢，仅地质公园占地就超过了 1000km^2。

哇，那么大的公园呀！我们怎么看呢？

神农架世界地质公园有 5 个园区：神农顶、官门山、天燕、大九湖和老君山。我们这次来主要看看神农顶和官门山两个园区。这两个园区的景点比较集中，看了这两个园区，对神农架世界地质公园就有一个大概的印象了。神农顶是 5 个园区中地势最高的。由于山太高了，气温会下降不少。咱们赶快把外套穿上吧。

爷爷，为什么山越高就会越冷呢？

这是一个很有趣的问题，还是留给你回去自己查查资料吧。

问题：

1. 中国有多少个省份？你知道每一个省的简称和省会所在城市吗？
2. 如果没有带指南针，你知道如何判断方向吗？
3. 神农架在什么地方？为什么神农架被叫作“华中屋脊”？
4. 你能系统地讲解一下中国地形地势的主要特征吗？
5. 为什么山越高就会越冷呢？

02

飞毯载着知识老人、小明和汪汪飞快地掠过田野山川，向着神农架的大山飞去，把一片片的白云甩在了后面。飞毯随着地势而逐渐上升着，太阳也越升越高，云层散开了，能见度越来越好。只见群山被森林覆盖，满眼郁郁葱葱，生机勃勃。

哇！神农架的树林好茂密呀！您看，这无边的森林随着地势的起伏，就像一片海洋，翻动着翠绿色的波浪。

可不是嘛。神农架的森林覆盖面积达到96%，具有当今世界中纬度地区唯一保存完好的亚热带森林生态系统。

什么是“中纬度”呀？

我先问你一个问题：在地球仪上你能看到些什么？

地球仪是仿造地球做成的一个模型，上面标示着大陆和海洋的位置，还有高山、平地、沙漠、河流，以及各个国家和地区的分布状况。

还有呢？

嗯……没有了。

没有了？你难道没有注意到地球仪上画的那些网格状的线条吗？

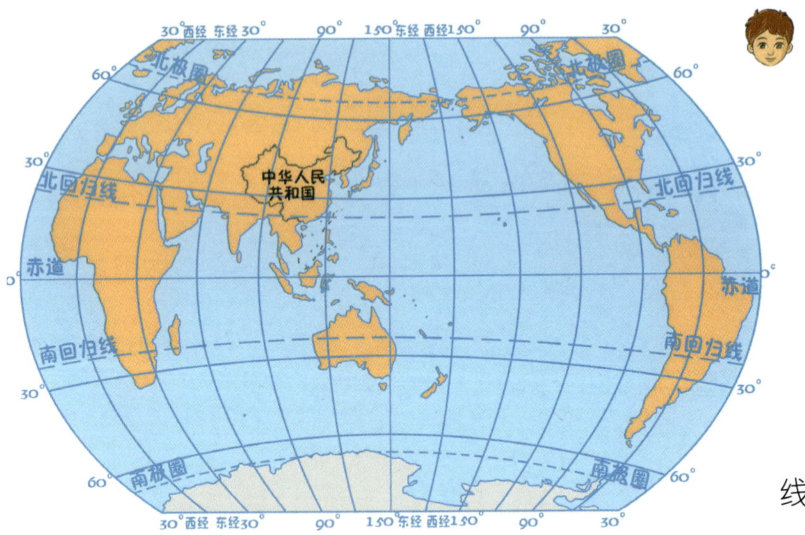

哦。那些线条是干嘛的？我看这些线条只不过把地球分成了许多小块而已。

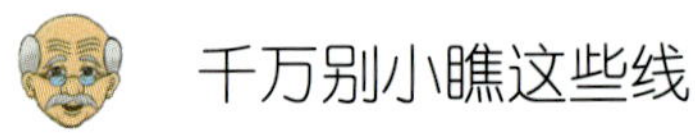

千万别小瞧这些线条，它们可重要了！那些连接南北两极的上下垂直的线条叫“经线”，任何一个经线圈都可以把地球对等地分成两个半球；那些横向的圈圈叫“纬线”，纬线与经线垂直相交。中间那条最长的“纬线”就叫作“赤道”。

哦，我懂了，我懂了。赤道把地球分成了南半球和北半球。

是的。地球的经纬是用“度”来标示的，全球的经线共计 360°，纬线以赤道为 0°，向南北极方向各划分 90°，它们构成了一个完整的坐标系统，按一定的规律把地球划分为不同的部分。全世界每一个具体的地点都可以通过“经度”和“纬度”精确地标示出来。

等一等，…… 我明白了。怪不得发生地震时，常常听广播电视说：震中位于东经多少度，北纬多少度，原来经线和纬线就起这个作用呀。

经线和纬线的作用远不止于此。例如，全球各地时间的划分就与经度有关，而地球上各个地方的气温则与所处的纬度密不可分。

那您刚才说的“中纬度”就是……

你已经知道赤道把地球分成了南、北两个半球。在靠近南极和北极的地方，有两个虚线画的圈（分别相当于南、北纬 66°34′的位置），它们标示出南极和北极的范围，叫作“南极圈”和“北极圈”；赤道的两侧也有两条虚线（分别位于南、北纬 23°26′），它们则被分别命名为“南回归线”和“北回归线”。

好奇怪的名字呀！画这么多线太麻烦了！谁弄的呀？

这是通过科学家的观察和研究确定的。看起来是有点复杂，但是给地理学的研究提供了很多方便。比如说吧，南、北回归线就是在一年时间之内，太阳直射地球所能到达的最南端和最北端的位置。直射的阳光到这里后不会再继续向两极移动了，而是朝着赤道方向返回，因此叫作“回归线”。

哦，原来“回归线”是这个意思呀。

北极
66° 34′
北寒带
北极圈
北温带
23° 26′
北回归线
热带
赤道
南回归线
23° 26′
南温带
南极圈
66° 34′
南寒带
南极

根据科学家的观测，直射地球的阳光总是反复在赤道和南、北回归线之间移动。当直射的阳光照射在赤道与北回归线之间的时候，就是北半球的夏天，而南半球则是冬天……

真的吗？！哇！这么说南半球和北半球的季节是相反的呀？

可不是嘛。当直射的阳光越过赤道，进入南半球的时候，咱们北半球的冬季就来到了。

哇，我们的地球太奇妙了！原来直射的太阳光老是以赤道为中心，一会儿向南、一会儿向北，来来回回地走动。我想这就是赤道地区整年都那么热的缘故了。

是的。赤道地带是地球上的热带；北回归线与北极圈之间的地带叫北温带，南边相对应的就是南温带。科学家把北极圈和南极圈之内的地区都叫作“高纬度地区”。

“高纬度地区”和别的地区有什么区别呢？等等，……您先别说！我知道了。

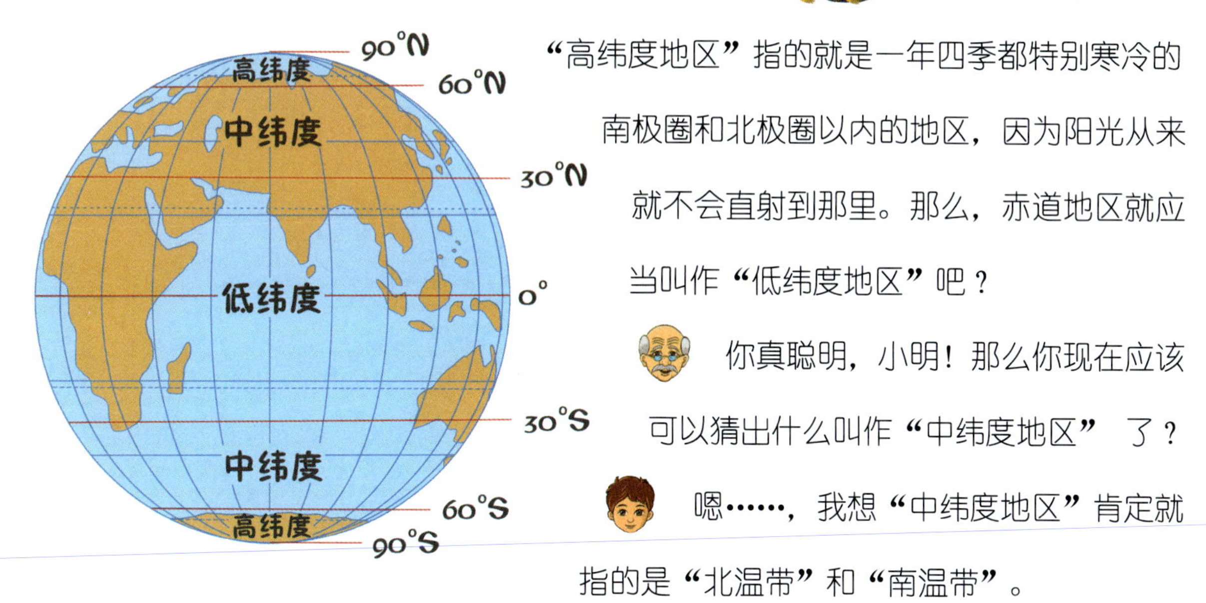

"高纬度地区"指的就是一年四季都特别寒冷的南极圈和北极圈以内的地区，因为阳光从来就不会直射到那里。那么，赤道地区就应当叫作"低纬度地区"吧？

你真聪明，小明！那么你现在应该可以猜出什么叫作"中纬度地区"了？

嗯……，我想"中纬度地区"肯定就指的是"北温带"和"南温带"。

大致是对的，但是不确切。"中纬度地区"确实分别处于南、北温带以内。但是科学上所说的"中纬度地区"，指的是南北纬 30°～60°之间的地带。

哇！干嘛搞得这么复杂呀。为什么要专门划出个"中纬度地区"呢？

因为中纬度地区是一个非常特别的地带。在这里，高纬度的冷气团与低纬度的湿热气团相互交叠，气旋活动频繁，气候变化非常复杂。尤其在北半球，因为这里集中分布了许多陆地，使得情况变得更加复杂。

哦！所以科学家把它专门划分出来进行研究，是吗？

是的。中纬度地区天气的非周期性变化和降水季节的变化都很复杂，加上北半球中纬度地带大陆面积较广大，海陆热力对比和复杂的陆地地形的影响，使得这个地区的气候更加变化多端、错综复杂。

什么是“海陆热力对比”呀，爷爷？

陆地和海洋对太阳热能的储存和释放速度存在很大的差异。海洋在白天储存的太阳热能，晚上会慢慢地释放出来，但是陆地则会很快地冷却下来。日积月累，陆地与海洋之间就会产生较大的热力差。这种热力差是造成季风的重要原因之一。

那什么是“季风”呢？

你是打破沙锅问到底呀。很好！孩子，现在我们快到神农顶了。至于什么是季风，就留给你回去自己好好查查资料吧。弄清楚什么是季风，是个很重要的问题。

问题：

1. 什么是“地球仪”？它有什么用途？你能说出地球仪上所有线条的名称吗？
2. 地球的经线和纬线是如何划分的？它们有什么作用？
3. 什么是“中纬度地区”？与地球上其他地区相比较它有些什么特点？
4. 你如何理解知识老人说的“全球各地时间的划分与经度有关，而地球上各个地方的气温则与所处的纬度密不可分”？
5. 为什么科学家要在地球上专门划分出“中纬度地区”？“中纬度地区”的气候如何受季风的影响？

03

飞毯迅速地爬升，穿过了一片片白云，向着神农架地区最高的山峰“神农顶”飞去。只见群山逶迤，绿浪滚滚；阳光普照，和风徐徐，一派生机勃勃的秀美景象。微风吹动着飞毯的边缘，发出轻微的“哗哗”声。飞毯离“神农顶”越来越近了。

爷爷，爷爷，您快看！那山顶上有个什么东西呀？黑乎乎的，不会是只大狗熊吧。

再近一点你就会看清楚了。那是一只大铜鼎，叫作“神农鼎”。

铜鼎？为什么要在山顶上放只大铜鼎呀？

“鼎”和神农顶的“顶”同音。人们把“神农鼎”放在神农架的最高峰上，表达对炎帝神农的崇拜和怀念，纪念他不畏艰苦造福于民的崇高精神。来神农架旅游一定要先了解神农是谁，他为人们做了些什么。这就是为什么出发前，我要求你一定要查询一下关于炎帝神农的资料。

我查了。炎帝神农是传说中中华民族农耕文明的创始者，他……

当心！我们马上就要着陆了。

好的。爷爷，就停在那个神农鼎旁边的平台上好吗？

好的。我们先转两圈，仔细看看这个大铜鼎，好吗？

好的。哇！这个铜鼎真大呀！有五六米高吧？

你眼睛看得还挺准的。这个“神农鼎”净高 5.9m。

爷爷您看，这鼎的上面还刻了好多字呢。

是的。鼎的中部以“羊”字造型，喻意“三阳开泰”；鼎耳为象形的“云”字，与神农顶吞云吐雾的景观相吻合。鼎身雕刻着东、西、南、北、天、地、山、川、金、木、水、火、土等象形文字，标志着传统中华文化的要素。

哇，顶天立地，好雄伟啊。真给力！

可不是嘛。人们编了一首打油诗：“神农顶，神农鼎，神农顶上神农鼎，顶上观日月，鼎下许宏愿，群山千骑出，白云天际来。不为浮尘遮望眼，只缘登上神农顶。”

这么大个鼎，怎么弄上来的呀？

你看，那边有一条路直通山下。人们先把大鼎分解开来，再沿着这条登山道，一部分一部分地运上来，最后再组装起来。

真不简单！哇，您看那登山道，像一条长蛇蜿蜒曲折，一会儿陡，一会儿缓。这可是一个了不起的大工程呀！

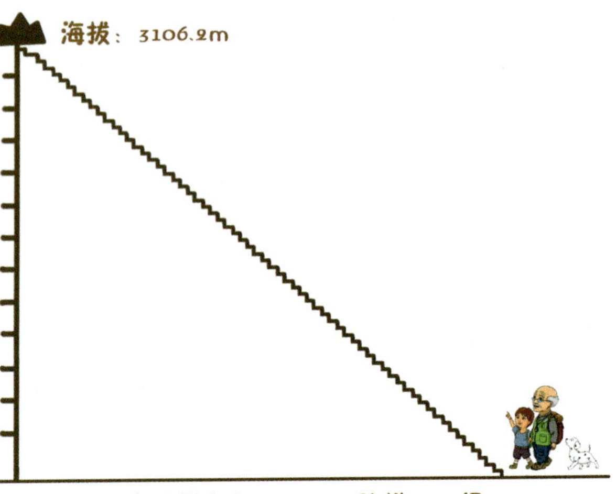

是的。这条登山步行道叫作“青云梯”，喻意“平步青云”，永远向上。它全长 1600m，共计 2999 级台阶。

神农架人真是了不起！

是的。青云梯的建设施工用了 310 天，用于铺砌阶梯的青石将近 30 000 条、沙料 $3000m^3$、水泥 625 吨。

哇，这么多的建筑材料，怎么弄上山的？

这些建筑材料绝大部分都是靠人工肩挑背扛运上来的。人们曾经动用过骡队运送沙和水泥，有 4 匹骡子因为劳累过度，“牺牲”在这里了。

哇，真伟大！神农架人付出艰苦的代价修这条登山道，真好，为游客做了件大好事。

来神农架旅游的人都以登临最高的神农顶为荣。神农架世界地质公园还为游客在青云梯的起点处修建了“神农营”。这是一个设备齐全的登山基地，为攀登

“华中屋脊”的游客提供许多相关的服务。

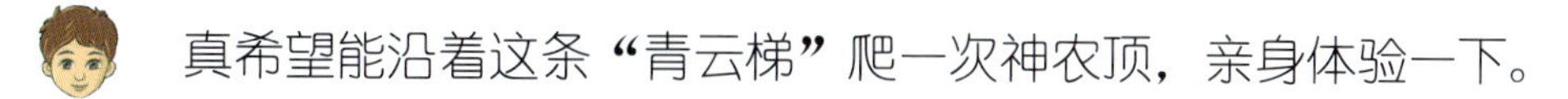

真希望能沿着这条“青云梯”爬一次神农顶，亲身体验一下。

你肯定会有机会的。每年神农营都会举行群众性的登山比赛，看谁最快爬上神农顶。

这是一项非常健康的体育运动。我很喜欢爬山，下次一定要来参加登山比赛，拿个第一名！

好呀！可是你要知道，现在登顶的冠军记录是 20 多分钟呢！

哇！这对我来讲可是一个巨大的挑战，看来我得好好锻炼锻炼。

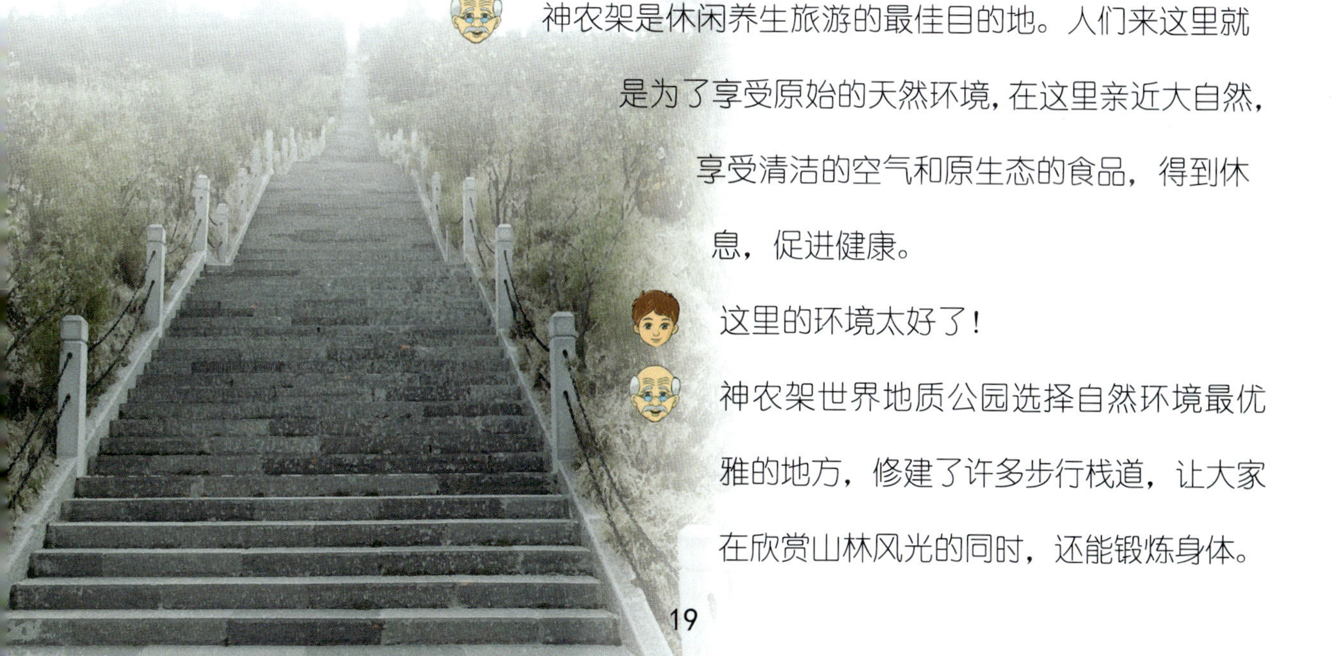

神农架是休闲养生旅游的最佳目的地。人们来这里就是为了享受原始的天然环境，在这里亲近大自然，享受清洁的空气和原生态的食品，得到休息，促进健康。

这里的环境太好了！

神农架世界地质公园选择自然环境最优雅的地方，修建了许多步行栈道，让大家在欣赏山林风光的同时，还能锻炼身体。

地质探秘神农架

地质公园想得真周到。

神农架环境优美，地域辽阔，地质公园每年都组织各种各样的登山越野活动。人们只要来参加过一次这样的活动，就会迷上神农架。尤其是暑期中小学生的夏令营，不仅能让孩子们通过野营和长途跋涉亲近大自然、享受大自然，得到野外生存的训练，并锻炼人的体力和意志，更能学到不少关于山林的知识。

哇！我一定要来神农架参加夏令营！神农架人为我们准备了这么好的旅游参观条件，向神农架人致敬！

我们确实应当怀着感恩之心，对那些为此做出贡献的建设者、管理者和组织者们致以衷心的感谢。

问题：

1. 谁是炎帝神农？他为人们做了些什么？有些什么历史功绩？
2. “鼎”是什么？它有什么用处？为什么要将“神农鼎”立在神农顶上？
3. 如果你攀登过神农顶，能向你的朋友详细描述一下登“青云梯”的感受吗？
4. 神农架人为什么要修建“神农营”和登神农顶的“青云梯”，以及许多的步行栈道？
5. 你知道野营或登山运动前需要做哪些准备工作吗？

04

知识老人驾驶着飞毯围绕着铜鼎缓慢地盘旋了两圈，轻轻地降落在神农鼎旁边的观景台上。汪汪跳下飞毯，兴奋地在观景台的木栈道上跑来跳去。小明站在山顶上张开双臂，深深地吸进了一口清新的空气，然后从背包里掏出望远镜，倚着木栏杆四下眺望。

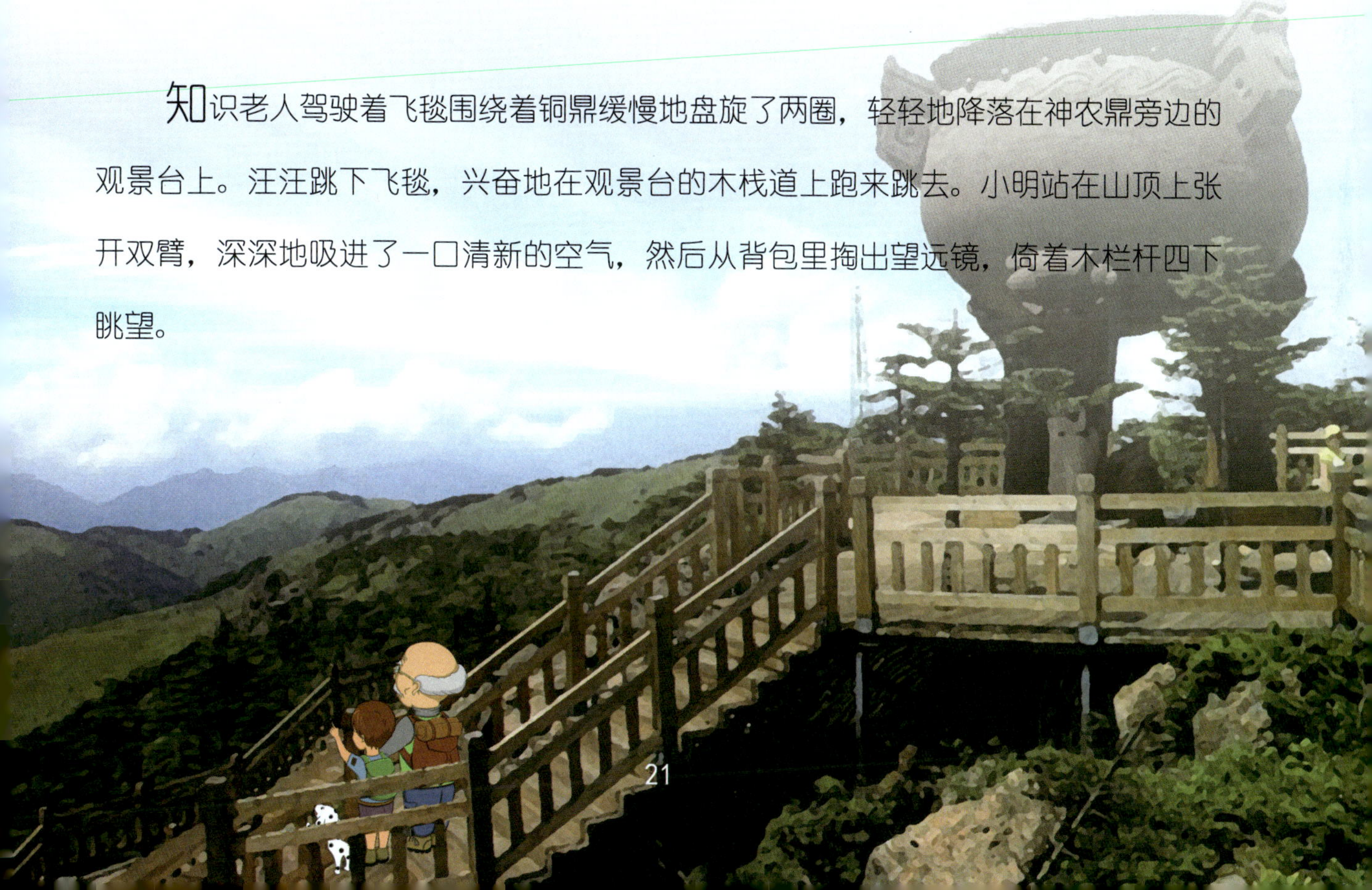

地质探秘神农架

爷爷，我们现在是站在“华中屋脊”上了吗？

可不是嘛！

可是我怎么觉得旁边有些山比我们这个山头还高些呀？

这是你视觉上产生的误差。不是有句话说“这山望着那山高”吗？当然，这句话还有别的含义。但是，当山的高度相差不是很大的时候，光凭肉眼，常常很难确定哪座山更高。

华中屋脊，海拔3106.2m
3000m
北岳恒山，海拔2017m
2000m
西岳华山，海拔1997m
安徽黄山，海拔1864m
东岳泰山，海拔1532m
南岳衡山，海拔1512m
中岳嵩山，海拔1440m
雁荡山，海拔1150m
1000m
0m

如果我没有记错，根据我查阅的资料，我们现在站立的神农顶海拔为 3106. 2m，是神农架最高的山峰，另外几座海拔超过 3000m 的山峰是大、小神农架，杉木尖，大窝坑和金猴岭。

你的记性真不错！的确，华中地区海拔超过 3000m 的 6 座山峰全部都聚集在神农架，所以这里才被叫作“华中屋脊”。

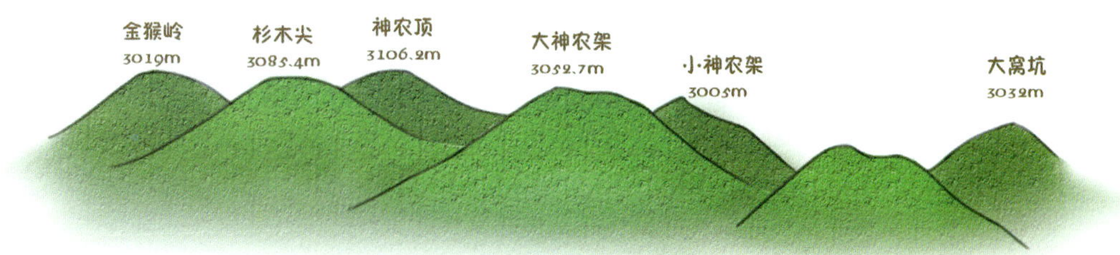

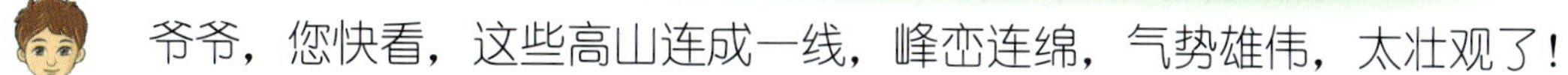

爷爷，您快看，这些高山连成一线，峰峦连绵，气势雄伟，太壮观了！

神农架的这些高山构成了一道天然的地理屏障，形成了长江和汉江的分水岭。

分水岭？是不是说山岭一边的水流入长江，另一边的水流入汉江？

对呀。高耸的神农架孕育了四大水系，即流入汉江的南河、堵河，以及流入长江的香溪河、沿渡河。

什么是水系呀？

河流连同它所有的支流一起，就构成这条河流的水系。

河流越大越长，水系就越复杂，是吗？

那当然呀。例如神农架的南河水系，它是汉江的一条支流，可是它又是由许多的支流组成的，例如关门河、古水河、洛溪河等。在神农架林区范围内，南河水系的流域面积高达 500 多平方千米呢。

爷爷，您说的流域面积是什么意思呀？

流域面积是河流与它所有支流所覆盖的地理面积。也就是说，在这个流域内的所有地表水都会汇集到这条河流中去。对了，小明，我来考考你。你能说说流域面积是根据什么来划分的吗？

让我想想。……支流的水都要汇集到主要的河流里，水是往低处流的，那分界就应当在山顶上，对了，是山！流域面积的分界就是山岭，也就是您刚才讲的“分水岭”！

很好。小明，观察大自然，就是要这样带着问题进行分析！

爷爷，您讲了神农架有四大水系，除了南河水系流入汉江之外，堵河水系也是汉江的支流，是吗？

是的。香溪河和沿渡河水系则是长江的支流。

长江支流的流域面积肯定更大些吧？

那倒不一定。流域面积是由具体的地形决定的。比如香溪河水系流域面积超过了 3000km^2，在神农架地区是最大的。可是沿渡河，又叫作神农溪，也是长江的一条支流，它的流域面积只有200多平方千米，比流入汉江的南河水系要小一半多。

长江流域水系图

真没想到神农架还是长江和汉江的重要水源地呢！

注意噢，我这里所说的“流域面积”，都只计算了神农架林区范围内的流域面积。至于整条河的流域面积，可能还要大很多呢！

问题：

1. 为什么神农架被誉为“华中屋脊”？神农架有哪几座山海拔超过了3000m？
2. 你能整理和讲述一下神农架地区的水系情况吗？
3. 什么叫作“分水岭”？从地形图上你能标画出河流的分水岭吗？
4. 什么叫作“水系”？查一下长江和黄河的水系分别包括了哪些支流。
5. 什么叫作“流域面积”？你能查出长江和黄河哪条河的流域面积更大些吗？

05

知识老人带着小明在神农顶的观景台上一边漫步，一边四下查看。小明一会儿趴在栏杆上用望远镜四处观察，一会儿举起照相机拍摄壮观秀丽的风景；汪汪则兴奋地跑前跑后，四下张望。这里视野开阔，居高临下，四周的山岭一望无余，真可谓“一览众山小”。面对着气势磅礴的群山，略显疲惫的小明坐在观景台阶梯上陷入了沉思。

爷爷，为什么神农架会这样高呢？

这很难一下子说清楚，得从头说起，我们先谈谈地球的构造吧。

地球不就是表面是石头，内部是岩浆吗？

事情可不是那么简单的。地球大致分为地壳、地幔和地核 3 个圈层。就好比一个煮熟了的鸡蛋：地壳，也就是地球最外的固体圈层，相当于鸡蛋的蛋壳。地幔则相当于鸡蛋的蛋白部分。

那地核就相当于蛋黄了！

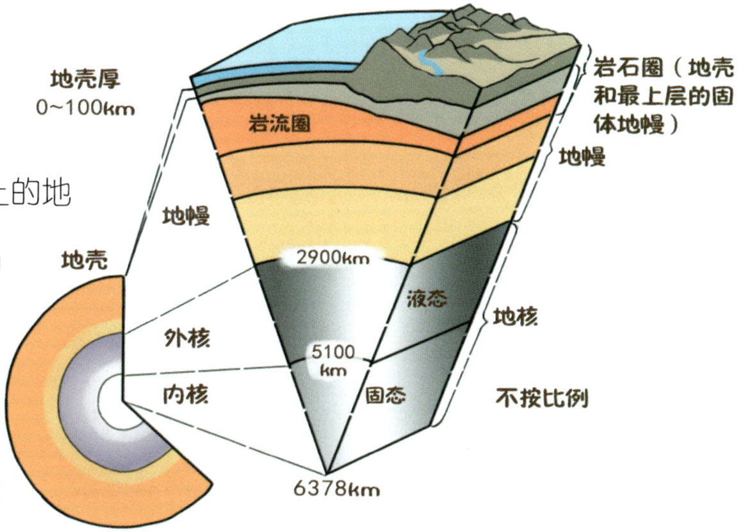

是的。全球地壳的平均厚度约 17km，其中大陆上的地壳，也就是“陆壳”平均厚度为 33km，高原或山区的地壳会更加厚些，最厚的地方甚至可以超过 70km。海洋中的地壳，就叫作“洋壳”，它远比陆壳薄，厚度只有几千米。

鸡蛋壳可没有那么厚。那地幔和地核是些什么东西呢？

可是鸡蛋远没有地球那么大呀。相对于地球的大小，几千米、十几千米厚的地壳确实微不足道。地幔位于地壳之下，是地球的中间层，主要由致密的造岩物质构成；地核则是地球的核心部分，主要由铁、镍元素组成，就像你刚才说的，地核就相当于蛋黄。

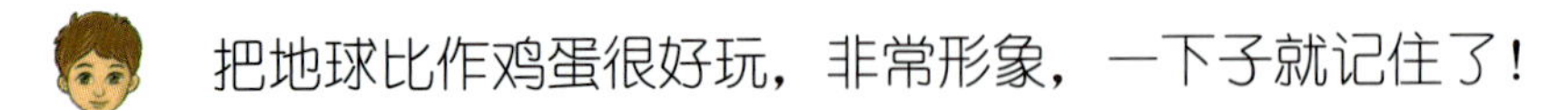

把地球比作鸡蛋很好玩，非常形象，一下子就记住了！

但是，实际上地球的构造比鸡蛋要复杂得多。在地壳和地幔中间还有一个软流圈。

软流圈？是软的吗？

是的。软流圈在全球基本上是连续分布的。也就是说，不管是陆壳还是洋壳，下面都有软流圈，而且它与岩石圈之间并没有明显的分界，呈现逐渐过渡的状态。

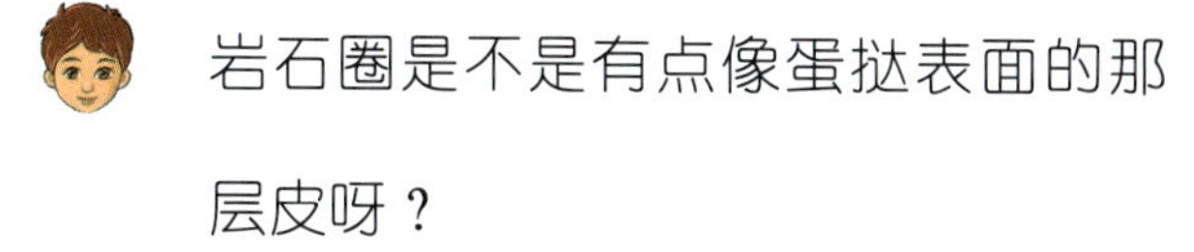

岩石圈是不是有点像蛋挞表面的那层皮呀？

倒还真有点像呢。科学家们认为软流圈是岩浆的主要发源地。

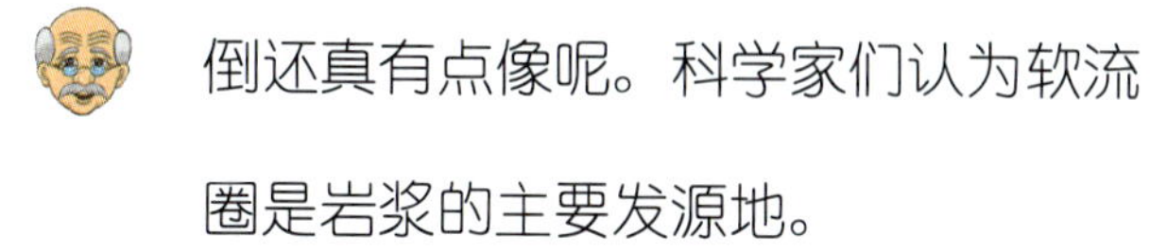

哇，看来地球确实比鸡蛋复杂多了！

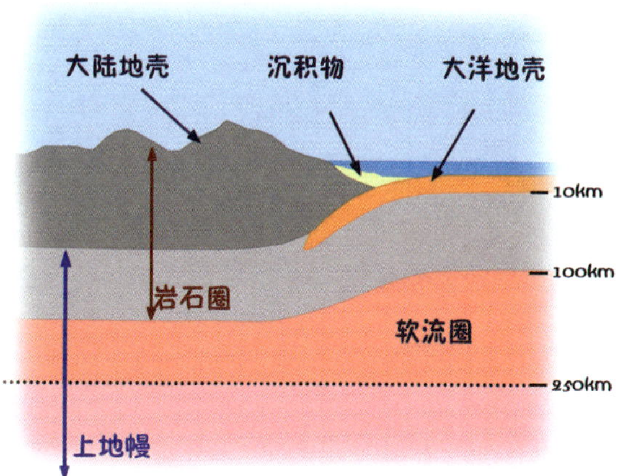

软流圈的分布具有明显的区域性差异，海洋下面软流圈的位置较高，而大陆之下它的位置则较深。

地球真的好神奇。

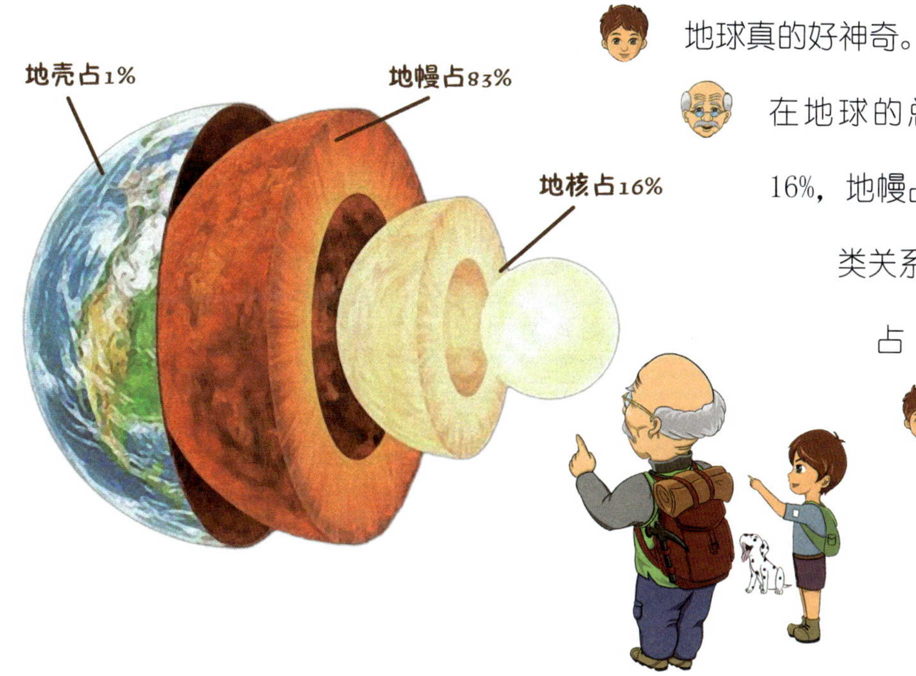

在地球的总质量中，地核占16%，地幔占83%，而与我们人类关系最密切的地壳，仅占1%而已。

与地球相比我们人类简直太渺小了。

是的。与宇宙和大自然相比较，人类确实微不足道。但是人类对大自然的破坏和影响却是不可忽视的。现在，由于工业的发展、社会的进步和人们对消费的追求，导致地球环境的破坏和污染已经到了极其危险的程度。

是的。现在很多地方的小河里没有鱼了，天空中没有鸟了；树林被砍伐光了，很多动物灭绝了。我们必须立刻制止人们对水、空气、森林、土地的污染和破坏！

人类若不保护自己生存的地球，后果将不堪设想。神农架是我国中部比较发达地区所剩下的最后一块世外桃源了。它是一颗绿色的珍珠，是华中地区的“生态肺”，担当着“吐故纳新”的重任，保障着人类有一个健康的生活环境。

问题：

1. 你能画出地球内部的结构图，并解说一下地球内部有些什么层圈吗？
2. 什么叫作“陆壳”？什么叫作“洋壳”？它们各自有什么特点？
3. 软流圈与陆壳、洋壳之间存在什么样的关系？
4. 你能举出一两个人类破坏地球环境的例子来吗？
5. “生态肺”是什么意思？为什么知识老人把神农架称为华中地区的“生态肺”？

06

小明听着知识老人的讲述，陷入了沉思。汪汪也懂事地蹲在小明脚边，一声不响。知识老人发表了一通关于环境保护的激昂感言后，突然想到小明的问题是：“神农架为什么会这样高”，于是转过身来，接着为小明解说。

好了，我们接着讲讲关于地球的结构。你听说过“板块构造”吗？

听说过，但是搞不太懂到底是怎么回事。

板块构造学说是在“大陆漂移”和“海底扩张”推论的基础上提出来的重要地质学理论。

大陆漂移？像船那样？

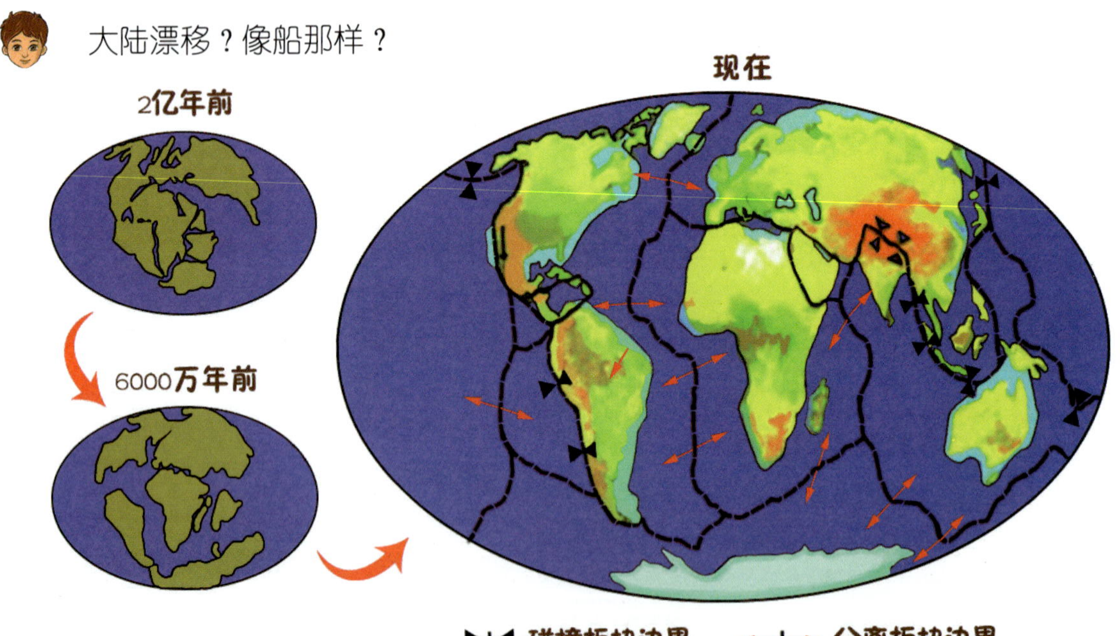

是的。德国科学家魏格纳发现大西洋两岸，特别是非洲和南美洲海岸的轮廓，好像可以相互拼接得天衣无缝。于是他做了大胆的推测，认为它们原先是连在一起，后来漂移分开的。

从地球仪上看，非洲和南美洲确实可以很完美地拼接在一起。

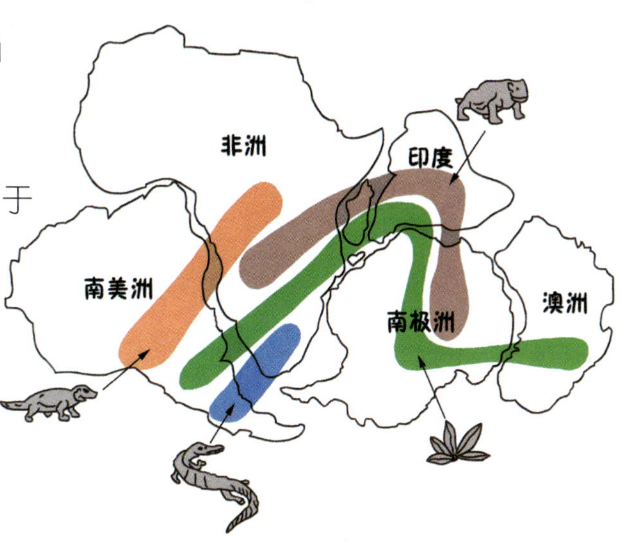

魏格纳做了许多的勘察和研究，于1921年在自己的著作《大陆和海洋的形成》中，首次提出了大陆漂移的推论，他认为现在地球上所有的大陆过去曾经是一个统一的巨大陆块，叫作“泛大陆”或“联合古陆”。这个“泛大陆”后来发生了分裂和漂移，四分五裂，逐渐形成了现在的地理形态。

哇，真有意思！不过我觉得也确实有点像这么回事。有点道理！

魏格纳还收集到了一些比较确凿的证据，比如同时代、相同种类的古生物化石，在如今被海洋分隔开的几块陆地上都有发现，而这些生物（包括植物）是没有能力漂洋过海的。唯一的可能性就是，在这些生物存活的时期内，这些大陆地块是彼此相连接的，这些生物是在同一块大地上迁徙扩散的。后来因大陆漂移，被海洋分隔开了。

对啊，这是个很有说服力的证据！

另外，冰川遗迹也提供了很好的证据。

古冰川示意图

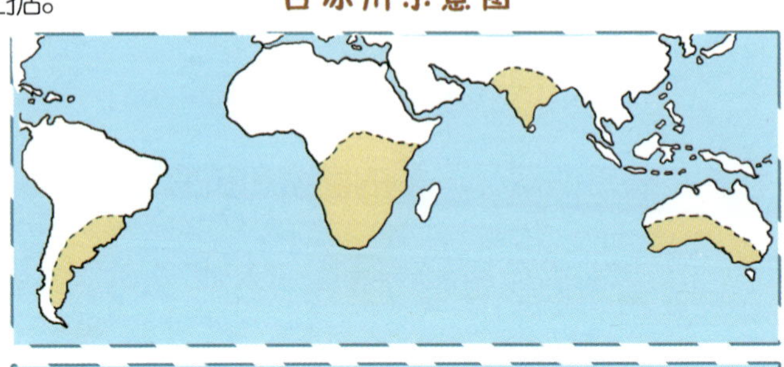

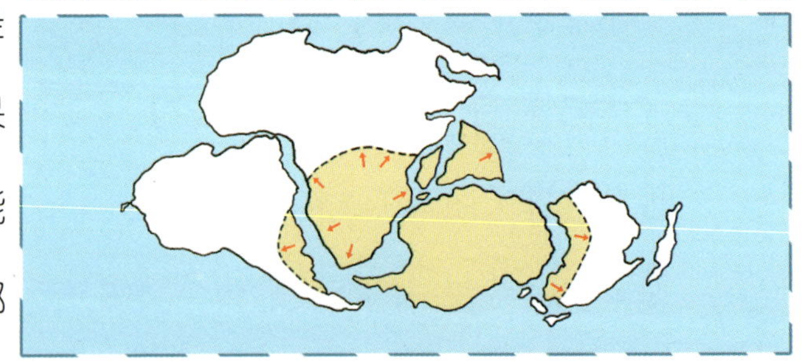

冰川怎么能证明大陆漂移呢？

在南美洲东南部、非洲南部、印度以及澳洲的南部都发现了同时期的冰川遗迹。这些地方现今是不可能出现冰川的。但是，如果把这几块大陆拼接在一起，就能发现这些出现冰川遗迹的地点都汇聚到了一起，而且冰川流动的方向似乎能指示古代的地极，就如同今天的南北极一样。

哇，太奇妙了！看来，在地球演化的历史中确实发生过大陆漂移的现象。

是的。但是你能说说，是什么力量促使大陆发生漂移呢？

是哦。是什么力量能推动大陆漂移呢？这可是要很大的力气的。

可不是嘛。因为这个问题没有解决，魏格纳提出的大陆漂移理论始终存在缺陷，不能被所有的科学家认同。

也是哦。科学家是最讲究真凭实据的，来不得半点马虎。

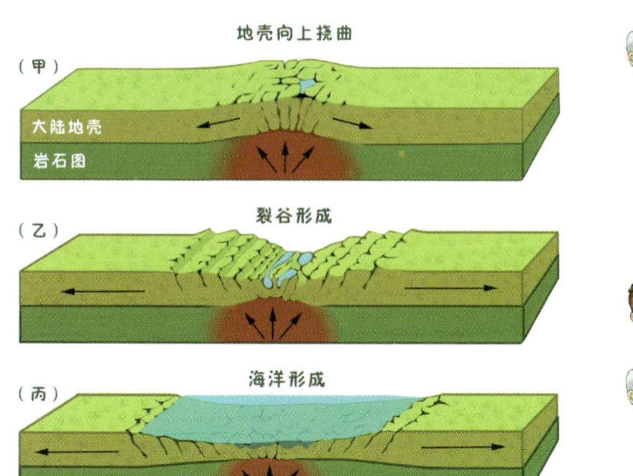

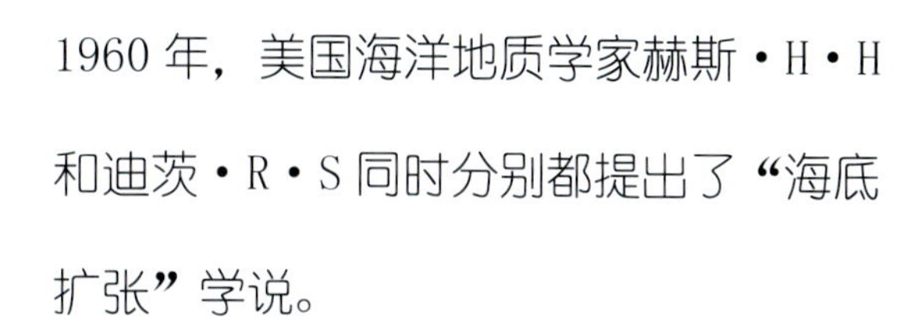

1960 年，美国海洋地质学家赫斯 • H • H 和迪茨 • R • S 同时分别都提出了“海底扩张”学说。

“海底扩张”？听起来好可怕呀！

海底扩张确实是非常可怕的地质现象。通过勘测，赫斯发现因地球内部存在热对流的缘故，地底高温的上升熔岩流会冲破大洋地壳，不断地涌出来，形成大洋中脊和新的洋壳，把早先形成的洋壳逐渐向两侧推移，不断地形成新的洋底地壳。

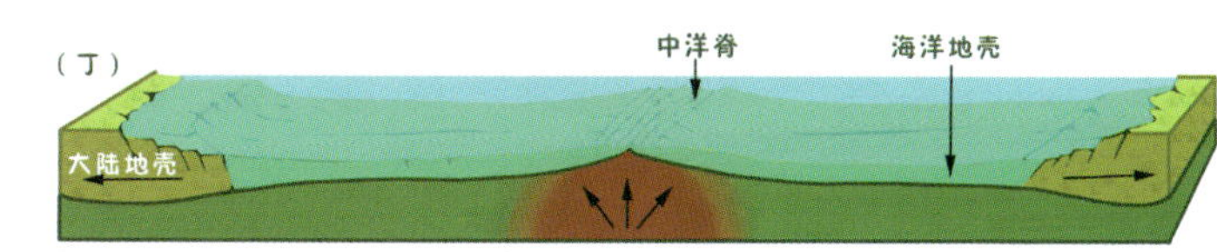

是不是像推折叠门那样呀？

是呀，聪明的孩子，你很善于想象！这是一个连续不断的过程，熔岩不停地涌出来，形成新洋壳，把早先形成的洋壳不断地向两侧推开。这样，两边的大陆就像浮在水上的冰块一样，越漂越远。

哦！原来大陆漂移是这样发生的呀！

海底扩张的理论解决了 40 多年前魏格纳大陆漂移学说无法解决的动力问题，

世界六大板块分布

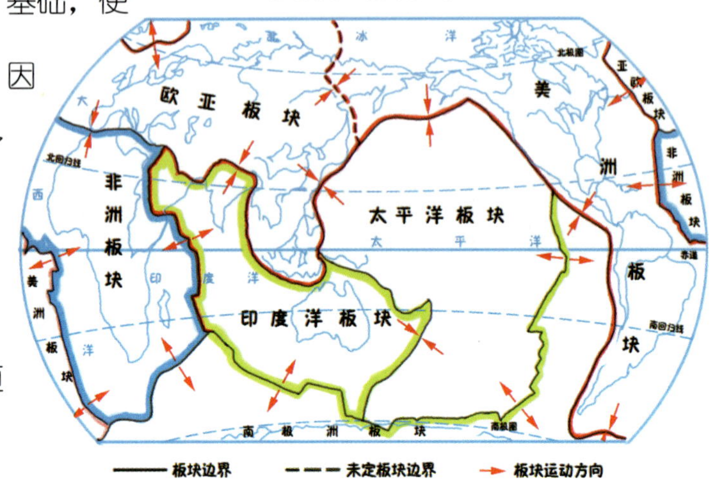

同时也为板块构造学奠定了基础，使它得到所有科学家的认同。因此，现在大陆漂移学说成了大家都认可的主流观点。

40 多年后魏格纳的推论才被大家认同。科学发展的道路真艰难呀！

科学研究和发现的道路确实是非常艰辛曲折的。

爷爷，您绕了这么大一个圈，但是还是没告诉我神农架的山为什么这样高呀？

“板块构造学说”就能解释为什么神农架的山这样高。

是吗？我听说过“板块构造”。据说地球上的五大洲四大洋都是一些板块拼成的，好像七巧板一样。

是的。板块构造学说是法国科学家勒比逊在“大陆漂移学说”和“海底扩张学说”的理论基础上，于 1968 年提出的。他认为地球的岩石圈不是整体一块，而是被分割成许多被叫作“板块”的单元。这些板块漂浮在“软流层”之上，它们因为地球内部的热对流，而处于不断运动之中：有时漂移开去，有时相互拼合。

那我们的房子会掉到板块下面去吗？

一般来说，板块内部比较稳定，板块之间的交界地区，是地壳的活动地带，不太稳定。若板块漂移分离，则形成大洋；若相互拼合，则发生碰撞。这时常常会因为挤压，或者会因为一个板块插到另一板块之下，使得它逐渐抬升形成高原和山系。

啊。我知道了，爷爷。听说我们的青藏高原和喜马拉雅山就是因为印度板块插到欧亚板块的下面，而被抬升起来，变成了“世界屋脊”。

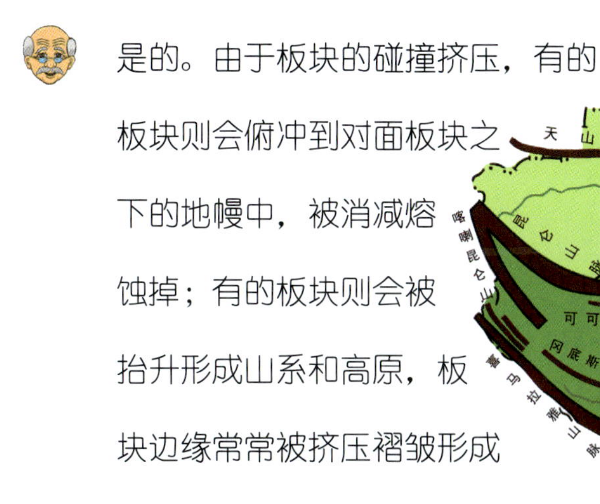

是的。由于板块的碰撞挤压，有的板块则会俯冲到对面板块之下的地幔中，被消减熔蚀掉；有的板块则会被抬升形成山系和高原，板块边缘常常被挤压褶皱形成沿一定方向延伸的山系。

我知道了，爷爷，神农架就是这样被挤压抬升的吧？

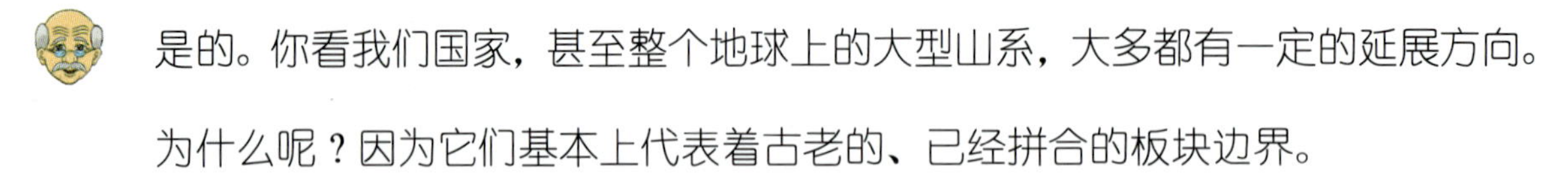

是的。你看我们国家，甚至整个地球上的大型山系，大多都有一定的延展方向。为什么呢？因为它们基本上代表着古老的、已经拼合的板块边界。

哇，太神奇了！科学家真了不起！我将来也要当地质学家。

神农架形成高山，成为“华中屋脊”，确实与板块构造运动相关联：有的科学家认为神农架属于一个大板块的边缘，由于板块拼合挤压褶皱而抬升；有的科学家却认为，神农架当时是一个夹在两个大板块之间的“微板块”，当两个大板块拼合时，它就被挤压抬升而成为“华中屋脊”了。

太神奇了！这一切是怎么发生的呢？

当时到底发生了什么事情，神农架是怎样一步一步成为高山区的，直到现在还没有完全弄清楚呢，还等着你这个未来的地质学家去探讨呢。

好呀！我长大了就专门研究神农架是怎样成为“华中屋脊”的！

问题：

1. 你能简单地叙述一下魏格纳的“大陆漂移学说”吗？为什么这个学说直到40多年后才被所有的科学家认同？
2. 什么是“海底扩张学说”？它有什么重要的意义？
3. 你能解释什么是“板块构造”吗？为什么地球上的大型山系都呈现一定的延展方向？
4. 当板块碰撞时会发生些什么现象？
5. 请查阅关于“东非大裂谷”的地质资料，并以它为实例，给你的朋友们讲解一下板块构造运动是怎么回事。

07

小明带着汪汪在神农顶的观景台上一边溜达，一边四处张望，陶醉在阳光与和风之中。突然，汪汪朝着观景台木栈道下面叫了起来，一只野兔一晃，飞快地钻进了旁边的乱石堆中。小明还想仔细看看野兔到底躲在什么地方，可一眨眼，野兔已在乱石的缝隙中消失得无影无踪了。

爷爷，周围怎么这么乱呀？干嘛把这些混凝土块堆在这里？

这些是混凝土吗？你再仔细看看！

我看就像混凝土。您看，这里面的碎石块都被水泥胶结在一起，成了一大块。

这可不是混凝土，而是一种角砾岩。

角砾岩？！

在自然环境中有很多不同种类的角砾岩。这是一种非常独特的火山角砾岩。

火山角砾岩！那这里有火山了？

你别害怕，这里没有火山。在地质学上，这种岩石叫作“隐爆火山角砾岩”。

为什么起这样怪里怪气的名字呀！

这个名字是有来历的：当地底的岩浆被挤压，沿着地下的裂隙上升到接近地表的地方，由于压力迅速减低，加之周围地下水受到岩浆高热的烘烤，瞬间被气化，产生了强烈的爆炸，把周围的岩石以及快要冷却石化的岩浆岩炸碎；这些岩石碎块又被不断涌上来的岩浆胶结，就形成了这种像混凝土一样的角砾岩。

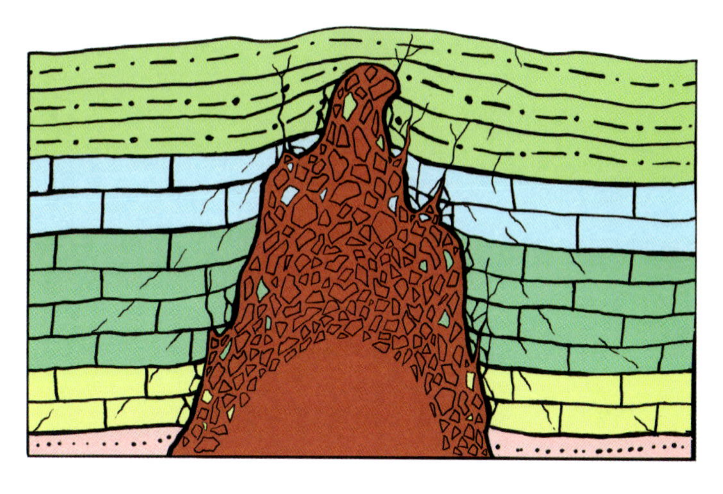

隐爆火山角砾岩示意图

哇！这个过程太可怕了！

地质作用常常伴随着破坏性极大的可怕现象。“隐爆火山角砾岩”形成的时候，虽然发生了强烈的爆炸，但并不是像火山喷发那样冒烟冒火。它是隐藏在地底下发生的，所以叫作“隐爆火山角砾岩”。

哈哈，这个名字还真贴切，名副其实！但是，它不是在地底下形成的吗？怎么现在都堆在这个山头上？谁挖出来的？

谁也没有挖它们。是大自然长年累月的风化剥蚀，把盖在它上面的石头和土层冲蚀掉了，才让它们“得见天日”。要知道，这种火山岩比周围其他的岩石都要坚硬，因此更难风化，所以在地势上它们常常形成高山的山顶。

可是这种“隐爆火山角砾岩”应当是一整块呀，怎么变成了这样的碎石堆呢？是谁把它们敲碎的呢？

这可不是人敲碎的，这又是大自然的杰作。

是吗？您怎么知道呢？谁看见了？大自然有那么大的力气，把一个山头都敲碎吗？谁相信呀？！

你别抬杠，小明。大自然的力气是不可比拟的，难道你不知道火山爆发、地震、山泥倾泻、滑坡这些地质事件，霎时间内就能释放出巨大的能量，给人类造成灾难吗？

是的，这些我都在电视上看到过，很容易懂。难道神农顶上的这些石头是地震震碎的吗？

那倒不是。我刚才说的火山爆发、地震、断层、褶皱、山体滑动等，都是由地球内部力量引发的，我们把这种作用叫作地球的“内营力作用”，也就是由地球内部的能量所引发的地质作用。其能量来自板块运动、火山爆发、地质构造运动等。

有“内营力作用”，那就肯定还有“外营力作用”吧？

对的。地球的“外营力作用”是指诸如大气、流水、太阳能，以及生物等外在因素对地球岩石发生的作用，例如风雨、河流、波浪、潮汐、冰川、风蚀、植物根的侵蚀等。

我觉得地球的“外营力”肯定没有“内营力”强大。

是的，这确实是事实。但是，你也别小看了地球的“外营力”。你知道“水滴石穿”是什么意思吗？

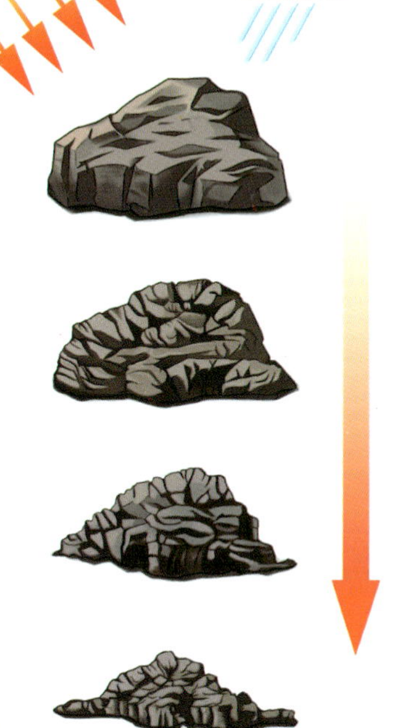

“水滴石穿”就是一滴水的力量虽然很小，但是一滴一滴长年累月不间断地滴在石头上，就能把石头穿一个洞。这个成语比喻做事情要有耐心和恒心，坚持到底就能成功。

说得很好！地球的“外营力”，比如阳光、风、流水等是最有耐心和恒心的。千百万年以来，它们永远不知疲倦地在修理和改造我们的地球。你看，地球的“内营力”造成高山，但“外营力”逐渐使高山变成小丘，河流携带着的泥沙沉淀，形成冲积平原。我们所说的“沧海桑田”就是地理地貌在“内、外营力”的作用下，历经千百万年发生的巨大变化。地球的“外营力”就好像是一个雕塑家，永远不间断地塑造着地球的外貌。

可是，您还是没有讲这山上的石头怎么被“外营力”敲碎的呀。我怎么都想不出阳光、风、流水是怎么样把一整个山头的石头都给敲碎的。

好吧，让我来告诉你：当这些石头裸露在山顶的时候，每天都要遭受风吹雨打、日晒雨淋。白天晒太阳受热膨胀，晚上则发生冷冻收缩，于是产生了裂隙。

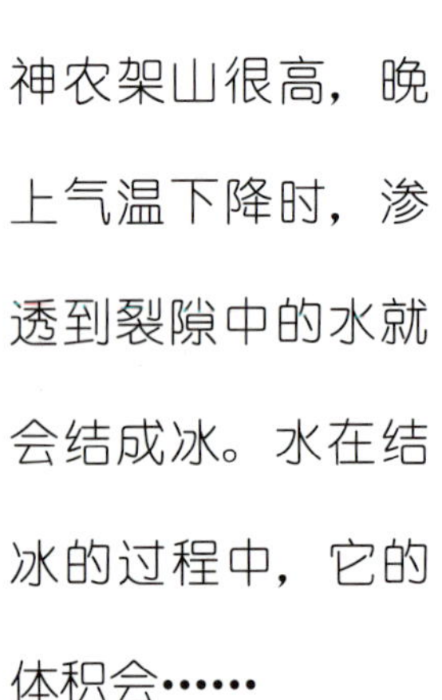

小小的裂隙怎么能把这一整个山头的石头都弄碎呢？

是的。小裂隙并不是致命的。可是，如果有水在里面掺合，渗透到裂隙中就麻烦了。

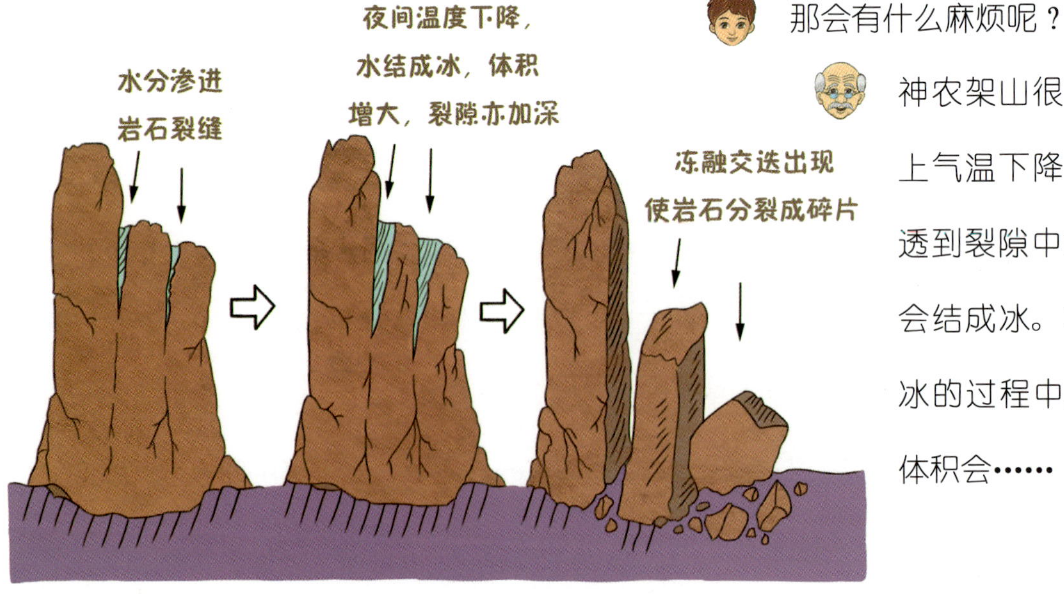
水分渗进
岩石裂缝
夜间温度下降，
水结成冰，体积
增大，裂隙亦加深
冻融交迭出现
使岩石分裂成碎片

那会有什么麻烦呢？

神农架山很高，晚上气温下降时，渗透到裂隙中的水就会结成冰。水在结冰的过程中，它的体积会……

哦，我知道了，我知道了！水一结冰，体积增大，就把裂隙撑开了。

不错。小明，你能在遇到问题的时候，把学过的知识应用起来，非常好！学习就是为了应用嘛。那你能不能讲讲，水是怎样具体地把这里的石头弄碎的呢？

好的，我试试吧。……嗯，其实也很简单：首先，石头的热胀冷缩产生了裂隙，雨水渗到裂隙中；当气温下降时，水就结冰，体积增大，把裂隙撑得更宽；这样就会有更多的水渗入到裂隙中，又结冰、又膨胀、又流进去更多的水……。就这样无限循环，渐渐地把坚硬的岩石劈开了、劈碎了。

你解释得很好！今后遇到有趣的地质地理现象，一定要先想一想，看看用自己以前学过的知识，能否进行解释。

好的，爷爷。我一定学以致用！

这些巨大岩石被劈成碎块的过程，地质学上把它叫作“冷冻风化”或者“冰劈作用”。其实，岩石发生风化还会受到很多其他因素的影响，风化作用也可以分成物理风化、化学风化和生物风化三大类型。

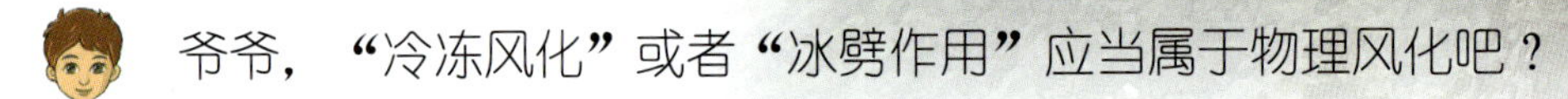

爷爷，“冷冻风化”或者“冰劈作用”应当属于物理风化吧？

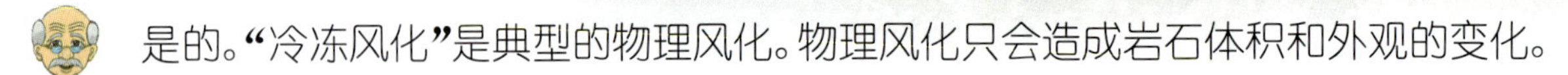

是的。“冷冻风化”是典型的物理风化。物理风化只会造成岩石体积和外观的变化。

那什么叫作化学风化和生物风化呢？

这个问题留待我们参观板壁岩和神农谷的时候再讨论吧。时间不早了，我们赶快去金猴岭看看吧。

好嘞！咱们走了，汪汪！

问题：

1. 什么是“隐爆火山角砾岩”？你还知道哪些其他类型的角砾岩？
2. 为什么“隐爆火山角砾岩”出露在神农架的山顶上？
3. 你能举出实例说明“内营力作用”对地球环境形成造成的影响吗？
4. 什么是地球的“外营力作用”？它在自然界有些什么具体的表现？
5. 你能叙述“冷冻风化”或者“冰劈作用”的具体过程吗？

08

知识老人展开飞毯，坐了上去，小明抱着汪汪也跳上了飞毯。飞毯平稳地缓缓升起，向着墨绿色的群山飞去。这次飞毯飞得很低，几乎是擦着树梢在飞行。小明仔细地察看着浓密的树林；汪汪一会儿站在飞毯的左边，一会儿又跳到右边，总想从下面的树林里发现点什么它感兴趣的东西，好对着它叫一叫。

爷爷，咱们这是去金猴岭吗？金猴岭一定有很多金丝猴吧？

是的。金猴岭是神农架海拔高于3000m 的 6 座高山之一，它以茂密的原始森林为特点。因为野生金丝猴常在这里出没而命名。

爷爷，我看过好多神农架金丝猴的照片，它们可漂亮了！

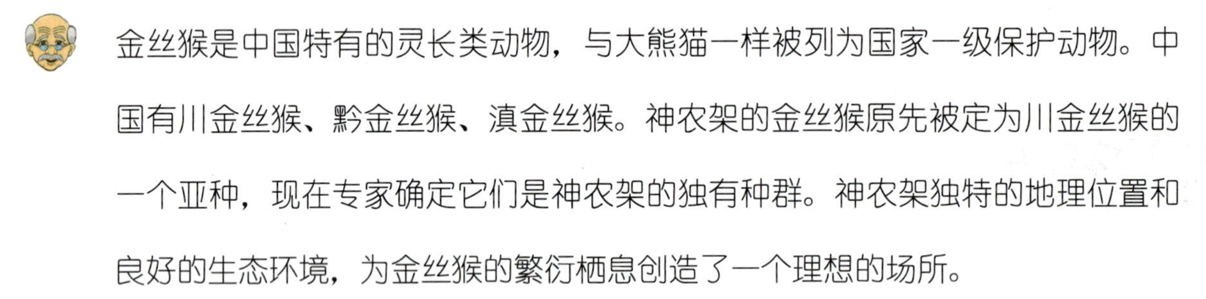

金丝猴是中国特有的灵长类动物，与大熊猫一样被列为国家一级保护动物。中国有川金丝猴、黔金丝猴、滇金丝猴。神农架的金丝猴原先被定为川金丝猴的一个亚种，现在专家确定它们是神农架的独有种群。神农架独特的地理位置和良好的生态环境，为金丝猴的繁衍栖息创造了一个理想的场所。

爷爷，我看过好多不同的金丝猴的照片和视频，但我觉得神农架的金丝猴是我看见过的最美的金丝猴。它们体态强壮，金发飘逸，目光炯炯，魅力无穷。

是哦，它们确实是非常美丽的动物，是神农架山野的精灵。神农架还有一个专门研究金丝猴的科研中心，设在大龙潭。它是全国首个建立金丝猴野外定点投食的研究机构。

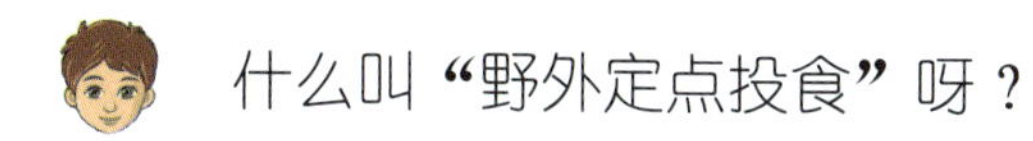

什么叫“野外定点投食”呀？

野外定点投食是研究野生金丝猴的一种手段。通过定点投食，使野生金丝猴逐渐形成习惯，到一个固定的地点去取食。这样，便于我们对金丝猴开展近距离的观察和研究。而且，当严冬来临，森林中食物剧减，金丝猴的生命受到饥饿威胁的时候，定点投食就起到保护金丝猴种群的作用。

是的哦。大森林里的野生动物冬天格外可怜，没有食物肯定会饿死的呀。

可不是嘛。但是，要让金丝猴养成到一个固定地点取食的习惯却是非常困难的。

我想也是的。它们听不懂我们的话，不知道我们这样做是为了它们好。

金丝猴原本就是警惕性非常高的动物，再加上过去盗猎者的偷猎和捕杀，使金丝猴对人类特别害怕和警惕，根本不让人接近，这给科研和保护金丝猴的工作带来了极大的困难。

那神农架人是怎样建立定点投食的呢？金丝猴怎么会听他们的话呢？

说来话长，为了摸清金丝猴的生活规律，神农架的科研人员整整三年冒着严寒，爬冰卧雪，吃尽了苦头，在茫茫林海中跟踪金丝猴的足迹，给它们投放食物。但是，金丝猴对人类非常警惕，它们情愿挨饿，也不去碰我们投放的苹果、胡萝卜、花生等。

它们可能没有吃过苹果和胡萝卜呢。不知道这些都是美味的食物吧？

我们的科研人员把金丝猴最爱吃的云雾草缠在苹果上，没想到金丝猴毫不领情，把云雾草吃了，苹果还是原封不动地丢弃了。

太傻了！那怎么办呀？

就在科研人员准备放弃的时候，有一只金丝猴大胆地进行了第一次尝试！那是在2005年12月28日，这可是一个重要的日子。那天和往常一样，科研人员投放食物的时候，一只金丝猴拿起一个苹果，一边端详一边犹豫着……

别犹豫了。快吃呀，快吃呀！真急死我了！

它对着苹果看了许久，可能因为实在是太饿了，终于咬了第一口，并且津津有味地吃下了整个苹果。这可是一个了不起的示范！

它真是一只勇敢的金丝猴，就像第一个吃螃蟹的人！它肯定马上告诉它的朋友们："苹果真好吃！快来吃吧，没事的！"

可不是嘛。科研人员给这只金丝猴起名叫"大胆"。榜样的力量是无穷的。自那以后，金丝猴开始信任我们的科研人员，逐渐养成了定点取食的习惯。

哇！太好了。它们终于懂得了我们的好心好意。

由于定点投食，保障了金丝猴种群的健康发展，神农架金丝猴的数量翻了1倍多。而且，由于我们的科研人员与金丝猴建立了相互信任的关系，所以能够较近距离地接触观察它们。神农架金丝猴研究中心因此取得了许多国际一流的科学研究成果。

神农架的科研工作者们真了不起。我要向他们致敬！

问题：

1. **你知道多少关于金丝猴的知识？你能向你的朋友们描述一下神农架的金丝猴吗？**
2. **为什么神农架的科研人员要建立金丝猴的"野外定点投食"？**
3. **神农架金丝猴的"野外定点投食"是如何建立起来的？**
4. **科学工作者为什么要研究金丝猴和其他的动物？**
5. **你觉得人类与野生动物之间应当建立一种什么样的关系？**

09

知识老人和小明乘坐飞毯，缓慢地盘旋在金猴岭的上空。只见山峦重叠、郁郁葱葱；流水叮咚、清泉长流。突然，汪汪不安地在飞毯上跳来跳去，朝着下面浓密的森林发出阵阵吼叫，显然它发现了什么。

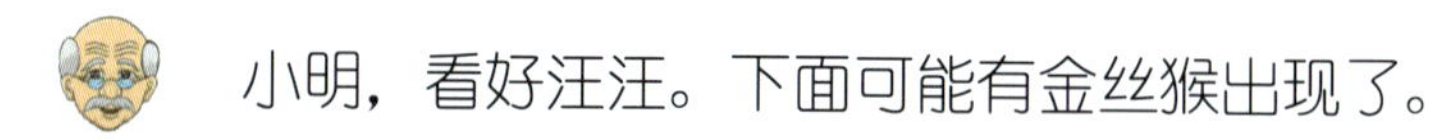

小明，看好汪汪。下面可能有金丝猴出现了。

是吗？汪汪，快过来，别大惊小怪的！让咱们好好地看看野生的金丝猴！

快看！下面有几十只金丝猴在树林间穿行呢！

哇！它们的身手好矫健呀！从一棵树攀到另一棵树，好像荡秋千一样。

金丝猴习惯于生活在海拔 1600~3000m 的地带。每一个种群都有它们自己活动的区域。

它们吃什么东西呀？

野果、嫩枝、树叶和草根都是它们的食物。它们也很爱吃原始森林中树枝树干上生长的松萝。

快看呀，爷爷。后面还有一大群呢！……哇，还有金丝猴 Baby 呢！

这不奇怪。前面刚过去的那一群金丝猴，我们称为“全雄单元”，后面的就是拉家带口子的一些金丝猴的家庭了。

什么？什么？什么是“全雄单元”？

金丝猴是群居动物，像我们人类一样，有着它们自己的“社会组织结构”。

肯定有一只“美猴王”统治它们，像花果山的孙悟空那样！

很多人都跟你一样，以为会有“猴王”。而事实上，每个金丝猴群体是按照“家庭”和“全雄单元”的组织结构来生活的。

这是什么意思呢？什么是金丝猴的家庭，什么是“全雄单元”？

每个金丝猴的群体一般包括 50~200 只的个体，它们组成一个“全雄单元”和若干个“家庭”。

什么是“全雄单元”呀？

从字面上你应该大概猜得出，所谓“全雄单元”就是这里面所有的金丝猴都是雄性的。

哇！金丝猴也分“男女界限”呀？

好像有那么点意思。“全雄单元”相当于金丝猴种群的“军队”，它们除了打前站，寻找食源，还要负责整个群落的安全保卫工作。

哈哈，真有意思。就像解放军那样？

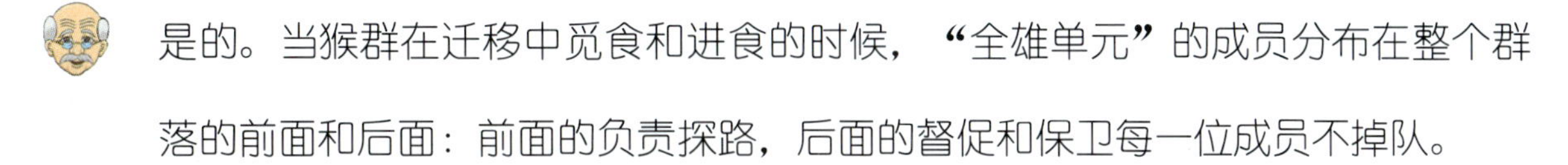

是的。当猴群在迁移中觅食和进食的时候，“全雄单元”的成员分布在整个群落的前面和后面：前面的负责探路，后面的督促和保卫每一位成员不掉队。

哇，组织纪律性很强嘛，真了不起！

猴群每天中午还要睡个午觉。这时候，“全雄单元”就负责放哨，肩负着保卫整个金丝猴群安全的重任。

哇！它们的责任还挺重的！那什么是金丝猴的“家庭”呢？

每个金丝猴的种群常常包括多个“家庭”。每个“家庭”都有一位成年雄猴，被称为“家长”，“家长”是最具权威的金丝猴，它可以有多位“妻子”，妻子们负责在“家长”的带领下和睦相处，抚养幼仔。

爷爷，爷爷，您快看呀，那儿有几个金丝猴 Baby，一身金黄色的绒毛，真可爱！有一个还在吃奶呢。

金丝猴幼崽是最可爱的小动物！它们天真无邪，充满好奇心。

真想抱抱它们。您看，连汪汪都不叫了，想跟它们玩呢。

在金丝猴的社会中，幼崽会得到整个群落的爱护。幼小的雌猴留在家庭里，一直到它们“出嫁”。而小雄猴，一旦成年，就要被“家长”也就是它们的父亲，赶到“全雄单元”去，就像去“参军”一样。

哇！它们的爸爸好狠心呀！

这就是大自然的规则。在金丝猴的社会中，有属于它们自己的“社会规则”。而这种社会规则是必不可少的，它可以保障金丝猴种群的延续和兴旺。

自然界真的好残酷。

是的。自然界的生物受到自然环境的制约，它们的生存确实很不容易。因此，我们更应当尽全力来保护它们。

是的，如果每一个人都把保护野生动物和自然环境当作自己的责任，我们的大自然将会更加和谐美丽。

问题：

1. 你能向你的朋友描述一下你在神农架见到的金丝猴吗？
2. 金丝猴的“全雄单元”与金丝猴的“家庭”是什么关系？
3. 为什么说金丝猴的这种“社会规则”能保障金丝猴种群的延续和兴旺？
4. 你知道还有哪些动物像金丝猴这样具有较严格的“社会规则”？
5. 为什么每一个人都应当把保护野生动物和自然环境当成自己的职责？

10

知识老人和小明乘坐在飞毯上，居高临下观察着金丝猴群的活动。金丝猴一边玩耍、一边进食，从一棵树飞跃到另一棵树，时不时传来它们欢快的叫声。只见森林里枝叶晃动，金光闪烁，一片喧哗。转眼间它们就消失在山脊的后面，喧哗声也渐渐地沉寂下来。飞毯在一个小广场上停了下来。汪汪一跳下飞毯就往山上树林里跑去……

汪汪，回来！ 别乱跑！

它想去找那些金丝猴玩呢。

汪汪快回来！金丝猴都走了。傻狗狗！

是的，它们都走了，别影响它们。我们到森林里去转转吧。

呀！这里还有泉水呢，“金猴溪”！好漂亮呀！

金猴岭这个景点最吸引人的就是茂密的原始森林和常年流淌不断的金猴溪。

哇，爷爷，这山上的树好高好密呀！

是的，这里树高林密，具有原始森林的主要特征。

原始森林有什么特征呀？不就是树老一些、粗一些罢了！

问题可没有那么简单。原始森林有很多独特的现象和特征。我们沿着这条小路向山上走一走，找一找原始森林的特征，好吗？

好的。哎，爷爷您快看，那儿有几朵花，粉红色的、白色的，好漂亮呀！

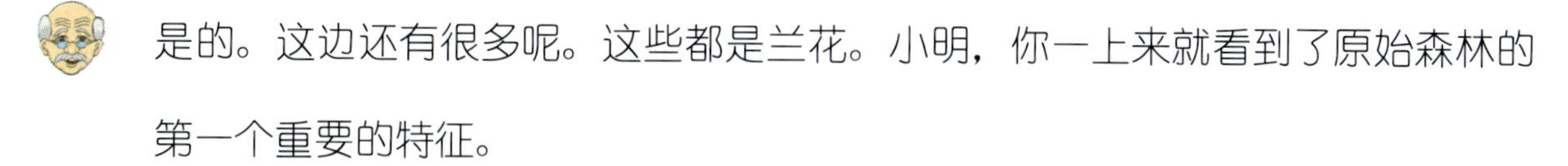

是的。这边还有很多呢。这些都是兰花。小明，你一上来就看到了原始森林的第一个重要的特征。

是吗？什么是原始森林的特征？兰花和原始森林有什么关系呀？

兰科植物和蕨类植物都是地球上比较古老的植物物种，大约在 3 亿年前就出现了蕨类植物。原始森林中必须要有野生的兰科植物和蕨类植物，才能证明它的原始性、古老性。这就是原始森林的第一个特征。

哦，原来是这样。那原始森林还有什么特征呢？

你看到前面有几棵倒在地上的大树干了吗？那边，远处还有一棵倒树呢。

是呀，它们都已经腐朽了，上面长满了青苔。呀！还长了几个小蘑菇呢！

小明，这就是原始森林的另外两个特征。原始森林要有自然倒伏的朽木：树木死亡后没有人为因素的干扰，自然倒伏在地上，直到最后全部腐烂，一切都呈现自然的状态。

那倒也是的，既然是原始森林，就不应当有人为的干扰。但是，是不是一定要有蘑菇才是原始森林呢？

倒不一定要有蘑菇。原始森林树高林密，森林的下部光照有限，因此树上会有附生的不需要光合作用的生物，如菌类、松萝及苔藓等植物，当然蘑菇也属于这类植物。你看很多树枝上都附着或吊着那些绿色的丝状物，那就是松萝。是金丝猴最喜爱的食物。

怪不得金丝猴常来这片树林。爷爷，原始森林还有什么特征呢？

还有。原始森林必定会展现植物的生长层次，即从下向上分别为青苔、草类、灌木及高大的乔木。

还有呢？

原始森林一般都有一定厚度的腐殖土。腐殖土是原始森林中植物落叶、死亡后所形成的，根据腐殖土的厚度可以推断森林的年代，证明这片森林的古老性。

嗯。很有趣，也很有道理！

根据刚才我讲的这些特征，你可以判断一下金猴岭到底是不是原始森林。

这些特征我们都找到了，毫无疑问，金猴岭是典型的原始森林。我一进到这树林里，就感到空气特别清新，特别舒服。

你知道为什么吗？原始森林是氧气聚集之处，是天然的负氧离子发生器……

“负氧离子”是什么？

负氧离子是指带负电荷的氧离子。它是无色无味的，被称为“空气中的维生素”。它对人体血液循环、增加血液氧含量和振作精神有非常积极的作用，因此具有镇静、催眠、镇咳、增进食欲及降血压等医疗效果。

哇，“负氧离子”有这么多功效呀！

是的。通常雷雨过后，空气的负离子就会增多，所以人们会感到心情特别舒畅。国际上对良好空气环境的指标是：每立方厘米的空气，含负氧离子 2000 个以上；而金猴岭原始森林空气中每立方厘米超过了 16 万个。所以这里被誉为“天然氧吧”。

怪不得呢，真舒服！金猴岭的原始森林真是太宝贵了！

是的。森林是地球上最宝贵的资源和财富。这里的每一棵树都是无价之宝。

无价之宝？！树木怎么能称得上是“无价之宝”呢？不就是能做建筑木材吗？

过去，人们砍伐森林获取木材，只计算森林产出的木材体积。但是，根据科学家计算，一棵生长了 50 年的树，按市场木材的价格计算，大约值 300 美元；但根据它的生态效益来进行计算，一棵生长 50 年的树，每年可以产生价值 31 250 美元的氧气；它还可以减轻大气污染、涵养水源、防止土壤流失；大树腐烂的树叶可以增加土壤的肥力……

还有，还有，森林是鸟类及其他动物居住的地方！

是的。森林养育了许多的生物物种。综合起来计算，一棵树的生态价值可高达 20 万美元以上。

哇！树木真的是无价之宝呀！我们一定要好好保护森林。

小明，你注意到刚才在我们身边流淌的小溪不见了吗？

是哦。看不见流水了。……但是我怎么还听得到水流的声音呢？水到哪里去了呢？

你伏在石头上听一听。

哇！流水的哗哗声好响呀，好像石头下面有一条河！

是的。这就是金猴岭神秘的地下河。虽然看不到流水，但却好像面对着一个大瀑布，听到“哗哗”的流水声。所以人们把这个地方叫作“金猴听瀑”。

太神奇了！这里怎么会有地下河呀？它是怎样形成的呢？

这是个好问题，留给你回去自己查查资料。我们现在向下走，沿路你再仔细地观察一下，可以看到清澈的山泉从岩隙中、从树根下涌出，汇集到山涧中奔腾而下，形成金猴溪的源头。

好的。汪汪，快回来！我们要走了。

问题：

1. 为什么说原始森林一定会出现菌类、松萝、苔藓等不需要很多光照的植物？
2. 为什么说根据腐殖土的厚度可以推断原始森林的年代？
3. 你能总结一下原始森林有些什么特征吗？
4. 为什么知识老人说每一棵树都是无价之宝？应当怎样理解森林对环境的影响？
5. 什么是地下河？它是如何形成的？

11

知识老人和小明乘坐飞毯，向金猴溪的下流飞去。只听见流水的“哗哗”声越来越响，涓涓细流的小溪渐渐地变成了一条在森林中奔腾穿行的山间小河，流水飞溅着向山下狂奔，如同脱缰的野马，势不可挡。突然，一面陡壁出现在眼前，湍急的溪流飞奔而下，形成了一条壮观的瀑布……

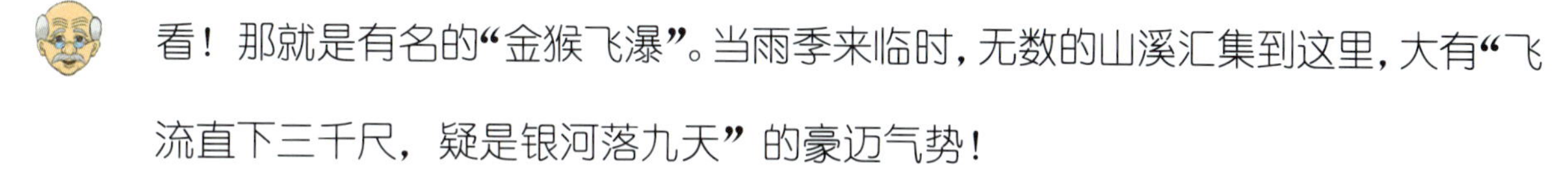

看！那就是有名的“金猴飞瀑”。当雨季来临时，无数的山溪汇集到这里，大有“飞流直下三千尺，疑是银河落九天”的豪迈气势！

太美了！我简直都不想离开金猴岭，不想离开这原始森林和飞泉瀑布了！

神农架的美景还多得很呢。我们马上就要到著名的“神农谷”了。

“神农谷”？那一定是神农架最大的山谷了。

倒不是最大的。只不过“神农谷”位于地质公园神农顶园区的中心位置。而它的形成和一个叫作“神农架背斜”的地质构造有密切的关系。

“神农架背斜”？什么意思呀？

“背斜”是一个地质术语。神农架的岩层，是10多亿年前在大海里形成的沉积岩。你还记得沉积岩与岩浆岩、变质岩的区别吗？来神农架之前我要你查过资料的。

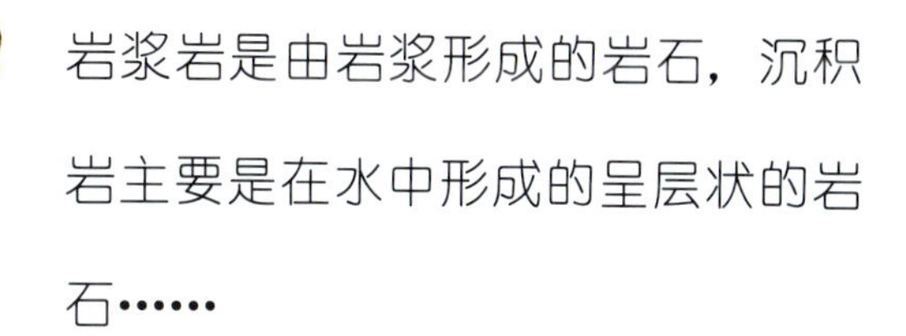

岩浆岩是由岩浆形成的岩石，沉积岩主要是在水中形成的呈层状的岩石……

是的。它们本应当是呈水平状态的，对吧？

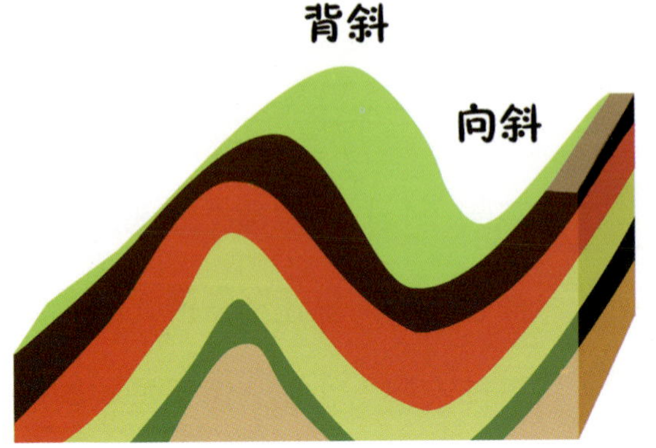

是呀。可是您看山上那些岩层，很少有水平的，都是些倾斜扭曲的。怎么会这样呢？

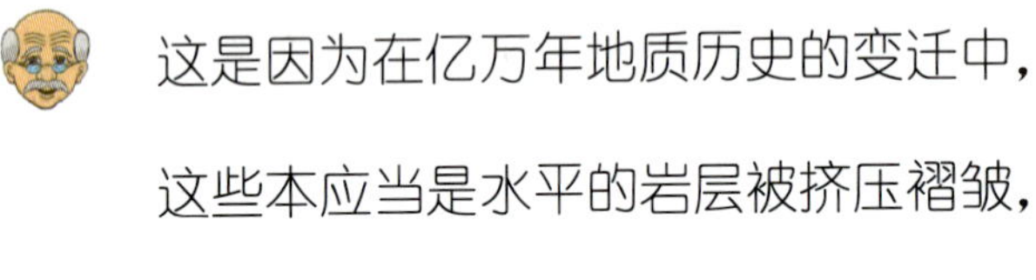

这是因为在亿万年地质历史的变迁中，这些本应当是水平的岩层被挤压褶皱，扭曲折断。当岩层发生褶皱的时候，一般会形成两种最重要的形态：向上拱曲，形成“背斜”；或者向下凹曲，形成“向斜”。

哦，背斜、向斜就是这么回事呀。

神农谷的形成与“神农架背斜”有着非常密切的关系。

……不对吧？爷爷，您不是说岩层向下凹曲形成向斜吗？山谷应当是向斜呀！怎么背斜反而变成神农谷了呢？

小明，我先考考你，你仔细思考一下：是背斜更容易形成山呢，还是向斜更容易形成山？

那当然是背斜容易形成山啰！它是往上拱的嘛。

先别急着回答。小明，你再想想。

…… 嗯，我再怎么想也想不出来。难道是向斜容易形成山？

在自然界，确实是向斜更容易形成山一些。

那为什么呀？向斜是向下凹的呀！它怎么能形成山呢？！

这要从地质作用的长期性来考虑！一般的人往往不懂得如何用“地质思维”的方法来考虑问题。

什么叫“地质思维”呀？

我们现在看到的所有地质现象，都是经历了数千万年甚至是上亿年漫长历史的产物。考察地质现象时，必须要以数千万年作为时间尺度来进行思考，这就是地质思维的方式。我们前面不是提到过“滴水穿石”吗？没有几千万年的时间，水能穿石吗？

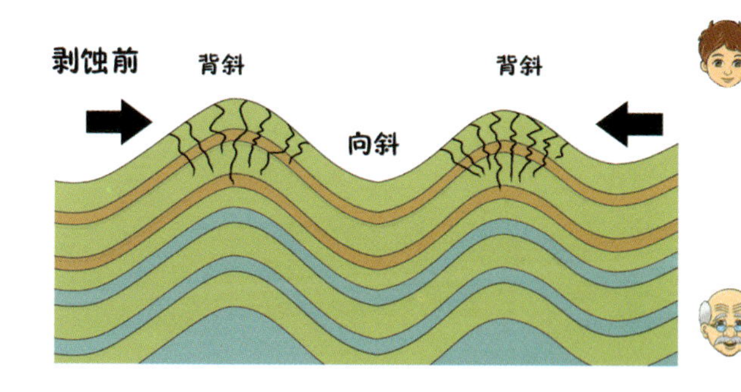

剥蚀后
向斜
背斜
背斜

肯定不能。但是，向下弯曲的向斜经历数千万年，也不可能比向上拱的背斜还要高呀。

那可不一定。让我们换一个角度来思考这个问题吧：岩层在发生褶皱的时候，形成背斜和向斜的岩层，所受力的方向是完全不相同的，所以岩石变形的性质也肯定存在很大的差异。

是吗？那您讲讲，背斜和向斜所受的力有什么区别呢？

神农架背斜构造示意图

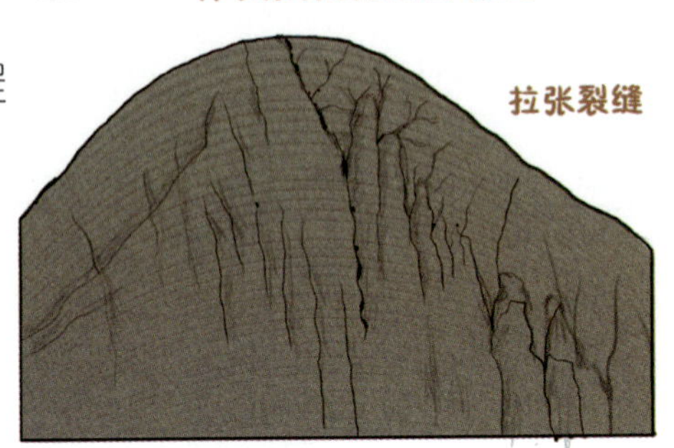

打个比方吧，当你把一根筷子弯曲到最大程度的时候，它就会被折断，是吗？

是呀，没错。

那筷子会从什么地方折断呢？

当然是从筷子的中间呀。

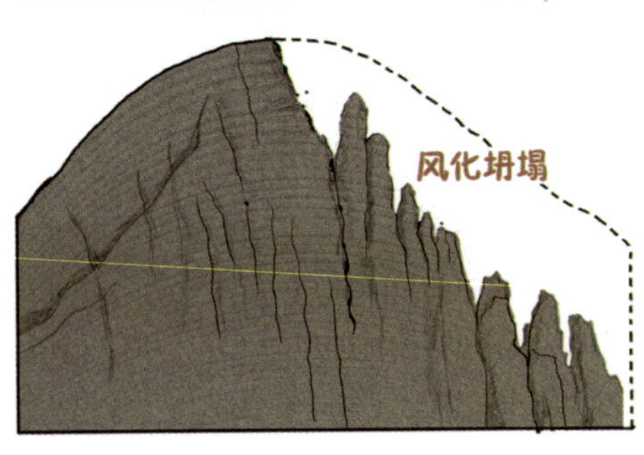

当地层发生褶皱的时候，所形成背斜的顶部，就相当于弯曲筷子的中部。

是呀。岩层在背斜的顶部就会被折断。

是的。岩层在这里不光会被折断，由于受到的是拉张的力量，岩层中会形成一系列的张性裂隙。我们把岩石中这些天然形成的裂隙叫作“节理”。由于产生了大量的张性节理，所以背斜部分岩石抗风化的能力就大大地减弱了……

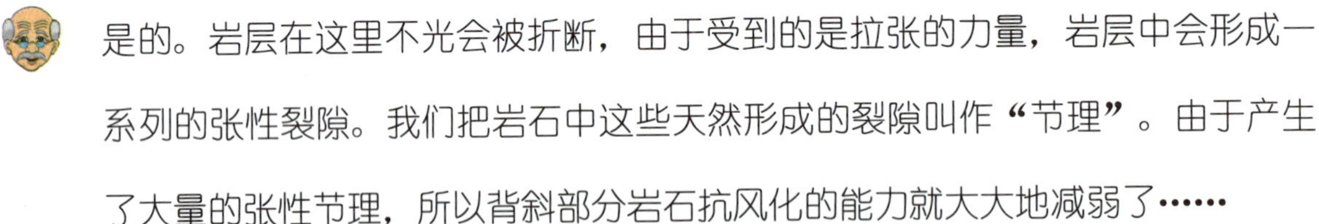

哦，我明白了。大量的节理使得岩石更容易遭受风化剥蚀。流水把风化剥蚀产生的泥土碎石冲走了，就逐渐形成了山谷。原来是这样。那向斜呢？

向斜部分在褶皱时受到的是挤压的力量，由于挤压使得岩石变得更加紧密，抗风化的能力反而更加强了，所以……

所以它们很难被风化剥蚀。当背斜被风化剥蚀的时候，向斜部分反而被保留下来，形成了山峰。哇！没想到真的是向斜更容易形成山一些！

这就是用地质思维的方式来分析问题！

太好了！我学会了地质思维！原来在大自然中，还存在这么多容易导致一般人产生误解的事实。

我们眼前的这些山山水水，都是亿万年地质作用的产物。当我们看到一些神奇的地质景观或现象的时候，都要用地质思维的方法去分析一下。

好的，爷爷。我觉得地质思维是一种非常有趣的分析地质现象的方法。

可不是嘛。自然界的一切都会在漫长的时间跨度中发生变化。用地质思维的眼光去看待我们面前的自然景观，可以锻炼我们的逻辑思维和推理能力，令我们在观察自然现象的时候充满想象。

确实很有趣。

用地质思维来观察我们面前的自然现象，能使我们在游山玩水的时候有更多的乐趣，边看边想，仔细领会大自然给我们的启发。大自然能够教给我们很多有用的知识。其实除了“向斜成山、背斜成谷”之外，还有一个很常见的地质构造也常常形成山谷。

那是什么地质构造呀？

那就是“断层”呀。你知道什么叫作“断层”吗？

岩石断开来了就是断层呗！

对了。但是从科学上讲这是不太确切的。地质学上给断层下的定义是：地壳岩层因受力而发生破裂，并沿破裂面有明显相对移动的地质构造称为断层。

哦，您是说断开两边的岩石发生了错动，才叫断层呀。

是的。光裂开来了只能叫作“节理”，两侧发生了错动才叫“断层”。

我知道了为什么您说断层也容易形成山谷。地层断开来，而且发生了错动，石头肯定都被搓碎了，因此很容易被风化，被流水冲蚀。常此下去自然就形成了山谷。

你分析得很对。地质学家在野外经常能根据地貌的变化来判断是否有断层，甚至能推断断层的性质。

断层的性质？断层有什么性质呢？

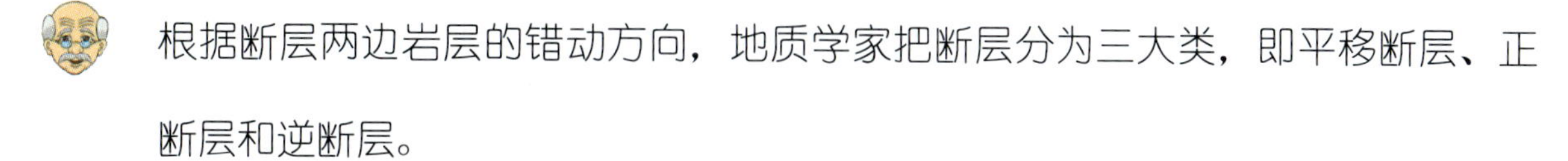

根据断层两边岩层的错动方向，地质学家把断层分为三大类，即平移断层、正断层和逆断层。

哎呦，断层还搞得这么复杂呀！它们是怎样划分的呢？

首先有几个术语得弄明白了，断层的断面，我们称为“断层面”，“断层面”往往是倾斜的。位于倾斜断层面之上的岩层我们叫作断层的上盘，下面的则叫作下盘。

平移断层

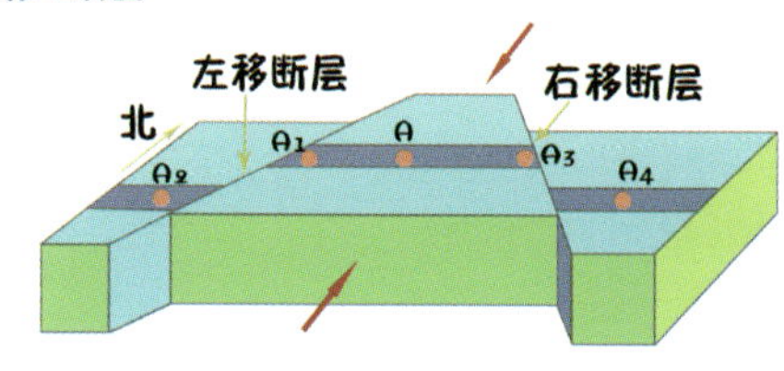

这个很容易懂。

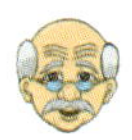

好。上盘沿着断层面向下滑动的断层，我们叫作“正断层”。这种断层是在张力作用下形成的，上盘在重力作用下自然地向下滑动，形成断层。

正断层

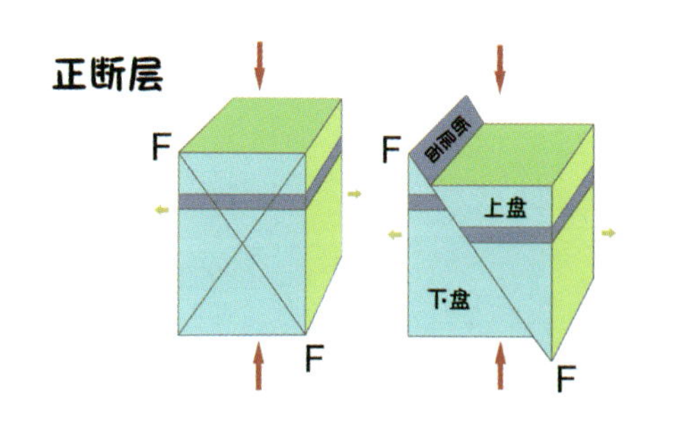

我猜“逆断层”肯定是在挤压力作用下形成的。

小明，你真聪明。“逆断层”确实是在挤压力作用下形成的。但是你能想象一下，当发生挤压时，断层的上、下两盘是怎样相对运动的呢？

逆断层

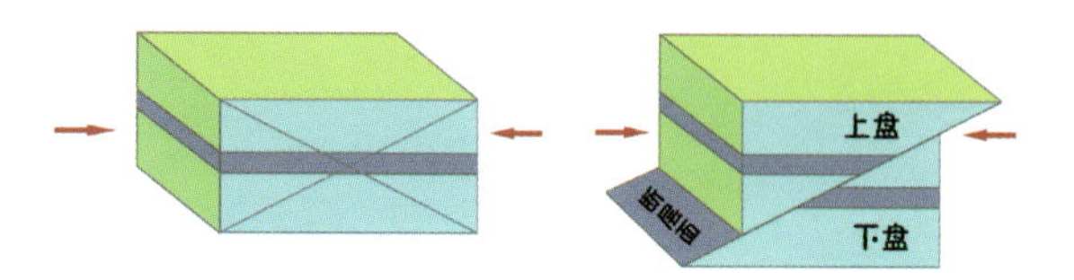

让我想想。……嗯，对了，这么一挤压，断层的上盘就会沿着断层面向上推举。是吗，爷爷？

对了！你的逻辑思维很好。

爷爷，那平移断层呢？平移断层是不是不上不下，断层的两盘发生水平方向的相对运动？

是的。小明，你归纳得很好。学习就是要这样善于思索、举一反三。好了，我们到神农谷了。

问题：

1. 从森林生态的角度，你能解释一下为什么金猴岭有那么多的泉水瀑布吗？
2. 什么是“背斜”构造和“向斜”构造？在野外还能看到些什么其他的地质构造？
3. 为什么“向斜”构造往往更容易形成山一些？
4. 什么叫作“地质思维”？如何用“地质思维”解释河口三角洲的形成？
5. “平移断层”“正断层”和“逆断层”的主要特点是什么？

12

知识老人把飞毯停在神农谷的入口处。下了飞毯后，带着小明和汪汪向神农谷入口的石头阶梯走去。一上到阶梯的顶部，面前展现的是一幅壮丽的“水墨画”：缕缕雾气从神农谷内冉冉升起，层层交叠的远山在虚无缥缈的薄雾中时隐时现；山谷一侧耸立着几根像巨人般的擎天石柱，云雾慢慢地飘过来，在石柱间缭绕……

地质探秘神农架

哇！太美了！真好像是仙境一样。

这是神农架世界地质公园最经典的观景点之一。

爷爷，这个峡谷就是神农谷吗？

是的。你好好看看这里的地质情况，然后分析一下神农谷到底是怎样形成的，好吗？

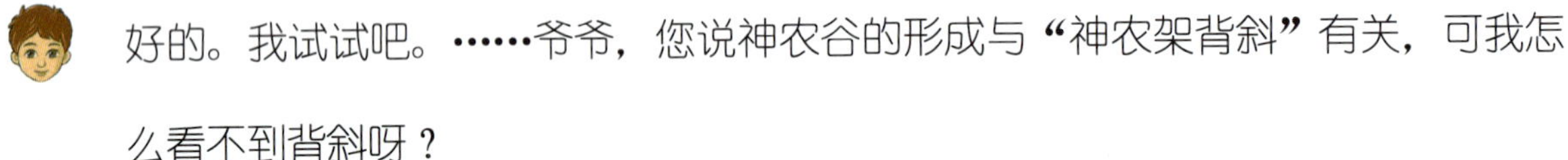

好的。我试试吧。……爷爷，您说神农谷的形成与“神农架背斜”有关，可我怎么看不到背斜呀？

要知道在野外，我们所说的那些地质构造往往都是非常庞大的，跨越和延伸往往达数十千米至数百千米。当你身在其中时，常常很难一眼看出来。所以，很多地质构造不可能像书上画的示意图那样清楚明白。

爷爷，那我们还是坐飞毯飞高点看吧！

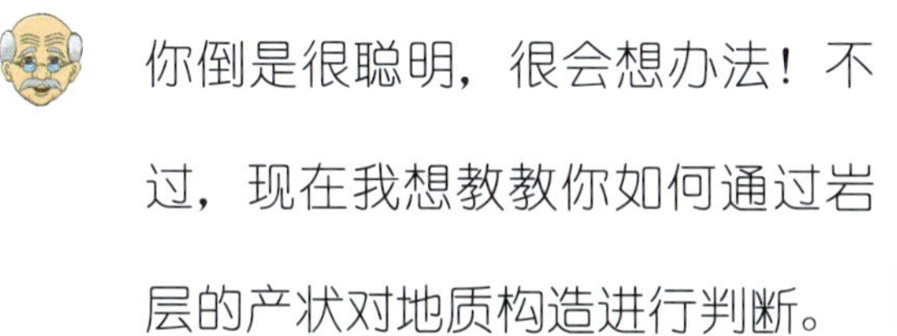

你倒是很聪明，很会想办法！不过，现在我想教教你如何通过岩层的产状对地质构造进行判断。

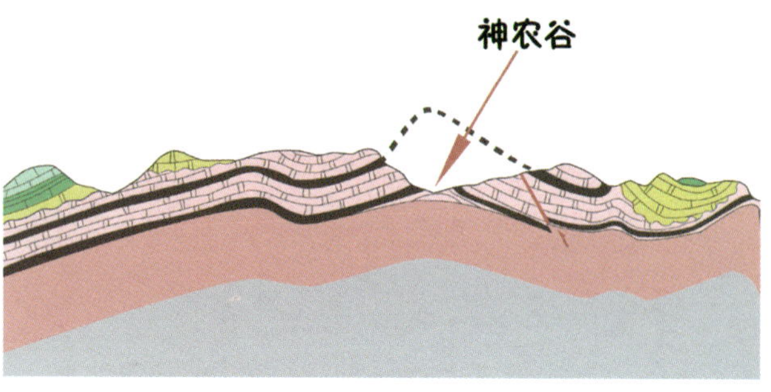

什么是岩层的“产状”呀？

“产状”是一个地质术语。简单地说，所谓“产状”就是岩层产出的状态，即岩层是水平的还是倾斜的、朝哪个方向倾斜、倾斜的角度多大以及与周围其他地质体的关系等。

哦。这很简单嘛。让我看看。…… 爷爷，我左手边的岩层是向左边倾斜的；右手边的岩层向右边倾斜…… 哇！爷爷，您看，这真是个背斜，真的是个背斜呀！

对了，如果两边的岩层都对着你发生倾斜，那就是向斜构造了。

背斜往往形成山谷，看来是真的。太神奇了！神农谷还真的是因为背斜构造形成的。

实践验证了我们通过地质思维得到的推理。那么，下一个问题是：你能解释一下为什么这里会有很多的石柱呢？

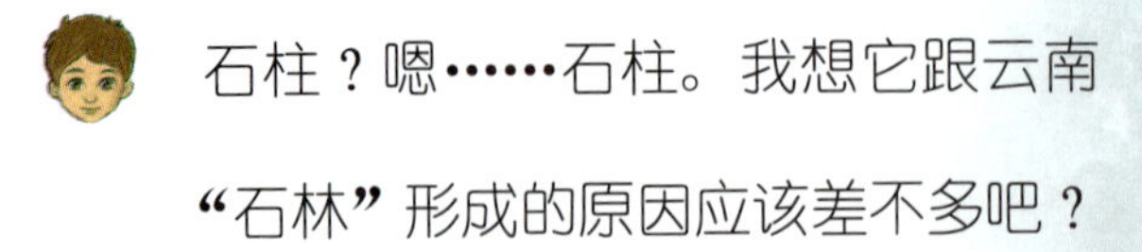

石柱？嗯……石柱。我想它跟云南“石林”形成的原因应该差不多吧？

是的，确实有点相像，但也有许多不同的控制因素。

比如说呢 ……

比如说石林往往发育在相当广大的一片石灰岩分布的地域内，而神农谷的石柱分布范围却是相当局部的。

确实是这样。爷爷，那是什么原因造成的呢？

你已经知道神农谷的形成与神农架背斜构造有关。那么，神农谷中出现的石柱肯定只限于神农架背斜的范围之内。

没错！但是，为什么只有这里才有石柱，旁边的那些岩石也属于神农架背斜，为什么它们没有形成石柱呢？

能否形成石柱，首先与岩层在发生褶皱弯曲的时候，岩石中出现裂隙的多少，也就是裂隙的密度以及裂隙发育的形态相关联。

是的。裂隙越多就越容易形成石柱。

是的。岩层在背斜形成的过程中，由于拉张作用，在某些局部的地带会产生大量的垂直裂隙，地质学上把这种自然形成的裂隙叫作“节理”。

是的，您说过，节理就是在岩石中自然产生的裂隙。

是的。大量直立的节理把岩石切割成了很多垂直的柱形方块。随后的风化剥蚀作用就会沿着这些垂直的节理进行。记住，要用“地质思维”的方式来思考风化剥蚀的过程……

我知道了！那就是说，经历了千百万年的风化剥蚀，加上垮塌，这些垂直的柱形方块周围被风化的岩石变成了碎石和泥沙被流水带走了，剩余的方块石柱就挺立起来了！

你理解得对极了！长期的日晒雨淋、风吹雨打，侵蚀加上重力垮塌，同时作用在节理密集的岩石上，就形成了这样雄伟壮观的石林。大自然的创造是多么不可思议呀！

大自然就好像是一个伟大的雕塑家，为我们塑造出各种各样的美丽风景。

问题：

1. 在野外如何观察向斜和背斜？
2. 什么叫作“节理”？你觉得自己有能力在野外发现“节理”吗？
3. “节理”在风化侵蚀和地貌形成上起到什么作用？
4. 你能向你的朋友们具体解说一下，神农谷是怎样由背斜构造逐渐地形成的吗？
5. 请详细解说一下神农谷的石柱是如何形成的，并解释它们与桂林石林的差异。

13

小明陶醉在神农谷如梦如画般的景色之中，时不时举起照相机，拍下一张又一张照片。汪汪在观景台上窜下跳地跑来跑去。当浓密的云雾升起、涌向观景台的时候，它如临大敌，昂着头，对着云雾汪汪地叫着；当云雾退去的时候，它又像一个胜利者，摇着小尾巴，高兴地俯视山谷。

小明，你注意到没有，不少的石柱都还比较圆滑呢，你知道为什么吗？

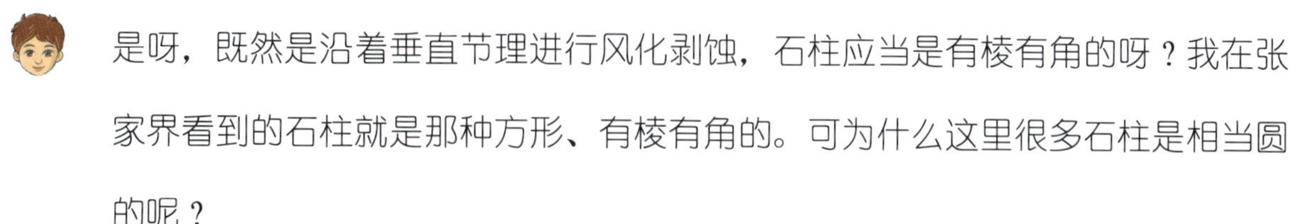

是呀，既然是沿着垂直节理进行风化剥蚀，石柱应当是有棱有角的呀？我在张家界看到的石柱就是那种方形、有棱有角的。可为什么这里很多石柱是相当圆的呢？

让我来告诉你吧。这与岩石本身的性质有关。神农架的岩石与张家界的岩石完全不同。张家界的石头是砂岩，而神农架的石头是10多亿年前在海洋里形成的一种叫作“白云岩”的沉积岩。

“白云岩”！多好听的名字呀！

白云岩就像石灰岩一样，也是属于碳酸盐类的岩石。碳酸盐类的岩石主要是由碳酸钙组成的，很容易被流水溶蚀，形成峰林、溶洞和天坑等。

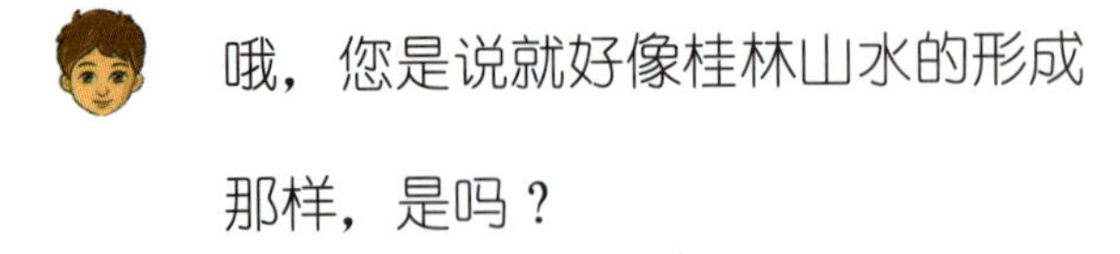

哦，您是说就好像桂林山水的形成那样，是吗？

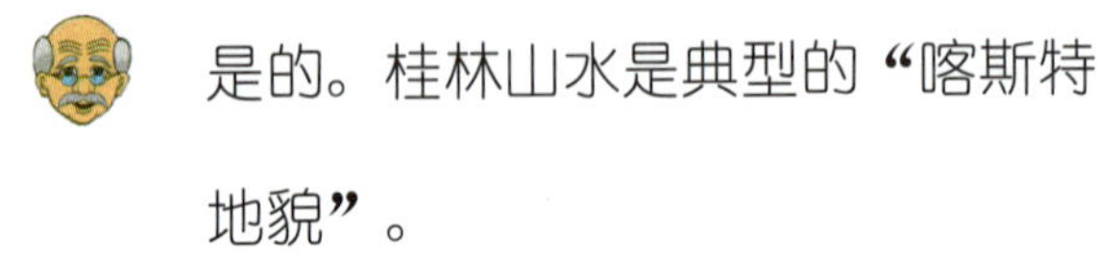

是的。桂林山水是典型的“喀斯特地貌”。

爷爷，什么叫“喀斯特”呀？

喀斯特是欧洲斯洛文尼亚的一个地名。

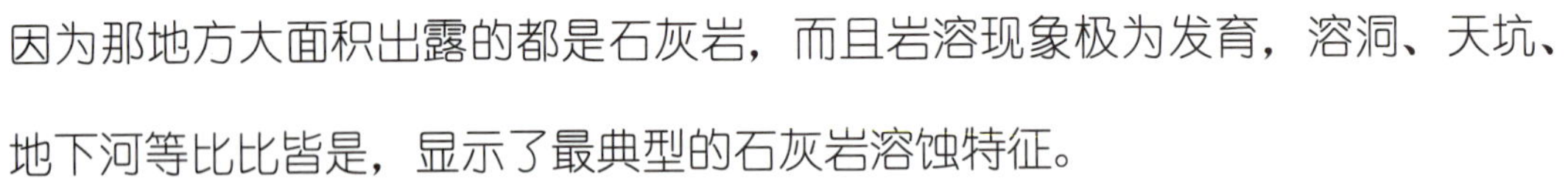

因为那地方大面积出露的都是石灰岩，而且岩溶现象极为发育，溶洞、天坑、地下河等比比皆是，显示了最典型的石灰岩溶蚀特征。

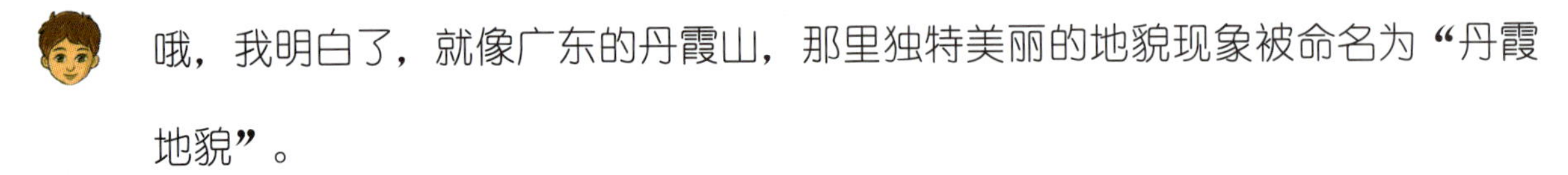

哦，我明白了，就像广东的丹霞山，那里独特美丽的地貌现象被命名为“丹霞地貌”。

是的。地质学上用“喀斯特”来命名碳酸盐岩的岩溶现象。“喀斯特”岩溶是一种以化学风化为主的“物理－化学风化”作用。

“化学风化”？就是岩石经过化学作用发生的风化？

是的。化学风化作用，就是岩石发生化学成分的改变或分解，使岩石硬度减弱、密度变小或体积发生变化，促使岩石分解后被剥蚀消溶。

您是说白云岩很容易发生化学风化，是吗？

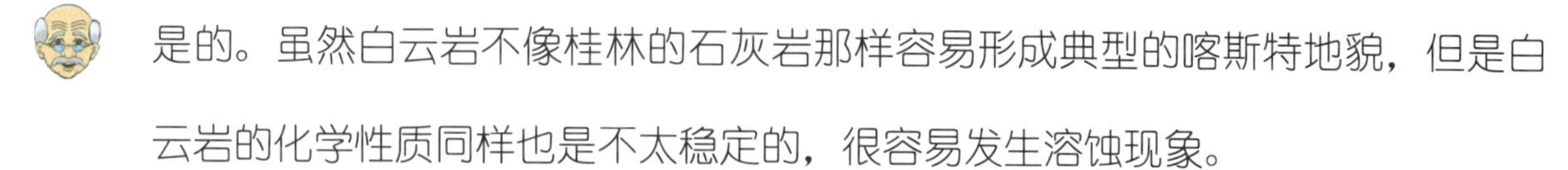

是的。虽然白云岩不像桂林的石灰岩那样容易形成典型的喀斯特地貌，但是白云岩的化学性质同样也是不太稳定的，很容易发生溶蚀现象。

从地貌上看，神农架的白云岩与桂林的石灰岩肯定还是有很大区别的。因为我看它们形成的地理地貌很不相同。但究竟是怎样不同呢？

小明，你观察得很仔细，也很会思考。你提的这个问题非常好。虽然白云岩和石灰岩一样都属于碳酸盐类岩石，但是，石灰岩的主要化学成分是碳酸钙，而白云岩的化学成分除了碳酸钙外，还含有碳酸镁。你知道碳酸钙和碳酸镁有些什么特点吗？

不知道。它们有很大的区别吗？

碳酸钙和碳酸镁在化学性质上有很大区别：碳酸镁的化学性质不如碳酸钙那样活跃。很多人只知道石灰岩和白云岩容易被溶蚀，发生岩溶现象。但是却不知道为什么它们会发生溶蚀。听着，我慢慢给你解释。

这个道理一定很深奥哦！

其实也并不难懂。首先，喀斯特溶蚀作

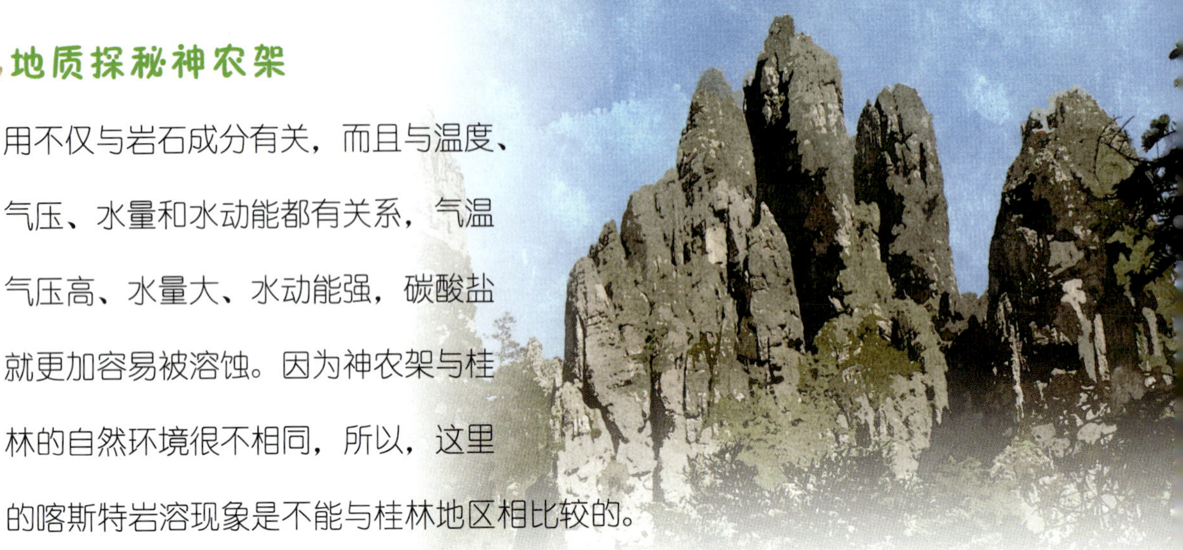

用不仅与岩石成分有关，而且与温度、气压、水量和水动能都有关系，气温气压高、水量大、水动能强，碳酸盐就更加容易被溶蚀。因为神农架与桂林的自然环境很不相同，所以，这里的喀斯特岩溶现象是不能与桂林地区相比较的。

怪不得这里看不到像桂林石林那样的景观。

另外，喀斯特溶蚀作用在很大程度上受到岩石成分的控制。刚才说过，石灰岩和白云岩抵抗风化溶蚀的能力有很大的差别。组成石灰岩的主要是碳酸钙，它们形成一种叫作方解石的矿物；构成白云岩的是碳酸钙和碳酸镁，它们形成的矿物叫作白云石。方解石的化学性质比白云石更加活跃。

也就是说，方解石比白云石更加容易被溶蚀，是吗？

是的。你知道地质学家在野外怎样鉴别石灰岩和白云岩吗？

他们用放大镜来看！

是的。石灰岩中常常可以看到细小的由碳酸钙组成的方解石的结晶，白云石的晶体往往不如方解石好。另外，他们还会用小刀刻一下石头……

为什么用小刀刻石头呀？

因为方解石的硬度很小，大大小于小刀的硬度。最后，他们还有一个鉴别方解石和白云石的“绝招”。

什么绝招呀？

他们会用随身带的5%的盐酸滴在岩石上，看石头会不会冒气泡。如果冒气泡，就可以断定岩石的主要成分是碳酸钙，也就是方解石，岩石应当属于石灰岩。

那为什么呀？

因为方解石，也就是碳酸钙遇到盐酸，就会发生剧烈的化学反应，形成氯化钙和二氧化碳。氯化钙是食盐的成分之一，它能溶解于水；二氧化碳则是气体，所以就会生成气泡。气泡冒得越剧烈，说明碳酸钙越多、越纯。地质学家就是这样来判别碳酸盐岩中碳酸钙含量高低的。

地质学家真聪明。那白云岩呢？

白云岩主要由白云石，即由碳酸钙和碳酸镁组成。碳酸镁不如碳酸钙那样容易与盐酸发生化学反应，滴盐酸时一般不产生气泡。但是地质学家也有办法，它们常常会刮下些岩石的粉末，然后滴盐酸在粉末上，若有气泡产生，就可能是碳酸镁，也就是白云岩。

地质学家还真有办法！可是……，可是为什么盐酸能使白云岩的粉末起泡呢？

从岩石上刮下些粉末再滴盐酸，是为了增加发生化学反应的接触面积，每一颗细小的粉末颗粒周围都能与盐酸接触，促使盐酸与碳酸镁进行化学反应。

哦，原来如此。但是，爷爷，我不明白，神农谷里面怎么会有盐酸呢？

其实，碳酸钙和碳酸镁遇到任何类型的酸都会发生化学反应，不一定非要是盐酸。举个例子说吧，如果你家的地板是大理岩的，就千万注意别把醋瓶子摔在地板上了。那会冒泡、毁坏你家的地板的。

大理岩？大理岩也是碳酸钙组成的吗？

是的。大理岩是由石灰岩在高温高压下形成的变质岩，它的主要成分还是跟石灰岩一样，是碳酸钙。碳酸钙遇到任何酸都会起化学反应。

哦。原来这样。但是神农谷里会有些什么酸呢？而且，怎么会有那么多的酸呢？

你知道“酸雨”吗？

听说过。但不太清楚酸雨到底是怎么回事。

酸雨是因为大气中的二氧化碳和工厂排放的二氧化硫，与自然界的水结合而生成的。它们的化学成分就是“碳酸”和“硫酸”。在自然界，二氧化碳和二氧化硫是无处不在的：动植物的呼吸、汽车排放的尾气、煤或其他燃料燃烧的过程等，都会产生大量的二氧化碳和二氧化硫。

我知道了！这些二氧化碳和二氧化硫上升到空中，与大气中的水汽结合，就形成了“酸雨”。……呀！那不是好可怕，我们淋到的雨都可能是酸雨？

是的。随着人类工业化的发展，酸雨的威胁愈来愈严重。神农谷的这些圆形石柱是经历了千百万年的溶蚀才形成的，这是化学风化长期作用的结果。

爷爷，那以后这些石柱是不是会变得更加圆滑些呢？

它们不仅会变得更加圆滑，而且会越来越短，最终会全部消失掉的。

不会吧？那多可惜呀！为什么呀？

你忘记了“地质思维”吗？我说的是千百万年以后。但是，如果我们注意保护环境，减少二氧化碳和二氧化硫的排放，这个过程将会大大地延长。

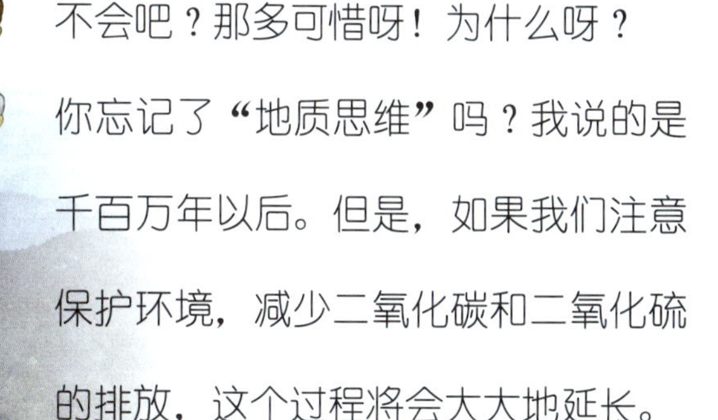

是的。我们应当尽量减少二氧化碳和二氧化硫的排放，保护我们的自然环境！

现在提倡“低碳减排”的生活就是这个道理。如果我们每一个人、每一个企业都真正做到了“低碳减排”，我们的环境就会大大改善，很多地质遗迹就可以保存更长的时间。所以，为了我们自己，更为了我们子孙万代，我们一定要用实际行动来保卫我们的地球！

问题：

1. 什么是白云岩？它与石灰岩有什么不同？
2. 什么是“喀斯特”？你能举出几个以喀斯特地貌为特征的旅游景点吗？
3. 根据你的理解，说说神农谷的石柱为什么比较圆滑。
4. “酸雨”是怎样形成的？它对大自然有什么影响？我们应该如何实行“低碳减排”？
5. 你能看懂下面的两个化学方程式吗？它们是喀斯特岩溶现象的理论基础：

   $CaCO_3 + H_2O + CO_2 = Ca(HCO_3)_2$ ，$Ca(HCO_3)_2 = CaCO_3\downarrow + H_2O + CO_2\uparrow$

14

知识老人带着小明和汪汪乘上飞毯继续旅行。飞毯沿着公路向西缓慢地飞翔着。突然，路边的石块上赫然出现了好几个庞大的脚印，小明不觉惊叫起来……

呀！爷爷，是谁在石头上留下了那么大的脚印呀？是神农架的“野人”吗？

不是的。这是神农谷的另外一个入口。但是这些脚印确实与神农架“野人”的传说有关。

那是怎么回事呢？

那些有脚印的石块是人们安放在那里的，脚印也是石匠雕刻出来的。

干嘛要刻些大脚印在这里呀？

你一定听说过神农架“野人”的传说吧。2011 年，中国和美国联合拍摄了一部叫作《大脚印》的传奇惊悚影片，主要的外景拍摄地就是神农谷。

这个电影肯定好看。一定是一个非常有趣的故事。

是的。影片以神农架著名的“野人”传说为基础，讲述了一个神奇动人的故事，展现了当地独特古朴的民俗和神农架人勤劳、勇敢、善良的民风。所以，地质公园把神农谷的这个入口更名为“大脚印峡谷”。

怪不得在这里雕刻了这么多大脚印。爷爷，您说真的有“野人”吗？

全世界各地都有关于“野人”的传说和记载。比如喜马拉雅山南麓的“耶提”、西伯利亚的“丘丘纳”、非洲的“切莫斯特”、澳洲的“约韦”等，美洲也曾有“大脚怪——沙斯夸支”的传说。几乎全球各个大陆都曾经有过关于“野人”踪迹的传说。

那到底有没有野人呢？

很多地方关于“野人”的传说都随着时间的流逝而渐渐地销声匿迹了，比如在云南、贵州和广西热带原始森林区曾经也有过“野人”目击报道，但都没有“后话”了。唯独神农架一直都不断地有关于“野人”的消息传来。

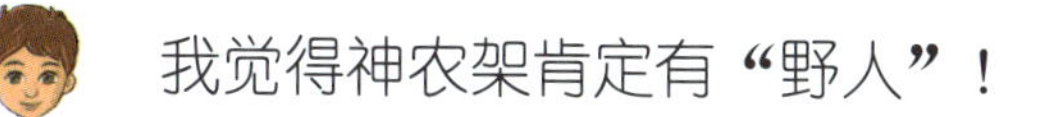

我觉得神农架肯定有“野人”！

很多迷恋“野人”的探索者来到神农架进行考察。中国科学院和一些民间组织，比如中国野生动植物考查协会、奇异珍稀动物专业委员会等都多次组织了对神农架“野人”的专门考察。

他们找到“野人”了吗？

很遗憾，没有。他们只发现了一些怀疑是“野人”的睡窝、脚印、毛发、粪便等间接的证据。但从未捕获或拍摄到一个真

正的活着的“野人”。因此无法证明“野人”的真实存在。

爷爷，我真的好想看看真实的“野人”呀！。

从理论上看，进化是一个完整连续的链条。如果确实有“野人”存在，从科学的角度，在目前人类活动范围空前广大的情况下，就一定会找到一些“野人”存在的实物证据。

什么样的东西才叫“实物证据”呢？

最能说明问题的实物证据就是它们的尸骨遗骸呀。根据自然规律，一个物种的生存和延续，必须具备两个基本条件：首先需要有一定的地域范围供它们生活和活动，而且在这个范围内要有充足的自然资源来支撑它们的生存；其次是这个物种必须是一个有一定数量的种群，这样它们才能得以繁衍延续。

可不是吗。没有一定的资源和食物，“野人”会饿死的；没有一定的数量，他们也会很快地绝灭的。

如果神农架的“野人”是具有一定数量的种群，就不可能不留下一定数量的确切实物证据。

神农架找到了这些实物证据吗？

很遗憾，目前还没有发现确凿的证据。不过，如今有没有“野人”这个问题已经不重要了，重要的是它已经在神农架形成了一种很独特的“野人文化”，而且对经济的发展起到了积极的促进作用。

可不是吗，只要一提到神农架，人们就会想到“野人”。

是呀，这就是神农架“野人文化”的迷人之处。它能引发人们的想象和思索，激发人们进行科学探索的兴趣。

是的。我总在想：如果“野人”还在神农架生活，他们每天会干些什么事呢？他们吃什么？他们住在哪里？他们会唱歌吗？他们碰见野猪黑熊会怎么办？如果我碰巧见到他们，尤其是碰到小“野人”，他会和我一起玩吗？他们听得懂我讲的话吗？…… 所以，我会更加仔细地观察神农架的自然环境，为他们找一个理想的生活地点，真希望能找到他们！

“野人文化”就是这样激发人们的想象和探索，很多优美的传说就从它而来，使得神农架更加魅力无穷。在神农架世界地质公园天燕景区有一个“人形动物博物馆”，我倒觉得这个“人形动物”的命名还比较科学确切。

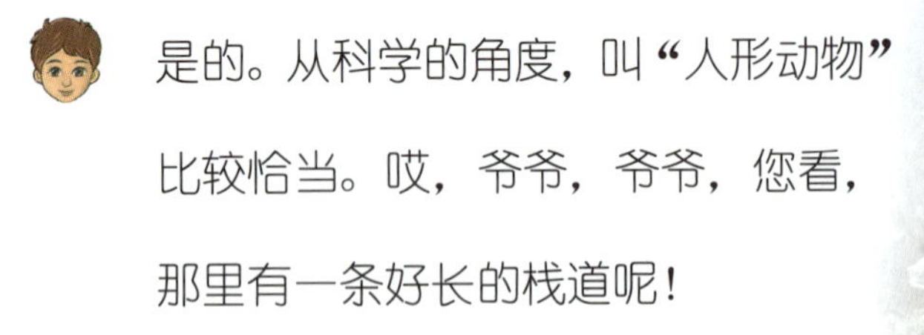

是的。从科学的角度，叫“人形动物”比较恰当。哎，爷爷，爷爷，您看，那里有一条好长的栈道呢！

那就是神农谷的旅游栈道。

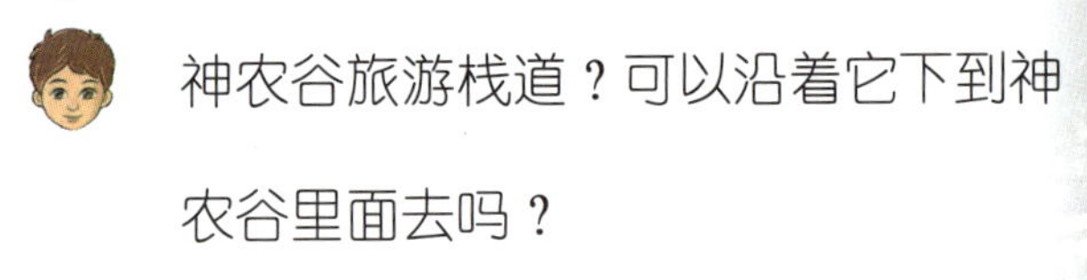

神农谷旅游栈道？可以沿着它下到神农谷里面去吗？

是的。为了让人们充分领略“大脚印峡谷”，也就是神农谷壮丽的自然风貌，地质公园修建了一条约 4km 长的“U”形观赏栈道。游客可以从“大脚印峡谷”的入口处，沿着栈道蜿蜒下到谷底，穿行于峡谷的原始森林之中，然后再回旋漫步攀升，从刚才我们停留的那个观景台附近回到山脊上来。

哇！这可是个相当浩大的工程呀！

确实是一个浩大的工程。入口处观景台的海拔高程为 2820m，从这里下到栈道最低处的垂直高差达到 400m。

400m！ 100 多层楼呀。太了不起了！

神农架世界地质公园为了让游客欣赏到山谷中绝妙的风光，克服了重重困难，耗费了大量人力、物力和财力，修建了这条栈道。你看，这条依山就势、蜿蜒曲折的栈道，就好像一条巨龙穿行于神农谷灵山秀水之中，逶迤盘桓于绝壁峰峦之间，与大自然融为一体，构成了一幅百看不厌的秀丽山水画。

可是爷爷，您看，下到神农谷里去的游客好像并不多呀？可能他们以为神农谷里没什么好玩的，光是上上下下地走路没意思吧？

确实如此。很多人的确不知道神农谷里到底有些什么可欣赏的。而且，下去走一趟起码需要3个小时，大多数游客往往没有准备，也没有事先安排好游神农谷的时间。

作为游客，来神农架一趟而没能下到神农谷去看看，实在是一种遗憾。

可不是吗。神农谷是由于张性断裂和垂直节理遭受长期风化、冰劈、剥蚀、坍塌以及喀斯特溶蚀等多种地质作用形成的。它以陡峭的岩壁和大量的峰林、石笋为特征，树林密布，气象万千、一步一景，可谓人间仙境，是外生地质营力造景的优美典型，不仅具有科学考察的意义，而且富有极高的美学景观价值。

爷爷，您快看，那栈道夹在窄窄的峡谷之间，好像天梯一样！

穿行于峡谷之间，能近距离观看谷中的美景，的确是一种美好的享受！你再看看那边，石峰林立、景象壮美，真是优美绝伦。

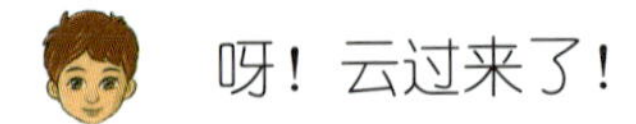

呀！云过来了！

神农谷的云景是旅游的亮点之一，流动的云雾更加增添了山谷的妩媚。在风云变幻的雾霭之中，挺拔多姿的峰林石笋参差不齐，错落有致；形态各异的陡崖石柱时隐时现，变化无穷。行走在蜿蜒盘旋的木质栈道上，可以尽情欣赏这一步一变、多姿多彩的自然景观。

爷爷，那我们下去沿着栈道走一走吧？

咱们今天时间不够了。下次来再走吧。我们还要去“板壁岩”呢。

“板壁岩”？那是个什么地方？

那是神农架最受游客欢迎的景点之一。那里的风景又是另外一种特色，到时候你一看就知道了。

好吧。那咱们快点儿去“板壁岩”吧！

问题：

1. 从科学的角度看，你觉得神农架真的有“野人”吗？
2. 动物种群生存和繁衍的基本条件是什么？
3. 为什么说云景是神农谷的一个“亮点”？
4. 你愿意顺着栈道下到神农谷去实地考察一下吗？为什么？
5. 你将会如何向你的朋友们介绍神农谷？

15

飞毯继续沿着公路缓缓飞翔。这条从神农谷向西一直延伸到太子垭的公路，基本上都是沿着山脊修建的。这里的山坡上林木稀少，高山草甸连绵不断，一丛丛的箭竹林使得山坡显得斑驳杂乱。虽然这里的自然景观相当单调，但沿途都可以居高临下，极目远眺。重重叠叠的群山气势恢弘，令人心旷神怡。山路上，载着游客的环保旅游车来来回回行驶着；不少背包客时而快步行走，时而驻足眺望，时不时还掏出照相机拍照留影，很显然，大家都陶醉在这壮丽的群山景致之中。

爷爷，这山顶上怎么没什么树了，都是些草呀？

我们现在看到的是高山草甸。神农架地区因为地理高差大，气温变化也大，所以出现了非常有趣的植被垂直分带现象……

什么叫“植被垂直分带”呀？

根据科学家的研究：海拔高度每上升 600m，植物种群就会发生较明显的变化。通常海拔 800~1200m 地带分布的是常绿阔叶林和落叶阔叶林。

那我们现在看到的草甸海拔高度是多少呢？

现在的海拔高度大约为 2400m。这里的植被以高山草甸和箭竹林为代表。再往西到太子垭，那里的海拔高度达到了 2600m 左右，除了草甸和箭竹之外，还出现了以巴山冷杉为代表的原始针叶林。所以高山草甸、箭竹和冷杉共同构成了神农架高寒地带植被的典型代表。

啊，山顶上还有原始森林呀？

是呀。神农架世界地质公园在太子垭原始冷杉林里面，修建了一条将近 2km 长的步行栈道和 6 个旅游观景台，人们能在这片静谧的原始森林中，尽情地欣赏树林和群山的美景，零距离地享受森林中独特的清新空气和花香鸟语。

“花香鸟语”？爷爷，您说得不对吧？山这么高、这么冷，原始森林里怎么还会有花香呢？

我说的没错。在冷杉、箭竹和草甸共同构成的高山植被中，还夹杂生长了不少高山海棠和高山杜鹃呢！

海棠和杜鹃？

可不是吗。海棠属蔷薇科苹果属，花朵红嫩娇妍，是一种非常优美的观赏植物。在太子垭的冷杉林中生长着好多海棠树，每到开花时节，小小的红花默默地装点着绿色的森林，有一首诗歌颂了海棠这种低调内敛的个性和不张扬、不与人攀比的风格，诗曰：“枝间新绿一重重，小蕾深藏数点红，爱惜芳心莫轻吐，且教桃李闹春风。”

哇，海棠花好有品位呀！那杜鹃呢？

神农架山区生长着不少的杜鹃。杜鹃的适应能力很强，既耐寒也耐热，四季常绿，被称为“绿色世界里的皇族”。据统计，在湖北省 58 种杜鹃中神农架生长了 35 种。它们分布在海拔 1800~3100m 一带的冷杉林或草甸山坡上。春天杜鹃开花的季节，山坡上一片片粉红亮白，景色美不胜收。

太令人向往了！什么季节来神农架看杜鹃花最好呢？我真想亲眼看看杜鹃花盛开的美景。

五月份是神农架看杜鹃花的最佳季节。到那时，在春天新绿的山坡上，红的、白的杜鹃花一簇簇、一片片，如火焰，似白云，将群山渲染得五彩斑斓，充满生机，美得令人窒息。

青翠的群山夹杂着一丛丛白色和粉红色的杜鹃花，这意境真像诗一样美！

杜鹃有“花中西施”之美誉。白居易有诗赞它：“回看桃李都无色，映得芙蓉不是花。”

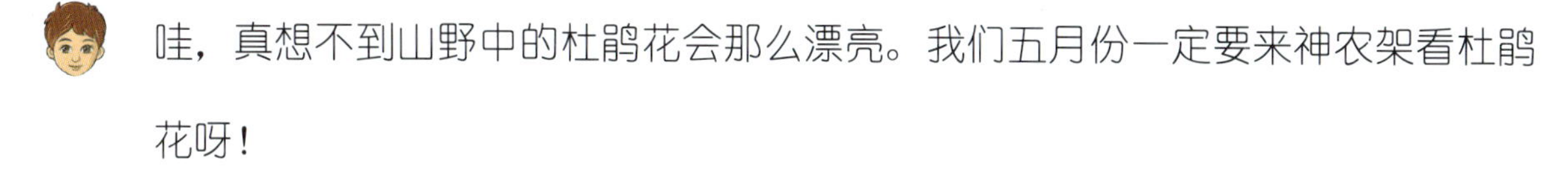

哇，真想不到山野中的杜鹃花会那么漂亮。我们五月份一定要来神农架看杜鹃花呀！

好的。春天一到，神农架就变成一个五彩斑斓的花花世界！

这里的宝贝植物还真不少呢！爷爷，还有什么新奇的植物呢？

呵，多的是呢！再给你介绍一种树吧：这里生长着一种比较罕见的红桦树。

红桦树？我经常听说白桦树，还有些歌里也唱到白桦树。好像还没听说过红桦树呢。

红桦树的树皮是粉红色的，它在绿树丛中显得格外醒目。它的树皮每年都会脱落一次。脱下的桦树皮很光滑，山里的年轻人爱拿红桦树皮来写情书传递爱情。因此在神农架，红桦树还有一个好听的名字，叫作“爱情树”！

爱情树，好浪漫呀！

红桦树还能分泌出一种汁液，叫红桦汁，味道非常甜美，含有十几种人体必需的氨基酸。过去山里人经常在树上钻孔取汁，制成非常珍贵的饮料。但自从神农架实行天然林保护措施以来，已经不允许采集红桦汁了。

地质探秘神农架

神农架山林的宝贝真多啊！

是呀。神农架保存着完整的亚热带森林生态系统，森林覆盖面积高达96%。现在已经发现高等植物3400多种，其中本地特有植物116种，包括33种新种植物；还有1800多种药用植物，其中有不少是名贵中草药，例如“头顶一颗珠”“江边一碗水”“文王一枝笔”和“七叶一枝花”等。相传炎帝神农曾在这里采药，为民治病，还编纂了著名的医书《神农本草经》呢。

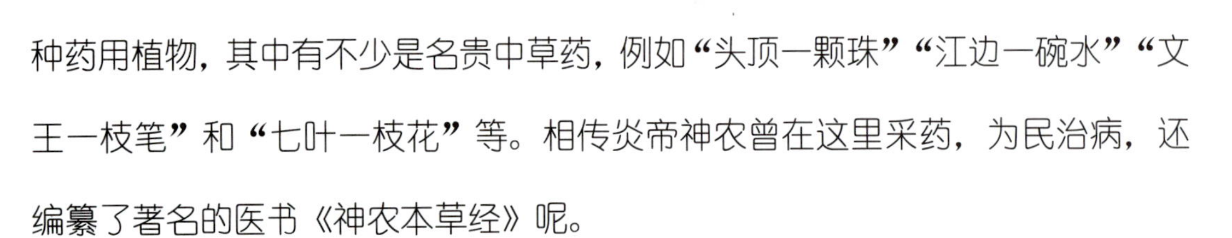

怪不得我看到书上把神农架称为“百草园”“绿色宝库”呢。

这些称谓都非常贴切。神农架优越的地理位置、茂密的森林、丰富的植被，确实是极为罕见的。

希望神农架能始终保持它的美丽和绿色。

神农架林区曾经是以采伐木材为目的而建立起来的我国第一个以林区命名的行政区。但是，2000年以后，神农架最终全面停止了木材采伐，而改为森林保护、植树造林。地质公园建立后，更加大了森林资源的保护力度。

多么巨大的变迁呀！

是的。神农架的历史见证了人们对大自然认识不断深化的发展过程。如今，人们越来越重视对山林自然环境的保护。相信神农架一定会永远保持绿色，越变越美。

问题：

1. 什么是植被垂直分带现象？
2. 神农架高寒地带的植被以什么植物为代表？有些什么特点？
3. 太子垭的冷杉原始森林中为什么还会有“花香鸟语”？
4. 你能介绍一下神农架的红桦树吗？
5. 神农架林区从砍伐森林木材转变为植树造林，森林保护意味着什么？

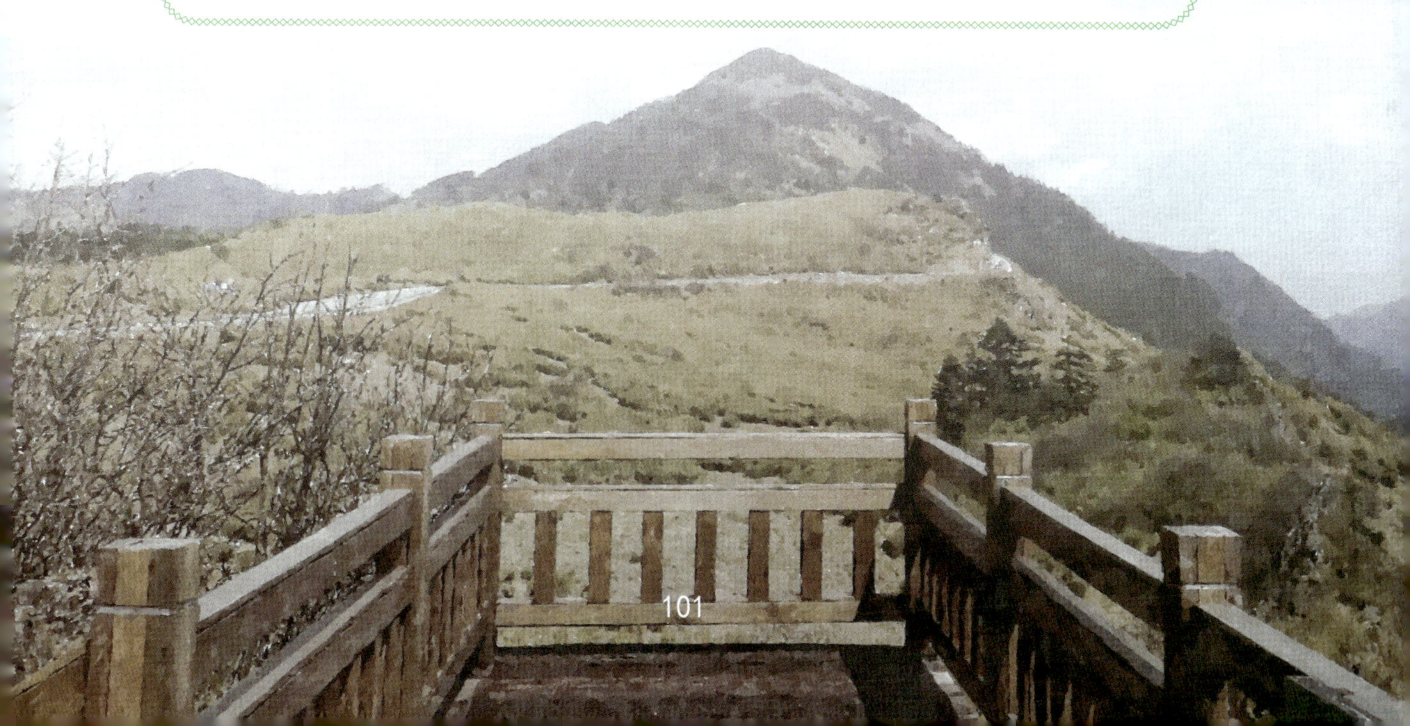

16

谈到森林和环境的保护，知识老人和小明都陷入了沉思。飞毯平稳地掠过青翠欲滴的群山。山坡被连绵的草甸和箭竹覆盖着，在灿烂的阳光下反射着养眼的绿色；微风轻轻地吹拂着大地，万籁静寂。远处传来了断断续续的鸟叫声。在蓝天白云之间，有两只雄鹰在翱翔。突然，随着一声呼啸，其中一只老鹰箭一般地俯冲下来，它掠过箭竹林，消失在一个小山包后面。汪汪盯着老鹰消失的地方，大声地吼叫起来。

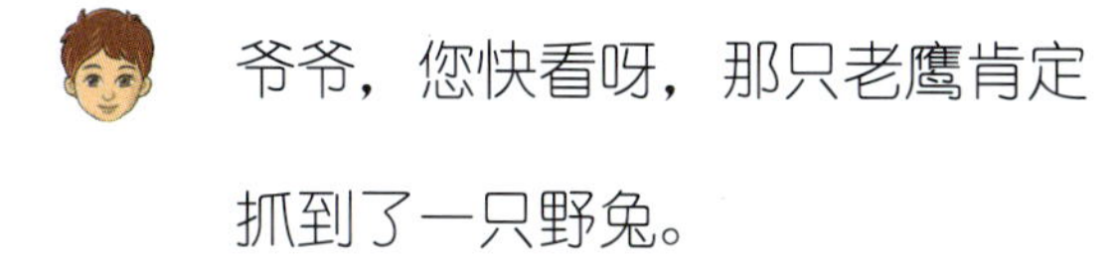

爷爷，您快看呀，那只老鹰肯定抓到了一只野兔。

是的。山里的小动物不少。

爷爷，除了金丝猴之外，神农架还有些什么有趣的动物呢？

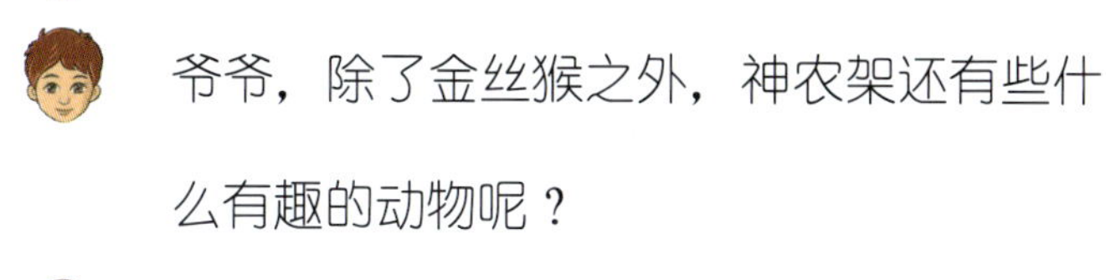

神农架不仅是一个植物和森林的宝库，也是各种野生动物的家园。

是呀，有森林就肯定会有各种动物。

神农架具有北半球中纬度地区极为罕见的生物多样性，早在 1990 年就被联合国教科文组织接纳为“人和生物圈”保护计划的成员。

什么是“人和生物圈”保护计划呀？

哦，这是联合国针对全球日益严峻的人口、资源和环境的挑战，以保护人类赖以生存的地球环境为宗旨，而制订的一个研究计划，简称 MAB（Man and the Biosphere）。

哦，这个 MAB 计划有什么作用呢？

它的主要目的，在于协调人与生物圈的关系。“生物圈保护区”是 MAB 的核心部分，具有保护、可持续发展，及提供科研、教学、培训、监测基地等多种功能。

神农架被联合国教科文组织确定为MAB的保护区，多年来一直发挥着这样的功能。

太了不起了！神农架真是一块宝地。

可不是吗，神农架被誉为“天然动物园”“物种基因库”，这里有79种国家重点保护的野生动物。其中国家一级保护动物有金丝猴、金钱豹、白鹳、金雕；74种国家二级保护动物包括了金猫、林麝、黄喉貂、秃鹫、大灵猫、大鲵等。

大鲵？就是娃娃鱼吧？

是呀。大鲵是世界上现存最大的，也是最珍贵的两栖动物。因为它的叫声很像婴儿的哭声，所以人们又把它叫作“娃娃鱼”。

我在动物园里见过。但是没看见过野生的。

神农架很多山沟溪流中都可以见到它们的身影。地质公园还专门建立了一个大鲵养殖研

究中心，饲养了两百多条大鲵呢。

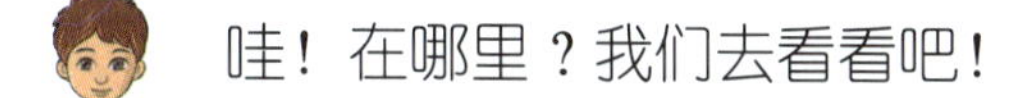

哇！在哪里？我们去看看吧！

在官门山景区。我们稍后会到那里去参观的。

太好了！真想看看野生的娃娃鱼！

复杂的气候和崎岖的地形使神农架成为一座野生动物的乐园，繁育了众多的珍稀物种，因此神农架已经成为濒危动物的避难所。

我想也是的。您听，虫鸣鸟叫，到处都充满了生机。

小明，你知道吗，神农架有野生哺乳类、鸟类、鱼类和两栖类动物共计 600 多种，其中受国家保护的有益或者有重要经济、科学研究价值的野生动物多达 259 种，这里的昆虫类更高达 4000 多种。

天啊！这么多！太神奇了！

更有趣的是，这里还出现了 30 多种白化动物。

白化动物？

是呀。有白林麝、白鬣羚、白蛇、白刺猬等。

白化动物是像北极熊、白天鹅、北极狐那样的动物吗？

不是的。北极熊、白天鹅、白鹭等并不是白化动物。它们的白色是显性基因的正常表现。而白化动物则是一对隐性基因纯合子的产物。

啊，说到基因，这可是个很复杂的科学问题，要弄懂得下功夫学习很多的资料。

但是这个问题确实很有趣，我回去一定要好好查查资料，研究一下。

爷爷就喜欢你这种爱钻研的优点！基因造成动物颜色变异的原因是十分复杂的。神农架出现较多的白化动物，除了基因之外，肯定与这里相对闭塞、局部隔离和多样化的地理生态环境有密切关系。

我想也是的。真是太神奇了！神农架确实是动植物的天堂！

岂止是天堂。如今，在地球环境严重退化的情况下，可以说神农架就像是圣经中所说的“方舟”一样，是生物物种的最后避难所。因此，保护神农架的自然环境具有极其重要的全球意义。

问题有那么严重吗？爷爷。

是的。情况确实很严重。就比如说酿造蜂蜜的“中华小蜜蜂”吧，这种曾经在我国到处都可以见到的勤劳可爱的小生灵，目前在大多数地方都已经绝灭了，只能在神农架和极其有限的几个相对闭塞的地方才能见到。

为什么“中华小蜜蜂”在很多地方都绝灭了呢？

因为引进的意大利蜂侵占了中华小蜜蜂的生存空间，排挤了我国土生土长的小蜜蜂。当然还有环境污染和气候变化等因素的影响。

为什么要引进意大利蜂呢？把它们赶走不行吗？

当初引进新的蜂种，是因为意大利蜂的产蜜量较高。但是，它们怕冷，不能适应高寒的气候。中华小蜜蜂却不怕冷，在气温为 4℃ 时就能出外采蜜了。神农架高山上的植物全靠中华小蜜蜂传授花粉才能得以繁衍。所以，如果没有了中华小蜜蜂，要不了多少年，神农架就将变成光秃秃的荒山野岭！

哇！那么严重呀。没想到小小的蜜蜂对自然环境的作用竟然这么大！

要知道，整个自然界是一个相辅相成、环环相扣的有机体。任何一个环节的缺失，都将造成难以想象的后果，甚至会引发一系列不可逆转的恶性变化，造成局部环境的崩溃。所以，保护自然环境、挽救濒临灭绝的物种，是地球人目前面临的最严峻的挑战之一。

看来我们人类确实面临着相当紧迫的任务，保护动植物和我们的地球环境刻不容缓。

为了我们的子孙后代，必须保护地球上所有的物种和它们赖以生存的环境。

问题：

1. 什么是“人和生物圈（MAB）”保护计划？它有什么作用？
2. 为什么说神农架是“天然动物园”“物种基因库”“濒危动物的避难所”？
3. 你见过白化动物吗？你觉得神农架为什么会出现较多的白化动物？
4. 为什么知识老人说“如果没有了中华小蜜蜂，要不了多少年，神农架就将变成光秃秃的荒山野岭”？
5. 你觉得应该如何保护我们的动植物和自然生态环境？我们每个人能做些什么事情呢？

17

飞毯继续沿着山脊上的公路飞行。缓缓起伏的山坡被高山草甸覆盖着，一丛丛比草甸长得高的箭竹，使本来应当平缓圆滑的山坡显得高低不一、有点紊乱斑杂；在箭竹连接成片的地方，山风吹拂的箭竹像大海一样起伏翻动。远处山坡上出露了一些奇形怪状的深灰色石芽。知识老人带着小明和汪汪在板壁岩宽大的停车场上停了下来，向着那块写着“板壁岩”三个大字的石头标牌走去。

爷爷，这个停车场可真大呀！

因为来这里游玩的人最多。尤其是节假日期间，这么大的停车场有时都还不够用呢。

为什么人们都喜欢来板壁岩玩呢？

因为这里有各种奇形怪状的象形石，有的像鸟、有的像兽、有的像人，栩栩如生、趣味盎然，就如同一座雕塑公园。

爷爷，为什么把这个地方叫作“板壁岩”呀？

你仔细地看看，这附近岩石有些什么特点？

让我看看。……哎，很多石头好像是从地里长出来的，一片片地直立着。爷爷，怎么会是这样呀？

这地方的岩石受到挤压变形，岩石呈现薄板状，沿一定的方向延伸，几乎直立地出露地表。也就是说，岩层的产状几乎是垂直的。当它们遭受风化剥蚀后，参差不齐，就像是从高山草甸中长出来的一丛丛石芽；有的高高地耸立着，看起来就像墙壁门板一般。因此人们给它起了一个相当形象的名称“板壁岩”。

哦，原来是这样的呀。……可是，爷爷，您快看，阶梯顶端的那一大块石头，像匍匐在地上的巨兽，它和旁边的那些石头都不一样，它的产状并不是直立的呀！

你很善于观察，并能发现问题。这个问题问得很好。但是，我想你只要再仔细观察一下，思考一下，就一定能自己来解释了。

让我看看。……哦！我看出来了，这一大块石头好像并不是在原地生长的，可能是从什么地方滚过来的。

对了！地质学把这种不是原地生长的石头叫“滚石”，而把原地出露的石头叫作“基岩”。你能说说滚石与基岩的区别吗？

嗯，我觉得基岩和它附近的滚石可能都属于同一种石头。但是，基岩保持着岩石原始的产出状态，所以我们可以从基岩看出岩层是向哪个方向延伸的。

完全正确！滚石只能为我们提供石头本身的性质和特征，比如它是什么种类的岩石、主要含些什么矿物等，但是无法从滚石追溯岩层的产出状态，也就是岩层是向哪个方向倾斜、朝哪个方向延伸的等。

哎呀！爷爷，您快看，那块大石头摇摇欲坠，马上就要滑下来了。谁把它放在那上面的呀？多危险呀。

那可不是什么人放在那里的。

难道又是地球的外营力干的？！

对呀！你仔细看看，好好琢磨一下，它应该是怎么形成呢。

嗯。这考不倒我。……您看，上面那块摇摇欲坠的石头，与它下面那块大石座的产状基本是一致的，原先肯定是连在一起的。哦，我明白了。因为风化剥蚀作用是沿着岩石层面和垂直于层面的裂缝进行的，风化剥蚀把这块石头周边都掏空了，虽然它还留在原来的位置，但它与下面的大石座已经分割开来了。

不错，你分析得很好！在野外每看到一种奇特的地质现象，就应当仔细想想、好好分析一下，找出这种现象产生的原因。这会给我们的野外旅行增加很多很多的乐趣。

爷爷，您说这块石头会滑下来吗？

你说呢？如果用“地质思维”的方式！

哦，我懂了！总有一天它会滑下来的。但是，如果我们努力实行“低碳减排”，保护环境，这一天就会大大地推迟。

说得很好！小明，你快看，远处那石头像什么？

像个鸡头！

对了。人们给它起了个好听的名字，叫作“金鸡报晓”。它就像一只雄鸡，英姿焕发、引吭高歌，简直惟妙惟肖。

您快看那边呀！爷爷，好像有一只石头小鸟呢！

是呀，像是一只刚刚孵出、嗷嗷待哺的小鸟，所以叫作“雏凤待哺”。

哇！这些石雕真有趣。

板壁岩是一个天然的雕塑公园。大自然在这里创造了许许多多精美的“作品”。人们给它们起的名字也惟妙惟肖，恰如其分，比如“悟空护驾”“孔雀石”“美女照镜”“生命之根”“骏马奔腾”等。

大自然真是个能工巧匠，它的创造简直太奇妙了。

大自然是无所不能的。我们今天看到的这崇山峻岭、悬崖陡壁、山岗沟壑都是大自然内

外营力作用造就的。更加奇妙的是，它们始终都处在一种变化的、动态的平衡之中。就比如刚才你问到那块看起来很危险、好像要滑落下来的大石块，总有一天它会真的滑下来。到那时，板壁岩的景观就会发生改变，说不定还会造就一个新的、更加美丽的雕塑呢！

那就太好了！可是，爷爷，我总是想不通，照说板壁岩这里的石头风化以后，都应该像是石碑、门板那样直立，或者像一丛丛竹笋那样呀。可是您看，这里的石头奇形怪状，尖的、平的、歪的、斜的，什么样的都有。这是为什么呢？

你真会提问题，这样很好，小明！说明你一边在观察，一边在思考。

凡事都要问一个为什么嘛，嘻嘻。为什么这里的石头会有这么多的变化呢？

这主要有两方面的原因。首先，岩石在漫长地质历史发展中会天然产生很多的裂隙，也就是地质学所说的“节理”。

是的。您说过，岩石中的天然裂缝就叫作节理。

节理在岩石中的发育是很不均匀的：有的地方特别密集，有些地方则比较稀疏。当风化剥蚀发生的时候，节理密集的部分很快就会被剥蚀掉。由于节理的延伸很不规则，风化剥蚀后保留下来的岩石就变得奇形怪状了。

哦。怪不得我们看见石头上布满了大大小小、长长短短、深浅不一的沟纹，风化剥蚀就是顺着这些节理进行的。

其次，岩石本身的成分也是不均匀的，有的地方或者层位方解石含量多些，有些地方白云石多些。方解石和白云石抗拒风化的能力有很大的差别。

哦，我知道了，方解石多的地方更容易被风化剥蚀掉，形成凹下去的坑洞或凹槽。

是的。所以风化后的石头常常是凸凹不平整的，地质学上把这种现象叫作“差异风化”。

“差异风化”？那就是说：由于不同的岩石抵抗风化的能力不同，所以岩石风化后，它的表面就会凸凹不平。

是的。你归纳得很好。这里的岩石中还常常含有硅质薄层、硅质条带、硅质团块等，它们就更难被风化了，常常凸起在岩石的表面。

是哦。难怪我看见好多凸凹相间的层理和形状不规则的凸起呢。

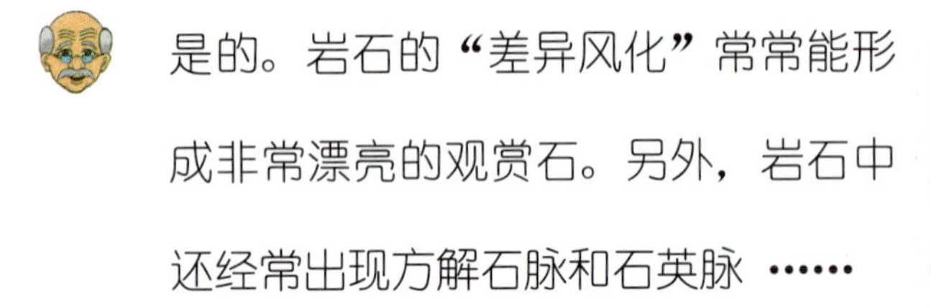

是的。岩石的“差异风化”常常能形成非常漂亮的观赏石。另外，岩石中还经常出现方解石脉和石英脉……

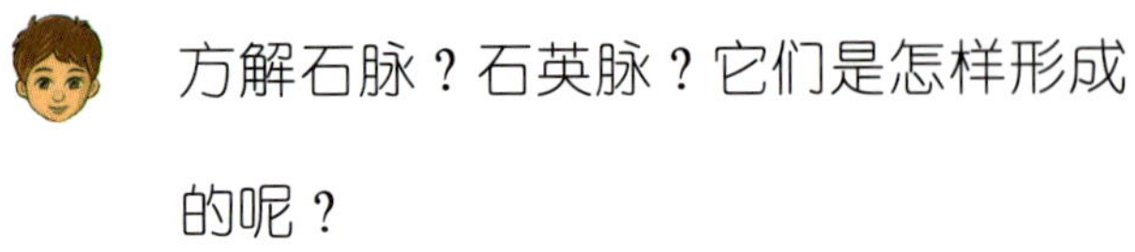

方解石脉？石英脉？它们是怎样形成的呢？

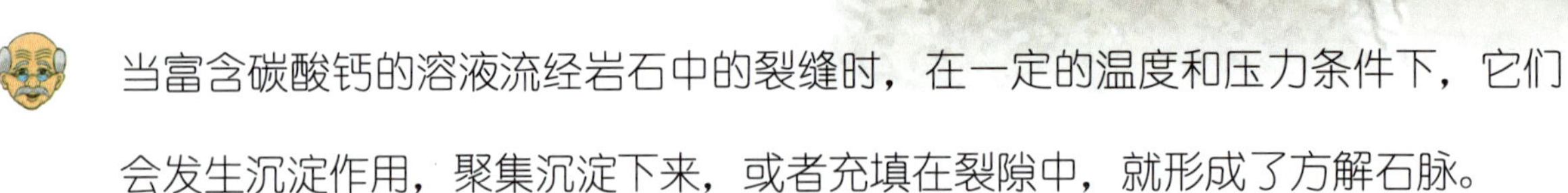

当富含碳酸钙的溶液流经岩石中的裂缝时，在一定的温度和压力条件下，它们会发生沉淀作用，聚集沉淀下来，或者充填在裂隙中，就形成了方解石脉。

方解石脉应当也是特别容易遭受风化剥蚀的，难怪石头上有些裂缝毫无规则，奇形怪状的。那石英脉呢？石英是从哪儿来的呢？

石英脉的成分主要是二氧化硅。它有两种形成的方式：一种是含有二氧化硅的地下水，在合适的温度、压力下沉淀在岩石的裂隙中形成的，与方解石脉形成的方式一样；另一种是含有二氧化硅的热液和汽水，从地底深处顺着岩石的裂隙上升到达地表岩石中……

您说的“含二氧化硅的热液和汽水从地底深处”是什么意思呀？

哦，它们往往与地底的岩浆活动有关。

哇！岩浆？这里有岩浆吗？那不是很危险？

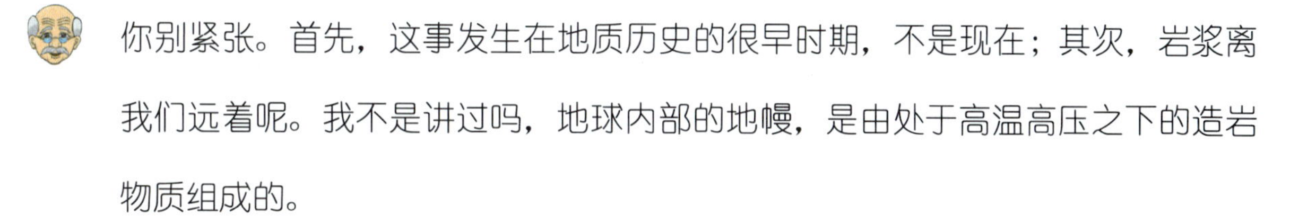

你别紧张。首先，这事发生在地质历史的很早时期，不是现在；其次，岩浆离我们远着呢。我不是讲过吗，地球内部的地幔，是由处于高温高压之下的造岩物质组成的。

什么是“造岩物质”呀？

“造岩物质”就是通常形成岩石的物质，主要含一些岩石中常见的矿物，例如长石、石英、角闪石、黑云母等。

哦，我知道，这些矿物常常出现在花岗岩中。

是的。地幔顶部出现的软流层富集了许多这样的矿物，同时软流层还富集了大量的放射性元素，它们发生蜕变，释放热量，使岩石在高温下软化，并局部熔融形成岩浆。在这个过程中，会产生大量含金属元素和非金属元素的热液和热气。

那这些热液、热气怎么会跑到板壁岩来的呢？

那是因为地质构造变动，使地球的岩石圈遍布断层和裂隙。这些含有各种元素的热液、热气顺着岩石中的裂缝不断向上运移，当温度和化学环境发生改变的时候，这些金属或非金属元素就沉淀下来，形成有用的矿产，例如铜、铁、磷、硼、氟、硒等。地质学上把这种原因形成的矿产通称为“热液矿床”。

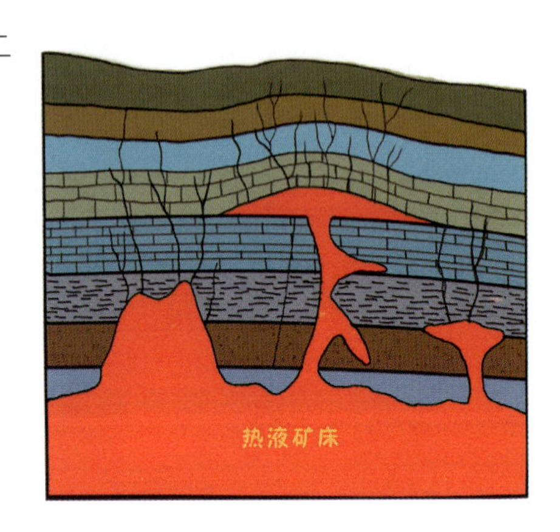

哦，我懂了！二氧化硅就是这样随着热液、热气来到板壁岩的，是吗？

是的。由于温度的降低和化学环境的改变，二氧化硅附着在岩石裂隙的表面，发生沉淀或结晶，甚至充填整个裂隙，就形成了石英脉。你知道吗，在这个过程中，有时还会形成非常美丽的石英晶洞。在特殊条件下，晶洞中巨大的石英晶体可达数米之长呢。

哇，那不是像童话世界一样美丽？

是呀，我们的地球到处都有童话般的美景。

咱们的地球真神奇！爷爷，我还想问一个问题：石英脉与硅质层的主要成分都是二氧化硅，那它们有什么区别呢？

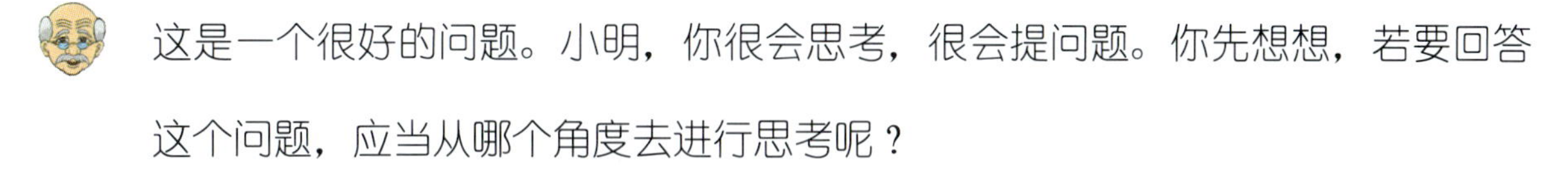

这是一个很好的问题。小明，你很会思考，很会提问题。你先想想，若要回答这个问题，应当从哪个角度去进行思考呢？

让我想想。……嗯，一个是脉，一个是层，……哦！我知道了，它们形成的原因和过程是不相同的！

你真是一个非常聪明的孩子！正是由于它们形成的方式和机制不同，因此造就了它们形态上的差异：硅质层是含硅质的沉积物沉淀形成的一个薄层，因此它与岩石层理的产状是一致的；而石英脉是沿着岩石裂隙充填沉淀形成的，所以它的形态是不规则的。

爷爷，有关板壁岩让我学到了很多东西，使我对大自然的变化和各种地质作用有了更深刻的理解。我现在敢说：板壁岩的任何一个象形雕塑，我都能说出它是如何形成的。

哦。看来你的进步真的很大！那你能说说这个“生命之根”是怎么形成的吗？

很容易。您看，这个“生命之根”就像根柱子一样竖立在这里。其实，它和两边陡壁上的岩石是一样的，而且原先它们应当是连成一体的。

但是，由于它的四周出现了许许多多垂直的节理，也就是裂隙裂缝。风化剥蚀作用沿着这些节理进行；既有化学风化，又有物理风化。破碎的岩石碎屑被流水带走，裂缝越掏越宽，加上坍塌，最后就剩下了这根柱子了。……我相信有一天，

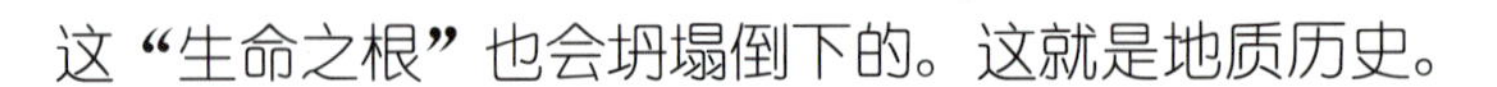

这“生命之根”也会坍塌倒下的。这就是地质历史。

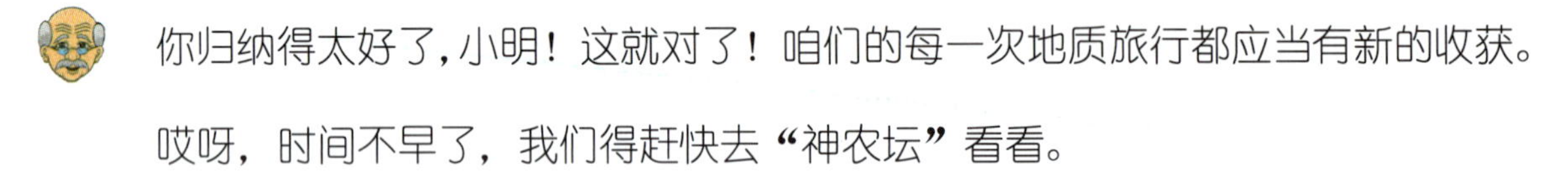

你归纳得太好了，小明！这就对了！咱们的每一次地质旅行都应当有新的收获。哎呀，时间不早了，我们得赶快去“神农坛”看看。

好吧，汪汪，咱们要走了。“板壁岩”，再见了！

问题：

1. 岩石发生风化剥蚀作用受到哪些因素的影响和控制？
2. 石英脉是如何形成的？硅质层又是如何形成的呢？它们有什么区别？
3. 什么叫作“差异风化”？你能在野外辨别“差异风化”现象吗？
4. 你能从风化剥蚀的角度，举例讲述一下板壁岩的象形石雕是如何形成的吗？
5. 请查询和归纳一下与岩浆活动相关联的“热液矿床”是如何形成的。

18

飞毯越过山脊向东飞去。山间漂浮着朵朵白云，山峦起伏，溪流蜿蜒；山坡上一垄垄墨绿色的茶田中，点缀着各色采茶人的衣裳；偶尔可见山中黑瓦白墙的民房和房前泛着蓝光的小池塘；这景色如同无比精致秀丽的盆景，带给人们世外桃源的遐想。飞毯沿着香溪谷很快地降低了高度，停在了一尊硕大雄伟的雕像旁边，这就是炎帝神农的标志性塑像。知识老人和小明带着汪汪下了飞毯，走到塑像前台阶的边缘。从这里居高临下，可以一览无余地看到插满彩旗的两条阶梯，向下延伸到每年举行神农祭祀活动的广场。

地质探秘神农架

爷爷，我们刚才好像经过了一个繁华的小镇。

是的，那个小镇叫木鱼镇，离这里只有 6km。木鱼镇是神农架世界地质公园最主要的旅游集散地。来神农架的游客大都住在这里，每天从这里出发去地质公园不同的景区参观游览。

木鱼镇，这个名字真好听，很有意思。我得回去查查看，这里面一定有个说法。

是的。听说关于木鱼镇还有一个非常优美的爱情传说呢！

是吗？一定非常有趣。爷爷，我们现在这是什么地方呀？

这个景区叫作“神农坛”。是神农架世界地质公园的南大门。这个景区以展现炎帝神农文化为特色。

难怪在这里修建了一座这么大的神农雕像，居高临下，好大气魄呀！

神农祭坛严格遵循了中国古代皇家园林的设计风格；上面神农氏塑像前设有“天坛”，是皇帝天子祭祀的地方；与其对应，下面设有“地坛”，是一般老百姓祭祀的场所。

我觉得这个山坡的地形太好了：面对开阔的河谷，后面有青翠的群山环绕。

这就是中国传统“风水”中所说的“背山面水”，是最吉祥的地势。可是，你知道吗，我们现在站立的山坡是一个巨大的滑坡体呢。

什么！滑坡？那不是好危险？

别怕。这是一个古老的滑坡体，现在它已经稳定下来了，暂时不会有任何的危险。

滑坡是怎样形成的呢？

滑坡是一种比较常见的地质灾害。简单地说，滑坡就是斜坡上的土和岩石受到地下水或地震等因素的影响，在重力作用下，沿着一定的软弱面或软弱带，整体顺坡向下滑动。它的破坏性相当大。

我们站立的这么大一块地方都是滑坡吗？这个滑坡发生的时候一定非常可怕！

是的，我们现在站立的地方，以及两边的坡地都属于这个巨大的滑坡体。这么大的滑坡发生的时候，肯定是地动山摇、惊天动地。

幸亏我们当时不在这里！不知道有没有人受伤。

据推测，这个滑坡发生在

很久以前，可能当时还没有人住在这里呢。但是，滑坡、泥石流、山泥倾泻等自然灾害，常常会给人们造成生命财产的损失。

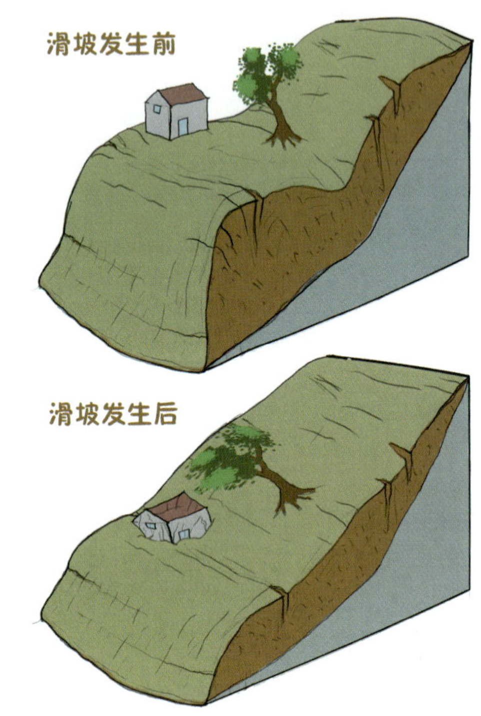

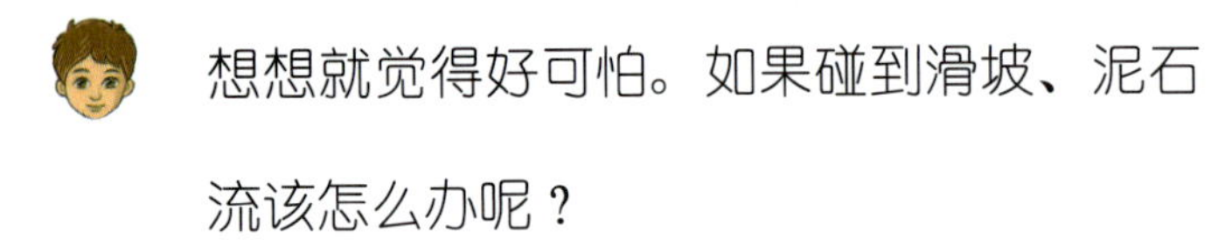

想想就觉得好可怕。如果碰到滑坡、泥石流该怎么办呢？

下雨天尽量不要在陡峭的山坡、山崖、山沟口等滑坡或泥石流高发的危险地带久留。万一碰到了滑坡或泥石流，要赶快向两侧的山坡上疏散。滑坡、泥石流、山泥倾泻都是危害性极大的地质灾害，会对我们的生命财产造成巨大的毁坏。

爷爷，滑坡之前这个地方的山应该很陡吧？

估计是比较陡峭的，因此很不稳定。

可是爷爷，您看现在，这里显得那样的静谧、平和。好像什么事也没有发生过似的，而且，您看，这山坡的地势相当平缓，咱们居高临下，俯瞰山谷，景色秀丽，还真挺不错的。

是呀，这就叫“坏事变成了好事”。如今人们充分利用滑坡天然形成的地势，依着山坡修建了这个神农坛，把神农的塑像建立在顶端，让人们仰望他，体现对这位农耕文明始祖的无限敬仰。

可不是嘛，好有气魄。

你看这座雕像，它高 21m，两侧宽达 35m，高宽相加为 56m，象征中国 56 个民族的大团结。

哦，真好！中国具有悠久的农耕文明史。当人们在广场上举行祭祀活动、传承优良的农耕文明的时候，仰望神农的塑像，心里充满了对他的热爱和崇拜，希望他像天神一样，保佑我们五谷丰收、国泰民安。

看来你做了不少的功课，对炎帝神农已经了解了很多，是吗？

是呀，来神农架游玩不了解炎帝神农那还行？

那你给爷爷简单地归纳一下炎帝神农的功绩吧。

好的。神农的功绩可多了，他是中国农耕文化的创始者；他分昼夜，定日月；教导人们使用工具，按季节播种五谷；他采药行医、烧陶绘画；教人们制作弓箭以狩猎、制造琴瑟以娱乐、练习舞蹈以健身，还教民众相互友爱、涵养智德。

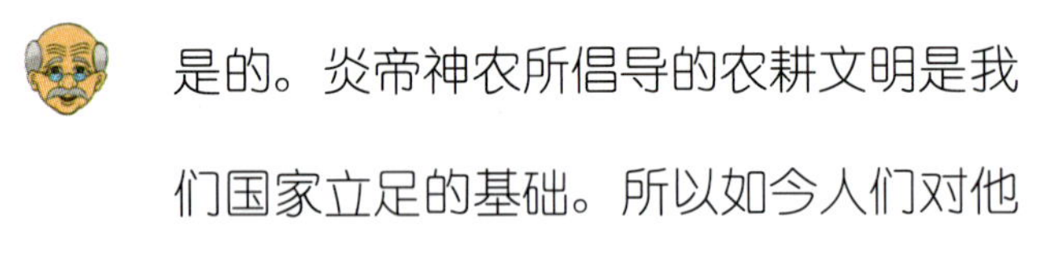

是的。炎帝神农所倡导的农耕文明是我们国家立足的基础。所以如今人们对他顶礼膜拜，将炎帝神农看成是丰收、兴旺、和平、安定的保障。

每年这里什么时候举办炎帝神农的祭祀活动呀？

农历 4 月 26 日是炎帝神农的诞辰日，所以地质公园每年都会在这一天在这里举办盛大隆重的祭祀大典。

那肯定非常热闹。我们一定要找机会来参加一次这样的祭祀活动！

是的，到神农架来，应当看看人们是如何敬拜炎帝神农的。除了每年的神农祭祀活动，神农坛这里还有许多与神农相关的东西呢。

是吗？快带我去看看。

你看到广场旁边那棵大树了吗？它高 48m，周长 9m，需要 6 个成年人才能将它合抱。人们把它叫作“千年杉王”。

哇，这棵树真高大！还挂满了红彩条呢。它真活了 1000 年吗？

它经历了唐、宋、元、明、清等朝代，至今已经有 1200 多年的历史了。虽然饱经风雨，历尽沧桑，但依然枝繁叶茂，昂首苍天。人们将它奉为“神树”。逢年过节人们都会前来祭拜求福；方圆百里的老百姓生疮害病，也常会前来求拜消灾。

哇，真是一棵宝树！

它之所以经千年而不衰，其原因之一可能是由于这个滑坡体的土石破碎，便于它的根部向大地深处伸展，能吸取更多的养料和水分。

这真是坏事变成好事！看来世间万物都是相辅相成、互相关联的。

是呀，我们应当尽量促使自然界的灾害转变为对人类有益的条件。你看到对面那边有好多圆形屋顶的小房子了吗？

是呀，很漂亮呢。那些房子是干嘛的？

那是供游客居住的客房，叫作“24 节气园”，每一间小屋都代表农历的一个节气。

就是阴历的 24 个节气吗？

是的。二十四节气是炎帝神农氏及华夏祖先，历经千百年的实践摸索发现的自然规律，是我国劳动人民长期对天文、气象、物候进行观测探索和总结的成果。是一份非常宝贵的科学遗产。

在这里修建“24 节气园”作为旅游客房确实很有创意！

是的。“千年杉王”后面还开辟了“神农百草园”呢。

哦，咱们快去看看。

你知道吗，神农架就是因为炎帝神农在这里搭架攀岩采药的传说而得名。《神农本草经》是我国现存最早的药学专著，记载了 365 种草药。

真了不起！

现在，在这个“神农百草园”里，栽种了 100 多种常用的中草药，主要用来培育和研究名贵的中草药在医疗中的应用。所以神农坛是继承和发扬传统中医药的重要实践场所。

哇，每一小块园地种有一种草药，还有解说牌呢！这里简直是一个学习中医和进行中草药实践的大课堂。

再往上走就是神农茶园了。说到茶叶，我想起了一个传说：相传神农品尝百草时，误食了一种罕见的毒草，昏倒在此地，幸亏有茶树上的露水滴到他的口中，他才慢慢地苏醒过来，由此他发现了茶树的药用价值，于是便教老百姓栽种茶树。

我看的资料中说，这里的高山“炒青”茶叶价廉物美，是很好的保健饮品，也是神农架著名的品牌。

是的。俗话说“高山云雾出好茶”，神农架海拔较高，没有农药化肥的污染；空气纯净，常年云雾环绕，土壤有机质含量丰富；茶树朝夕饱受雾露的滋润，白天光照充足，光合作用制造了丰富的有机物，而夜晚温度较低，细胞呼吸作用减弱，降低了这些有机物的消耗，所以神农架的茶叶芽叶肥壮，叶质柔软，属于茶中之极品。

地质探秘神农架

哎，这茶园里怎么还有房子呀？

前面就到“神农农耕文化园”了，它以茶园为主，分为农耕文化、采茶制茶和农趣体验3个部分。我们先看看功能展示室吧。

呀，这里展示了好多古时候的用具呀，这是以前人们用的灯吧？……哎，这是打渔用的吧？

是的。在手工作坊区还专门介绍了传统的榨油、做豆腐等的生产程序，让人们了解我们勤劳聪明的祖先是如何劳作、如何生活的。

这里简直是个人类生活百科博物馆！

哦。说到博物馆，差点忘了，咱们该走了，赶快去官门山吧，那里有四五个博物馆呢！

是吗？那我们赶快去看看吧，我最喜欢看博物馆了！

问题：

1. 什么是滑坡？应当怎样避免滑坡造成的灾害？
2. 为什么神农架每年都要举办盛大隆重的祭祀大典来纪念炎帝神农？你能详细介绍一下炎帝神农的功绩吗？
3. 你知道24节气的名称和含义吗？它们是如何与农业生产活动相互对应的？
4. 为什么说神农架的茶叶是“茶中极品”？
5. 神农坛内有些什么与炎帝神农相关的设施？参观游览神农坛的最大收获是什么？

19

飞毯越过了一条葱翠的山脊后，沿着山沟缓慢地向上游方向飞行。山沟里流水“哗啦啦”地响着，清澈的激流冲击着河边的石头，溅起白色的浪花。在较宽阔的一段河面上，流水平稳地泛着涟漪向下游流去；沿河修建的公路蜿蜒曲折，平坦的路上时不时有公园专用的电瓶车载着游客来回奔忙。看来这里也是游客非常喜爱的地方。

一条修建在山边的深色木质栈道，时隐时现地沿着河谷延伸着。河流两边岩壁陡立，时宽时窄。河谷在一个“S”形的大拐弯处，陡然显得格外宽敞起来。绿树丛中，几栋造型美观的建筑物出现在眼前。飞毯停在一座有宽大阶梯的建筑物——“地貌厅”的前面。

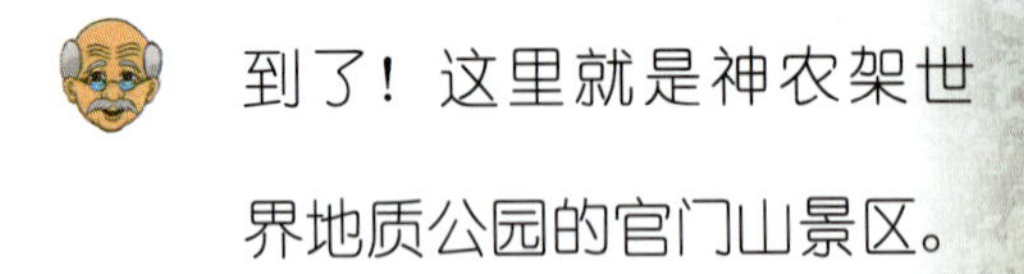

到了！这里就是神农架世界地质公园的官门山景区。

这河里的水好清亮呀！真是山清水秀，好地方呀！

这条河叫作石槽河。官门山景区就是沿着这条河修建的。

这里好像离木鱼镇更近些，是吗？

是的。住在木鱼镇可以很轻松地步行到这里来游玩。

太好了，太方便了！

这个园区是以“探索·发现”为主题的科研、科考、生态旅游体验区和最佳的科普教学现场。

哦，这里有些什么科研科普设施呢？

这里集中了好几个博物馆，比如地质馆、奇石馆、动物馆、植物馆、科考馆、民俗馆，还有一座4D影院和许多动植物生态培育养殖园。

哇，这简直是一个博物馆群呀！

可不是吗。它们统称“神农架自然展览馆”，始建于2008年，2013年建成投入使用。占地面积近 3 万平方米，建筑面积将近 1.5 万平方米，投入了 6500 万元资金，将这个山谷变成了一个科学普及的殿堂。

哇，神农架世界地质公园在这些旅游和科普设施上做了这么多的工作。我最喜欢参观博物馆了，这回可要大开眼界了。让我们一个个慢慢地参观吧！

好吧。来，我们先看看这个“地貌厅”。

“地貌厅”？我们从空中不是已经把地貌看得很清楚了吗？

那可不一样。“地貌厅”里有一个 1∶1.2 万比例尺的地形地貌沙盘模型，可以让参观者对神农架全区一目了然。另外，“地貌厅”里还设有咨询、投诉、救援、残障人士服务和休息的设施。

哇！这个“地貌厅”占地面积不大，但是设施倒真周全。

小明，你过来看。从这个沙盘模型上可以清楚地看到神农架地区的河流水系，从而建立一个整体的概念。你看，以神农顶为首的山系在地势上确实构成了长江和汉江的分水岭，每一条河的流域范围都可以看得一清二楚。

是的，一目了然。哎，那儿还有个飞机场呢！

神农架机场是我国内地海拔最高的飞机场！海拔高度为 2580m。机场与地质公园景区之间地理位置的关系，在这里也一目了然。

这个沙盘模型真好。爷爷，虽然我们已经参观了不少地方，可是一看这沙盘模型，我才发现我们看过的地方实在是太少了！

是呀，神农架是个大型地质公园，园区面积超过 $1000km^2$ 呢！我们现在看过的只不过是很小的一部分。好了，咱们该去参观一下博物馆了。

哎，这个女孩子怎么骑在豹子身上？

这座雕像讲的可是一个真实的故事。那还是在 1957 年的夏天，一个家住神农架林区叫作陈传香的 20 岁姑娘，为了从豹子口中解救一个 3 岁的孩子，不顾危险，勇敢地与豹子搏斗，将豹子打死了。

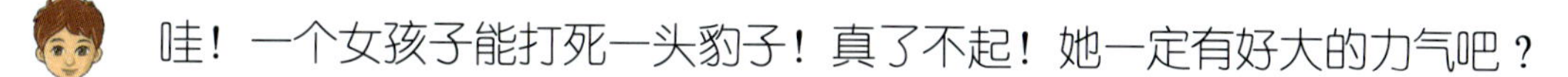

哇！一个女孩子能打死一头豹子！真了不起！她一定有好大的力气吧？

陈传香在与豹子搏斗的时候，突然想起老人们曾说过，豹子是“铜头铁尾麻杆腰”，于是很机智地骑上豹子的背，用力向下一坐，将豹子的腰压断了。

真机智，真英勇！

从这里也可以看出神农架林区勤劳勇敢的人们对森林中的动物非常了解。今天，我们正好可以利用这方面的知识更好地保护它们。

哎，到“地质馆”了。

“地质馆”是神农架自然展览馆的主题馆之一。

这个展览馆布置得很奇特，深蓝色的穹顶上繁星点点，如同浩瀚的太空，令人遐想；四壁的展板和照片，对应于玻璃柜中陈列的岩石矿物标本，述说着神农架亿万年的地质历史，令人仿佛经历着时空的穿越。很有趣。

这里通过陈列的标本、照片和解说，形象化地介绍了神农架的地质概况，告诉人们在长达 10 多亿年的漫长地质历史中，神农架地区如何经历沧海桑田、海陆变迁而上升成为“华中屋脊”的曲折过程。楼下还有个“奇石馆”，更是游客极喜爱参观的地方。

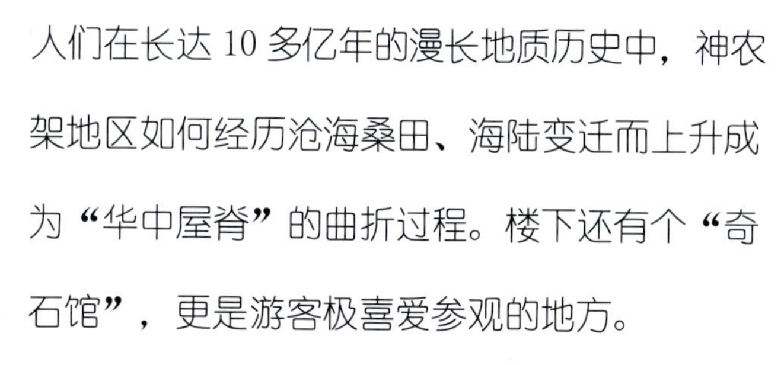

爷爷，为什么会有这么多奇形怪状、色彩绚丽的石头呀？

大自然是一个神奇的创造者。我们所说的奇石，就是指天然形成的形状或花色奇特的石头，它们因具有不同寻常的材质、造型、色彩或者花纹，能够满足人们猎奇或审美观赏的欲望，可供人们收藏、观赏和把玩。

大自然的创造实在是太神奇了。哇，爷爷，您看，这石头上花纹简直就像是一幅水墨山水画！它是怎么形成的呀？

你是说这个吗？这是典型的大理岩观赏石。我不是说过吗，大理岩是石灰岩遭受后期高温高压变质而形成的。

哦。石灰岩变质成大理岩会发生什么样的改变呢？

石灰岩中的方解石常常会发生重结晶，就是由原来细小的晶粒逐渐再次结晶成为较大颗粒的晶体，变成结晶状的石头。白色结晶状的大理岩晶莹剔透，被人们称为“汉白玉”，是一种上好的建筑装潢石料。天安门广场上的华表和金水桥的栏杆都是用汉白玉造的。

“汉白玉”确实很漂亮。岩石的变质作用就是矿物发生重结晶，是吗？

重结晶只是变质作用的一种。另外，岩石在变质的过程中常常会形成一些新的矿物，我们称之为“变质矿物”。遭受变质作用的岩石本身所含的物质，在不同的温度压力条件下，会形成不同的变质矿物。因此，地质学家常常根据岩石中变质矿物的组合，来判断岩石所遭受到的变质作用的强度。

哦。地质学家真了不起！但是，这块大理岩那么细腻光滑，根本看不到重结晶，也不知道它里面有没有变质矿物。

这就是我下面要讲的第三种变质作用的结果，也就是岩石结构构造的改变。岩石受到高温高压的作用，会发生结构构造上的变化，造成岩石中矿物的碎裂或发生定向的排列，形成片理状或片麻状。如果在显微镜下就可以看得更加清楚了。但是，也有部分岩石在高温高压下，变得像面团一样柔软，这块大理岩的原岩应当是夹有暗色含碳薄层的石灰岩，在遭受高温高压变质的过程中，它像面团一样被搓揉、扭曲、揉皱，就形成了像山水画一样的纹饰。

哦，就好像是蒸花卷那样吧？

是呀。其实很好理解，比如打铁的时候，把铁块烧红了就可以随心所欲地敲打成各种形状了。

这比喻很贴切！

石灰岩在发生变质作用时，原来岩石中黑色的含碳薄层在塑性状态下被扭曲、变形、揉皱，形成了山水画一样的美丽图形。这种类似中国山水画的大理岩，是观赏石的一个重要类别。

哦。大自然真是个能工巧匠。呀！爷爷您快来看呀，这块石头上画了好多的竹叶呢！真漂亮！

这是典型的竹叶灰岩。

竹叶灰岩？竹叶怎么会跑到灰岩里面去的呢？

这些并不是真正的竹叶。只不过形态像是竹叶，所以这样命名的。

那它是怎么形成的呢？比画家画得还像，而且还是立体的！

竹叶灰岩是一种比较常见的“同生砾岩”。

“同生砾岩”？什么意思呀？

我们通常所说的砾岩，是指那山上的碎石块，被山溪河流带到湖泊或海洋中堆积下来，在漫长的地质历史中逐渐地被泥沙或其他化学物质胶结固化形成的岩石。从沉积到最后被胶结成岩石之间经历了较长的时间跨度。而“同生砾岩”则不同……

怎么不同呢？难道碎石块堆到一起马上就被胶结成岩石了？我想不通。

那倒没有那么快。我们只能说“同生砾岩”的砾石和胶结固化作用大致是同时期发生的。这个过程是这样的：首先，浅海里沉积的薄层灰岩，在还没有完全固化的时候，被海浪打碎、冲刷、磨蚀，变成像竹叶般的大小和形态，然后又被后来沉积的石灰质沉积物胶结、固化，最后就变成了这种漂亮的竹叶灰岩。

太神奇了！真想不到我们的大自然还会按照一定的“程序”来制造这些巧夺天工的美丽“艺术品”供我们欣赏！

大自然确实是一个神奇的创造者！从她的所造之物，我们可以追溯这些奇石形成的过程和故事。小明，你能分析一下竹叶灰岩形成时候环境的变化吗？

我试试吧。……嗯，我觉得，起初，大海应当是比较平静的，所以沉积了一些薄层的石灰岩，这些石灰岩还没来得及固结变成石头，就起了很大的风浪，可能是暴风或者台风吧。

也有可能是海啸，是吗？

是的。风浪把这些还没有变成石头的薄层石灰岩打碎了，这些碎块在海底随着海浪翻滚摩擦，把棱角都磨掉了，变成了像竹叶那样的薄片。当大海又逐渐归于平静后，新的沉淀物把这些竹叶状的灰岩碎块胶结起来，又过了数千万年，它们固结石化，变成了竹叶灰岩。

很好，小明，看来你是真正地理解了竹叶灰岩的形成过程。我再问你一个问题：如果还没有固化成岩的沉积层被波浪打碎后，还没来得及翻滚摩擦，马上就被后来的沉淀物充填掩埋起来，然后固化变成了石头。这种石头应当是什么样子的呢？

打碎后马上就胶结形成岩石，嗯……那就不是像竹叶那样边缘被磨圆了，而是棱角状的。是吗，爷爷？

对了。这时候，那些砾石就是有棱有角的。我们把这种沉积岩叫作沉积角砾岩。

真有趣！光看看石头就能说出它们是怎样形成的。地质学真是一门非常有趣的科学！

地质学家就是按照这样的方法，来分析判断地质历史中发生的事件和环境变化的。你现在学会了一种非常重要的而且很常用的地质分析方法，那就是“将今论古”的方法。

“将今论古”？什么意思呀？

将今论古就是根据现今我们所看到的、所了解的自然现象，以及它们形成的原理，来分析地质历史中类似的现象和事件，推测可能的原因。

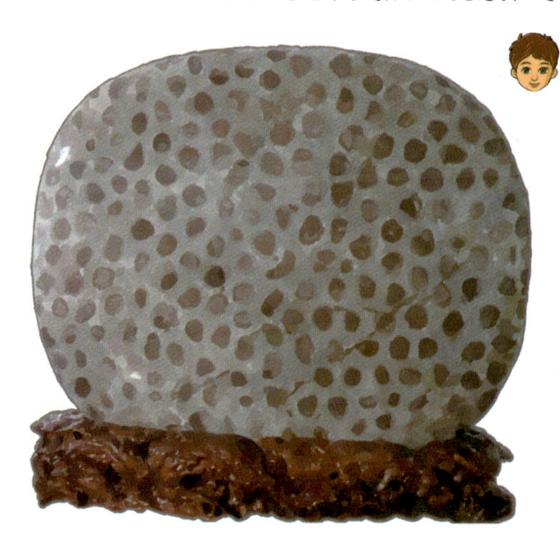

哇，太有趣了！我觉得“将今论古”真有意思。我一定要好好学习这个方法，今后不管到哪里、不管看到什么地质现象，都能够用“将今论古”的方法讲出一个有趣的故事！

是的，我们对周边的自然现象了解得越多，运用“将今论古”就会越得心应手，分析的结论就会越正确。

是的。今后我一定要更加仔细地观察研究我们身边的自然现象，积累尽可能多的知识。这样，以后碰到任何神奇的地质现象，都可以分析它形成的过程和原因，讲出它背后的神奇故事。

大自然中有太多神奇的事物等待着我们去研究探索、去分析解释。

是呀，每块石头都有一个有趣的故事。我以后想当个专门讲述石头故事的人：随便捡一块石头，就能讲出它的来历和经历。哇，那多有趣呀！

只要你现在注意学习和积累，你一定会成为一个很好的专讲岩石故事的人！

问题：

1. 你觉得神农架地质公园官门山是一个什么样的景区？你会怎样向你的朋友介绍这个景区？
2. 大理岩中像山水画一样的花纹是如何形成的？
3. 岩石发生变质作用时会产生些什么样的变化？
4. 你能讲述一下“竹叶灰岩”是如何形成的吗？它与沉积角砾岩有什么区别？
5. 什么是“将今论古”方法？你能运用这个方法分析一下为什么有的河边是砾石滩，有些则是沙滩吗？

20

知识老人和小明边聊边参观。小狗汪汪跑前跑后，但由于展柜太高，它什么都看不见，显得有点无聊。只是在进入动物馆的时候，才兴奋起来。一下子面对那么多各种各样的野生动物，它高兴得又跳又叫，还不时趴在地上对着那些动物低声“咕噜咕噜”地说着什么。最后，当它意识到这些呆呆地站在那里的动物根本不可能和它一起玩耍时，便懒洋洋地走到门口打起瞌睡来。

从博物馆出来后，知识老人带着小明越过一座小桥，走上了沿着河边修建的木质步游栈道。栈道的一侧悬崖陡立，另一侧则是“哗哗”流淌的清澈溪流。树林里吹来的微风凉爽清新，令人精神振奋。和煦的阳光透过密林，在栈道上洒下晃动着的光点。小狗汪汪追逐着太阳的光点或飘落的树叶，高兴地跑来跳去，对它来讲这里有趣多了。只要有通向河边的阶梯，它就会跳下去，在河边的石头滩上闻来闻去，寻找它感兴趣的东西。

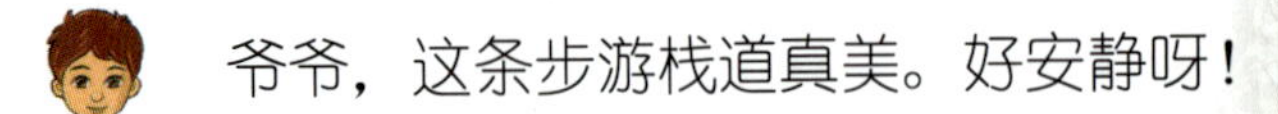

爷爷，这条步游栈道真美。好安静呀！

是的。这条步游栈道从进入官门山景区的大门开始，沿河而筑，全长 8km，是一条让人们亲近大自然，与山水树林、岩石花草零距离接触的绿色通道。

没错。走在这里，绿茵环绕，溪流欢唱，花香鸟语，令人神清气爽。我觉得心境格外平和，非常享受！

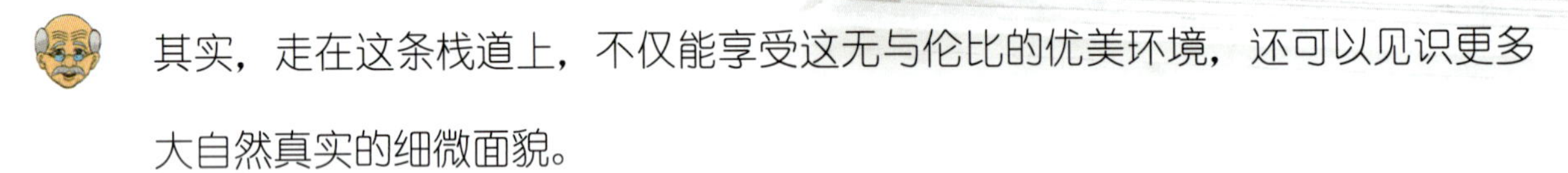

其实，走在这条栈道上，不仅能享受这无与伦比的优美环境，还可以见识更多大自然真实的细微面貌。

爷爷，您刚才说的我没听懂。这栈道不就是一条路吗？怎么能使我们见识到大自然的细微面貌呢？

来，我讲给你听。地质公园设计修建这条栈道，是为了更好地向游客展示神农架地区无比丰富的生物多样性和地质遗迹，以及优美的自然环境。因此，这条栈道不仅仅是一条路，而且是一条零距离亲近大自然、亲身体验大自然的长廊。

哦，那在这条步游栈道上能看到些什么呢？

沿着这条栈道，地质公园修建了茶园、药园、兰园、腊梅园、红枫园、杜鹃园、百果园、沧桑园、名人名树园、国际园、蜜蜂园，以及全国自然保护区联盟林、中国世界地质公园发展林等。在龙头寨下的河边，正在修建熊猫馆，准备让国宝熊猫在神农架安家落户呢。

哇，这么多林园呀！一整天都逛不完。

不止是各种林园，这条栈道沿途还有好多有趣的地质遗迹呢！

是吗？爷爷，有些什么好玩的地质遗迹？

比如说地下暗河、叠层石化石、重要地层单元命名地、奇风口……

奇风口是什么呀？听起来怪有意思的。

那是一个很奇怪的溶洞口。这个溶洞深不见底，但是一年四季都有呼呼的风穿进穿出。

是吗？怎么会有这种奇怪现象呢？

其实道理也很简单：溶洞深入到地下，洞中的空气温度常年都保持着相对稳定的状态；但是，洞外的温度却随着季节而发生变化。因此，洞外与洞内的空气温度，始终都存在一定的差别，于是洞内、洞外就会自然而然地发生空气的循环流动了。

是哦，道理确实很简单，但是在自然界出现这种现象却是比较罕见的呀。

是的。这可能与这个溶洞的结构有关。目前因为洞口太小，无法探寻洞内具体情况，因而一般人讲不清楚到底是怎么回事。所以人们给它起了个名字叫“奇风口”。

大自然总是为我们创造一些发人深思的奇迹。

是呀。这条栈道沿途有很多有趣的东西供人们探讨。可是现在来官门山园区旅游的人们，大多都不了解，甚至不知道有这么一条趣味盎然的步游栈道，还以为官门山只有博物馆和电影院呢。

其实，博物馆、电影院与步游栈道是相辅相成的：在博物馆里看到的是标本，而沿着栈道看到的是实物，是纯粹的自然！应当更加有趣和给人以享受。

你说得很对。在“国际园”附近河对岸的山沟里，建有一个“伐木场再现”的特殊园地。

“伐木场再现”？那是个砍树的现场吗？

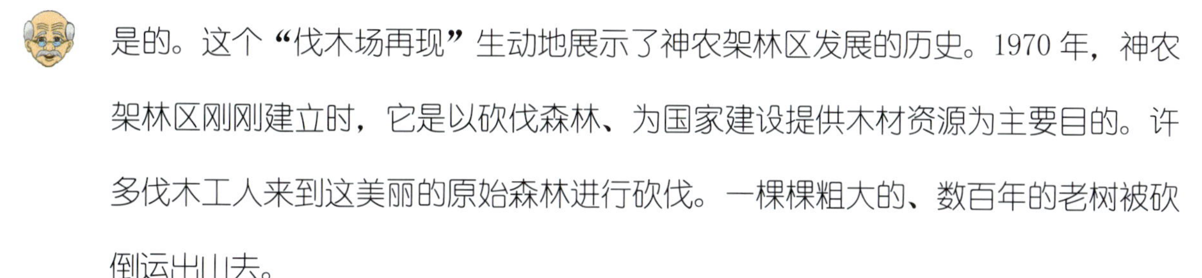

是的。这个“伐木场再现”生动地展示了神农架林区发展的历史。1970 年，神农架林区刚刚建立时，它是以砍伐森林、为国家建设提供木材资源为主要目的。许多伐木工人来到这美丽的原始森林进行砍伐。一棵棵粗大的、数百年的老树被砍倒运出山去。

太可惜了！您不是讲过，每一棵树都是无价之宝吗！

是呀，可是当时人们并没有意识到树的宝贵价值，更不理解森林对人类生活环境的重大影响和意义。

那怎么办呀？把那么多树都砍了，真急死人了！

随着保护环境的呼声越来越高，21 世纪以来，神农架已经完全停止了森林砍伐，而转变成为一个自然资源的保护区，并开展了持续性大规模的植树造林活动。

太好了！

从砍树到植树，充分反映了人们思想观念的巨大变化，体现了我国对环境和资源进行保护的战略性逆转。所以这个“伐木场再现”

富有极其重要的教育意义和旅游价值，它揭示了我们曾经走过的弯路，激发人们对森林、对大自然的热爱和保护环境的自觉性。

“伐木场再现”确实很有教育意义。

沿着这条亲近大自然的体验长廊进行游览，可以获得很多的知识。天气晴好的日子，听着小溪流水淙淙，闻着花草阵阵清香，漫步在这里实在是一种很好的享受。

嘘，爷爷，您看那小鸟，好漂亮呀！

这里草深林密，昆虫众多，又有流水，常常有很多美丽的小鸟在这里觅食生活。这里可以说是观鸟者的天堂。

真是鸟语花香。走在这条栈道上太舒服了！还能看到各种不同的小鸟和花草植物，长不少知识，真好！

是啊。就拿杜鹃园来说吧，修建于2008年，占地面积超过6000m²，收集了神农架和周边地区杜鹃科植物及其伴生树种达 50 余种。

哇，这么多种杜鹃呀！开花季节一定非常美丽。

可不是嘛。杜鹃是世界三大高山观赏花卉之一，自古就有“人间美西施，花中唯杜鹃”的说法，神农架的杜鹃有红、黄、紫、白等多种颜色，每年四五月份是杜鹃的开花季节，盛开的杜鹃花把神农架装点得五彩斑斓，美不胜收。

真令人向往！爷爷，我们一定要在杜鹃花开的季节来这里赏赏花！

好的。你想看花？现在就能看！前面就是“兰花园”了。

是看兰花吗？太好了！我最喜欢兰花了。

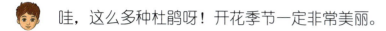

中国人历来把兰花看作是高洁典雅的象征，并与“梅、竹、菊”并列，合称“四君子”，被评为中国十大名花之一。人们常借兰花来表达纯洁爱情的专一，有诗句说“气如兰兮长不改，心若兰兮终不移”。

兰花确实有一种高洁的气概。神农架的兰花得天地之灵气，肯定格外的美丽幽香。

官门山的“兰花园”建于2007年8月，占地面积16 000多平方米，位于海拔1280多米的山坡上。这里集中栽植了神农架兰科植物36属，90余种，1万余蔸。

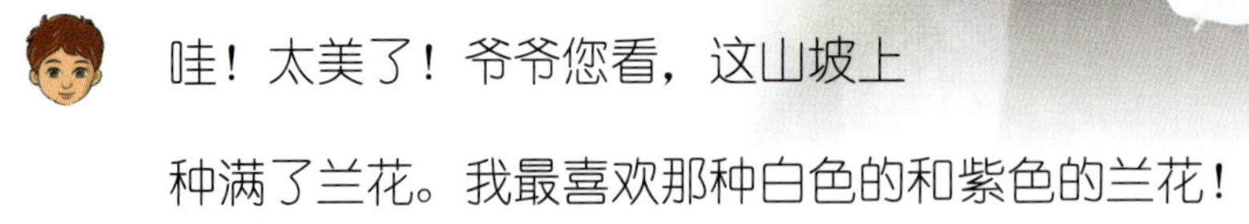

哇！太美了！爷爷您看，这山坡上种满了兰花。我最喜欢那种白色的和紫色的兰花！

这种黄色的和粉色的兰花也很好看呀。对了，你不要光顾着看山坡上的兰花，我们脚下铺路的石板上也“刻”了一些非常美丽的“花”呢！

哦，刻的是什么花呀？

我现在不告诉你，看你自己能不能找到。

让我仔细找找。……哎，爷爷，是这个吗？一圈圈的，好精细呀！这是什么东西呀？

这是一种化石，叫作“叠层石”。

“叠层石”？这名字起得很好，确实是一层层地堆叠在一起！哇！这里好多呢，路上石板上有，路边的石头上也有。

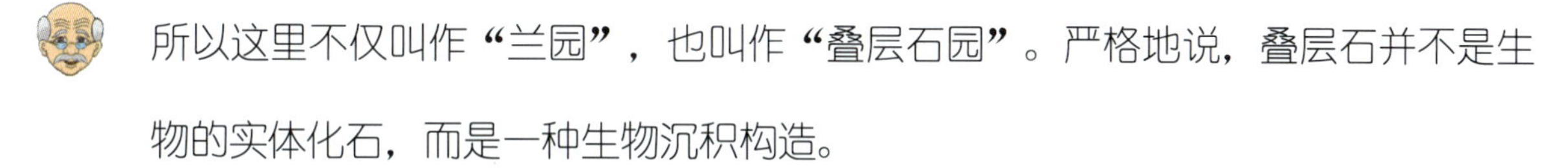

所以这里不仅叫作“兰园”，也叫作“叠层石园”。严格地说，叠层石并不是生物的实体化石，而是一种生物沉积构造。

这是什么意思呢？叠层石到底是什么东西呀？它是怎么形成的呢？

叠层石是一种海洋中原始的单细胞藻类生物在生活过程中形成的一种特殊的沉积构造。神农架的这些叠层石生活在十三四亿年之前。形成叠层石的这些藻类是地球上最古老的生命之一。

哇，那么古老呀！我们今天能看到它真是个奇迹！

是的。而且更重要的是：形成叠层石的藻类本身，对于我们的地球来说，就是一个奇迹的创造者。

是吗？它创造了什么奇迹呀？

对于我们这颗蓝色星球上众多生命的诞生和繁衍，这些藻类可以说功不可没，起到了举足轻重的决定性作用。

这些藻类真的那么重要吗，爷爷？

一点不假！早期我们地球大气中充满着二氧化碳、甲烷、氮、硫化氢和氨，没有任何生命。

作为地球上最先出现的原始藻类，它通过光合作用，制造了大量的氧气，经过数千万年的积累，逐渐地改变了地球大气的组成，为后来生物的繁衍创造了一个必不可少的富氧环境，如今的地球才会有如此丰富的生物多样性。

哇，这些藻类确实为我们的地球立了一个大功！爷爷，它们怎么会长成这种一层层、一圈圈的样子呀？

形成叠层石的藻类是附着在岩石表面生长的。它们一代又一代，老的死去，新的又在上面生长出来，在生长的过程中，它们不断地吸附着海水中微小的沉积物颗粒，一层层地构筑了这些看起来像同心状的化石结构。

哦，原来它是这样长成的呀。真有趣！

更有趣的是，根据叠层石纹层的厚度变化和其他特征，科学家可以推断出当时季节，以及昼夜长短和气候变化的状况。

真太奇妙了！科学家真是伟大，他们怎么

能根据叠层石的化石推断出 10 多亿年前的气候环境变化呢？

基本原理就像是根据树的年轮判断树的年龄，研究它生长的环境那样。

哦，这很简单嘛，我数过树的年轮。

这可比数年轮要复杂多了。根据化石推断亿万年前的气候，需要考虑很多因素，并测试大量数据。这是一个相当复杂的科学研究课题。

爷爷，在博物馆里我还看到过恐龙的足迹，他们说也是一种化石。

是的。化石可以分为实体化石和遗迹化石两大类。我们看到的恐龙骨架、贝壳、鱼虾等都属于实体化石。你刚才提到的恐龙足迹则属于遗迹化石。

遗迹化石？就是生物留下来的痕迹，是吗？

是的。遗迹化石包括足迹、爬痕、掘穴、钻孔，甚至生物的排泄物，比如粪便等。

哦，这些东西都能成为化石？对了，爷爷，我还见过琥珀包裹着的蚂蚁呢，可漂亮了！它们也是化石吗？是怎么形成的呢？

是的。琥珀里保存的昆虫也是化石。你知道琥珀是怎样形成的吗？那是植物的树脂，就是类似于“松香”那样的物质，在流动的过程中把一些昆虫包裹进去，经历了千百万年的地下埋藏，发生氧化、固结，逐渐形成为琥珀。我们称之为特殊的有机化石。

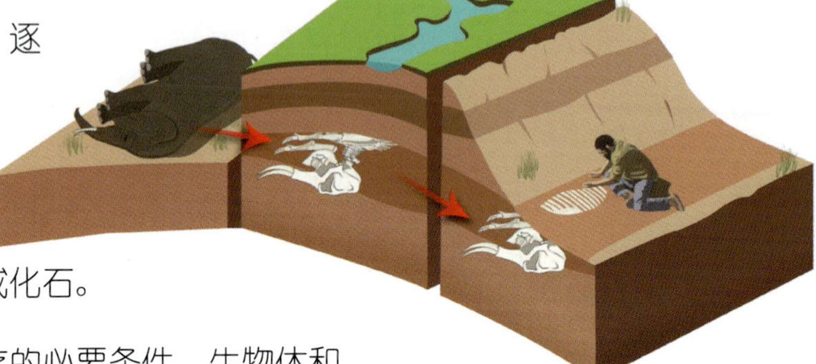

看来任何动植物都能形成化石。

基本如此。只要具备保存的必要条件，生物体和它们留下来的痕迹都有可能被保存为化石。

什么是化石保存的必要条件呢？

动植物死后，它们的遗体或遗迹若能够被迅速地掩埋，与空气隔绝，那么在石化的过程中，它的外壳或骨骼被矿物质交代而保留下来，就成了实体化石；它们生活过程中留下的痕迹就成为遗迹化石。

我特别喜欢看化石。可以了解到远古时代那些长得稀奇古怪的生物。

化石还能告诉我们许多地质历史中的故事。通过研究化石，科学家可以认识遥远的过去生物的形态、结构、类别，可以推测出亿万年来生物起源、演化、发展的过程，还可以恢复漫长的地质历史时期各个阶段地球的生态环境。

是呀，化石背后的故事可能比石头的故事更加有趣。

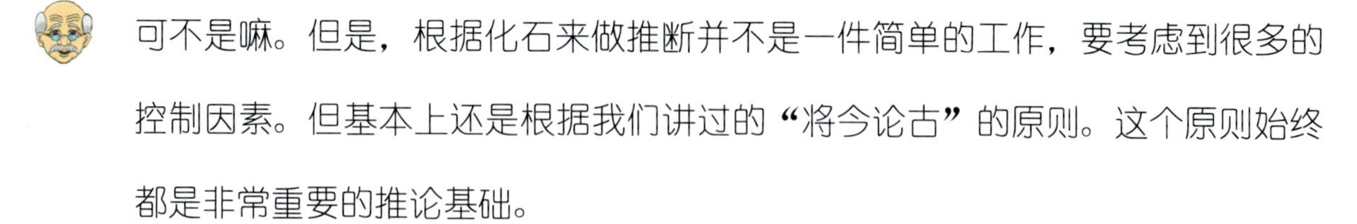

可不是嘛。但是，根据化石来做推断并不是一件简单的工作，要考虑到很多的控制因素。但基本上还是根据我们讲过的“将今论古”的原则。这个原则始终都是非常重要的推论基础。

我想也是的，因为自然变化的法则是不会轻易改变的，改变的只不过是当时具体的自然条件。

是的。运用已经研究清楚了的自然法则，去分析远古时代的环境或地球上发生过的地质事件，是地质学家经常使用的有效方法。

问题：

1. 请解释一下官门山“奇风口”神奇现象发生的大致原因。
2. 通过参观“伐木场再现”，你能说说神农架发生了什么样翻天覆地的变化吗？
3. 为什么神农架会具有如此多样化的生物群落？
4. 什么是叠层石？它在地质历史发展中起到过什么样的作用？
5. 关于化石你知道些什么？什么是“实体化石”、什么是“遗迹化石”？

21

知识老人和小明沿着步游栈道边走边聊，小狗汪汪兴奋地跑跳着。在一片草坪上，它发现了一只美丽的黄蝴蝶，于是它蹑手蹑脚、悄悄地向蝴蝶靠近。狡猾的蝴蝶假装没有注意到汪汪的接近，仍然不慌不忙地从一朵花飞到另一朵花。当小狗汪汪举起前爪，正准备抓住它的时候，蝴蝶突然向侧边一闪，飞快地绕过附近的一块石碑，消失了踪迹。汪汪懊恼地对着石碑叫了起来。

知识老人和小明循着汪汪的叫声，来到石碑的前面。

汪汪，你怎么了？谁欺负你了！

也许它发现了一只小田鼠吧。

哦，这里是“国际园”呀，爷爷！

是的。以前叫作“国际友谊园”。

为什么在这里建一个“国际友谊园”呢？

这是为了纪念2007年第三届世界植物园大会在我国武汉市召开而建立的，当时神农架作为分会场。在会议期间，有来自10个国家31名中外专家、学者到神农架考察，并在这里植树作为纪念。

这确实具有很好的纪念意义！

“国际园”位于海拔1290m处，占地面积近2万平方米，分为亚洲区、北美区、南美区、欧洲区、非洲区、大洋洲区。真可谓：“世界植物群英荟萃，神农架珍稀植物成园。”

哇，看来神农架早就走向世界了！

“国际园”还栽植了我国的珍稀树种红豆杉、珙桐、银杏、小勾儿茶等200余种不同品种的植物呢。

这里真是个植物大观园呀！

是呀。官门山这里的每一个园林和植树基地都有自己的故事和发展历史呢。

真是个户外的植物大课堂！

这里不仅植物丰茂，还有蛇类爬行动物馆，大鲵（娃娃鱼）、红腹锦鸡、梅花鹿等动物养殖研究中心和蜜蜂园呢。

真是一个集科研、科考、生态旅游、科普教学、养生休闲为一体的好地方！

官门山的科学普及教育宣传和公园设施非常适合于家庭旅游，老少皆宜，一家人可以在这里悠哉游哉、轻轻松松地玩一整天，享受美好的大自然：老人可在这里休闲观赏，孩子们可在这里亲近山水和大自然，学到很多课堂里学不到的关于动植物的知识。

的确是的。我看见沿途有不少的标牌，对照着景物进行解说。这样很好，真可以学到不少知识呢。

可不是嘛。这些解说牌就是为了帮助人们理解大自然的各种现象而设立的。就比如我们现在站立的这个地方就是一个很重要的地质点。

“地质点”？什么意思呀？

 地质点就是有重要地质现象出现的地点。

 那我怎么看不出这里与别的地方有什么不同呢？

 你看看这个解说标牌就明白了。这里是“神农架群石槽河组”命名的地方。

 让我看看。“石槽河组”距今约 13 亿年。哇，这么古老呀！什么是“群”、什么是“组”呀？

 “群”和“组”都是地质学的专业术语。我们看到的岩层大多都成百上千米厚，为了便于研究，并揭示岩石形成的环境和过程，必须将数百上千米巨厚的岩层进行划分。“群”和“组”就是地质工作者对岩石地层进行划分时使用的单位。

 怎么分呢？是分成厚度一样的几部分吗？

 不是。地质学家根据岩石的成分和岩性的变化，推断岩石形成的过程和环境，来对巨厚的地层进行划分。由大至小，按照“群、组、段、层”4 个级别加以分类和归纳。

 哇！这么复杂呀！

 是比较复杂。但是，把复杂的地层分别划归到不同的地层单位中之后，事情就会变得简

单些了。“群”是最大的地层单位，它往往反映这一套岩石形成的整个地质阶段和过程。比如这里所说的“神农架群”就记录了一段长达数亿年的海洋沉积历史。

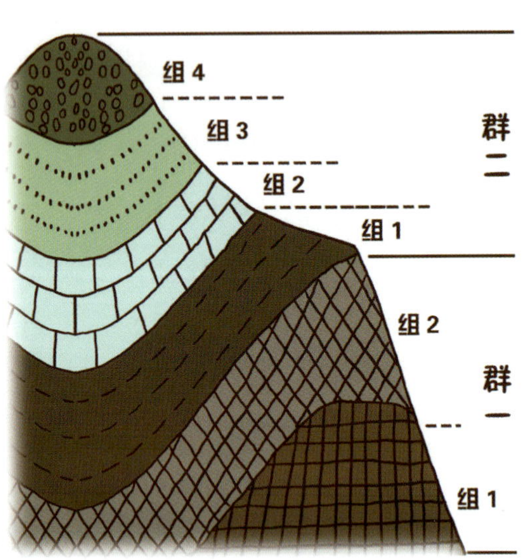

哦！这很有意思呀！

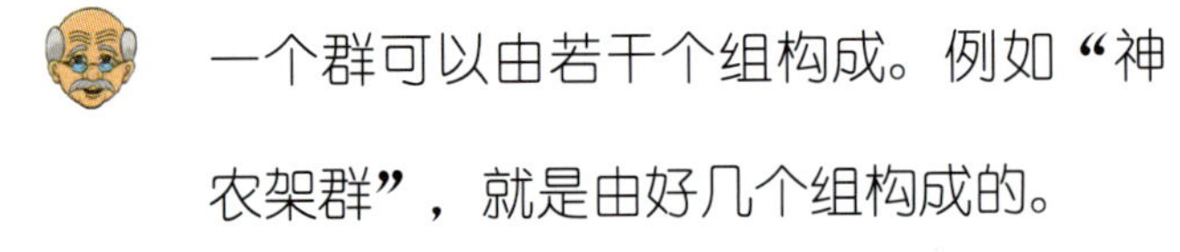

一个群可以由若干个组构成。例如“神农架群”，就是由好几个组构成的。

那“组”又是怎么分出来的呢？

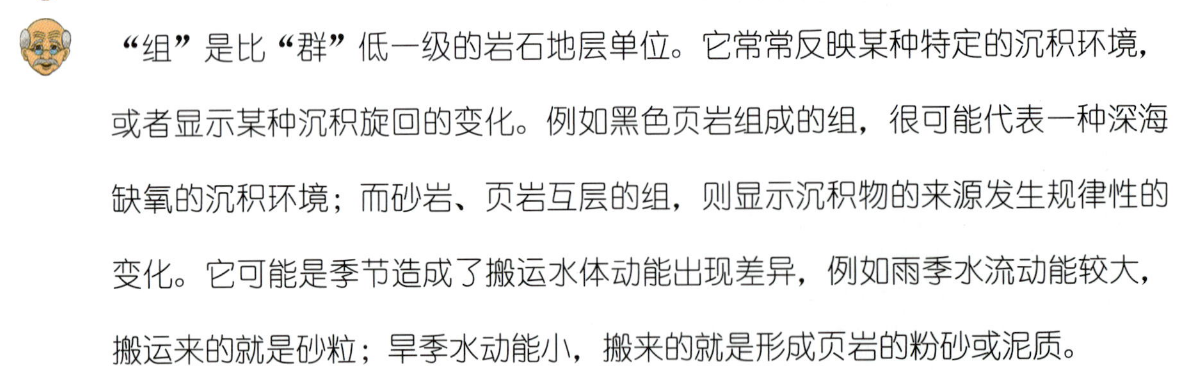

“组”是比“群”低一级的岩石地层单位。它常常反映某种特定的沉积环境，或者显示某种沉积旋回的变化。例如黑色页岩组成的组，很可能代表一种深海缺氧的沉积环境；而砂岩、页岩互层的组，则显示沉积物的来源发生规律性的变化。它可能是季节造成了搬运水体动能出现差异，例如雨季水流动能较大，搬运来的就是砂粒；旱季水动能小，搬来的就是形成页岩的粉砂或泥质。

真神奇！从石头的组成就能分析出它们形成的自然环境。那神农架最常见的白云岩是什么环境下形成的呢？

白云岩是海洋中发生的化学作用形成的。海水是咸的，不是吗？那是因为海水

溶解了大量各种各样的盐类化合物。碳酸钙和碳酸镁是海洋中最常见的成分。

这些盐类化合物是从哪里来的呢？

它们有可能是流入海洋的河流带来的，也有可能是海洋中的生物，尤其是那些带壳的生物死后，它们的骨骼和外壳被海水溶解后形成的。

既然这些碳酸钙和碳酸镁能被海水溶解，怎么又会在海洋中沉积形成白云岩和石灰岩呢？

这就是大自然的奇妙之处。很多化学反应都是可逆的，它们随着控制化学反应条件的变化而改变。比如当温度较高的时候，碳酸钙是很容易被溶解的，但当温度降低、压力加大的时候，它又很容易沉淀下来。所以，在特定的温度、压力条件下，溶解在海洋中的碳酸钙、碳酸镁就会沉淀下来形成石灰岩或白云岩。

大自然真是太奇妙了。那我们是不是可以根据岩石的成分和特征来推断自然环境的变化呢？

是呀。这正是地质学、沉积学要解决的问题。也是我们对地层进行划分，并进一步找寻有用矿产的目的。

哦！地层划分还能找矿？大海里有什么矿呢？

地质学最重要的目的就是找寻有用的矿产。大海里形成的矿产很多。我们国家著名的“鞍山式铁矿”就与地质历史中的沉积作用密切相关。

是吗？它怎么跟沉积作用相关联呢？

地质学家把“鞍山式铁矿”归纳为“变质火山沉积矿床”。它的基本成矿步骤是：首先，在大海里沉积了巨厚的含铁质的火山沉积物；然后，在成岩的过程中，或者成岩之后，遭受了多期的变质作用，使铁质成分富集起来，就形成了有用的矿床。

啊。我想那些铁质成分是火山沉积物带来的吧？

大部分应该是的，但也有可能是后期变质作用产生的。沉积作用能形成许多不同的矿床，它是指在地表条件下，成矿物质被水或风、冰川、生物等搬运到水体内沉淀聚积而形成的矿床，我们称为“沉积矿床”。煤、石油和天然气就是最典型的代表。我国北方煤矿比较丰富，这与当时那里的气候环境有密切的联系：大片的森林覆盖为煤的形成提供了物质基础。另外，铁、锰、铝、磷等很多元素也能通过沉积作用形成矿床。

地质学简直太伟大了。我以后一定要当个地质学家，走遍山山水水去为国家找矿。

侏罗系
白垩系
深湖-半深湖相沉积
远岸浅湖相沉积

找矿并不是没有目标地漫山遍野地去瞎碰。一个地区有没有矿、可能会有什么样的矿，首先要根据这个地区的基本地质条件进行分析。所以，基础地质研究是最重要的步骤。地层学的研究就是最基本的地质勘查手段。

为什么说地层学的研究是最基本的呢？

因为地层学研究的成果能告诉你这个地区的地质发展历史，告诉你这里过去是海洋还是陆地、是大洋深海还是陆棚浅海、是否曾经有过火山活动、经历过什么样的地质构造变动等。

地层学怎样获得这些信息的呢？

哦，这个问题就相当专业了，而且很复杂。我们说，一个地区的地质历史主要就是根据它的地层层序来复原的，而地层层序建立的过程，实际上就是对沉积环境进行研究，并推断环境的演变历程、综合分析该地区地质历史发展的各个阶段和其特点。因此，地层学研究的成果必然构成对地区地质历史的总结。

问题：

1. 为什么说官门山景区的科普设施非常适合于家庭旅游，老少皆宜？
2. 从地质公园“地质点”的科学解说牌能学到些什么？
3. 为什么地质工作者要将地层划分为“群”“组”等单位？
4. 请查阅相关的资料，讲讲“沉积矿床”形成的基本原理。
5. 了解一下我国有哪些重要的沉积矿床。

22

知识老人滔滔不绝地讲解着地层学研究的科学道理，小明认真地听着，还不时提出一些新的问题。小狗汪汪跑前跑后，显然对知识老人的讲解毫无兴趣。它感兴趣的是栈道边草丛中的蝴蝶。

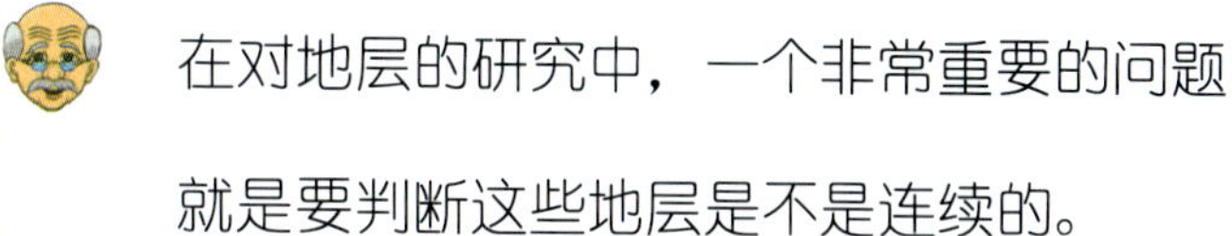

在对地层的研究中，一个非常重要的问题就是要判断这些地层是不是连续的。

那怎么判断呀？

地质学家是根据岩层之间的接触关系来判断的。岩层与岩层之间的接触关系大致有 3 种类型，即“整合接触”“平行不整合”和“角度不整合”。

我想，“整合接触”就是连续的沉积吧？

你说得非常对。“整合接触”，就是岩层之间没有明显的间断。它所反映的是连续的沉积作用。

“平行不整合”和“角度不整合”呢？

首先，“不整合”就意味着岩层之间存在着沉积间断。“角度不整合”是上下两套地层的产状不一致，它们以一定的角度相交。

哦，我明白了：早先的地层形成之后，肯定发生过很大的地壳变动，使它们不再是水平的了。

是的。你现在已经学会推论了。很好！“角度不整合”说明两套地层的时代是不连续的，两者之间存在时间上的间隔，

角度不整合的地层接触

甚至出现构造褶皱运动，有的还会留下一个明显的“剥蚀面”。

什么是“剥蚀面”？

“剥蚀面”是岩层从水中出露到地表后，遭受风化侵蚀，岩石被破碎氧化，留下一层砾石或砂土，这就是“剥蚀面”，或者叫作“古风化壳”。

“古风化壳”？这个名称挺有意思。一听就知道以前在这里曾经发生过风化作用。

是的。如果大面积地区内出现层位稳定的砾岩层，我们就要特别注意了。因为，它们有可能是“古风化壳”的残迹。

那说明整个地区都曾经经历过风化剥蚀。

说得对。岩石露出水面被破碎氧化后，就会留下一些砾石，当再一次接受沉积的时候，往往就形成砾石层。地质学家把它叫作“底砾岩”。

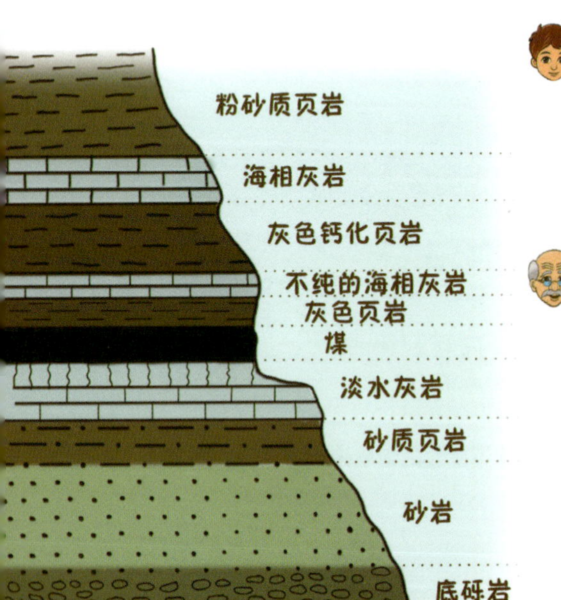

“底砾岩”，这个术语很好懂。这些砾石可能会被后来的沉积物覆盖。因为它在最底部，所以叫作“底砾岩”，是吗？

你真聪明，小明。我们可以把“底砾岩”看作是一个“标志层”。它显示另一个沉积旋回的开始。在神农架地区也有这种叫作“底砾岩”的砾石层，当地的人们把它叫作“宝石砾岩”。

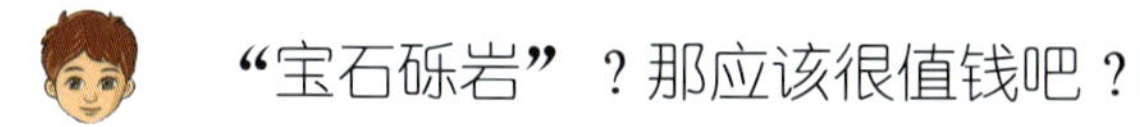

“宝石砾岩”？那应该很值钱吧？

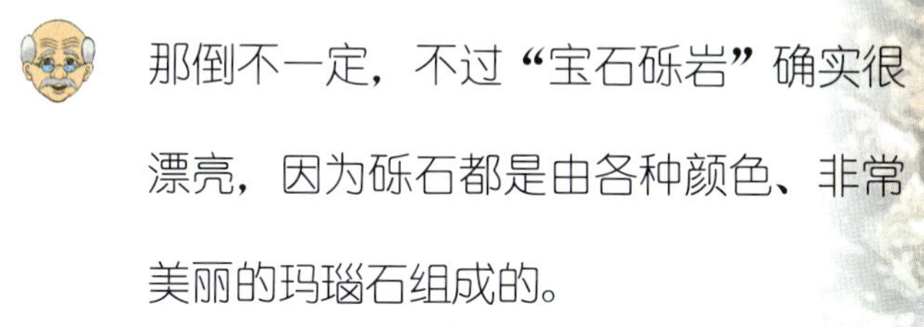

那倒不一定，不过“宝石砾岩”确实很漂亮，因为砾石都是由各种颜色、非常美丽的玛瑙石组成的。

什么是玛瑙呀，我经常听说玛瑙，是一种宝石吗？

玛瑙的主要成分是二氧化硅，它属于水晶家族的成员。只不过水晶是结晶状态的二氧化硅，而玛瑙则是隐晶质的，也就是说根本看不到矿物的颗粒。颜色或花纹漂亮的玛瑙具有较高的观赏性和收藏价值，例如非常著名的南京“雨花石”。

我见过雨花石，确实非常美丽。大自然真是很奇妙，为我们创造了这么多赏心悦目的奇迹！爷爷，什么是“平行不整合”呢？

“平行不整合”，又叫作“假整合”，指的是上、下两套岩层的产状基本上保持一致，但它们在形成的时间上存在着一段差距，也就是说，这两套地层之间有较长期的沉积间断。

那我们怎么知道有没有沉积间断呢？

我们可以根据上、下两种岩石性质的变化进行判断，但更重要的证据，应当是上、下岩层中所含的化石在时间上是否连续。

是的。化石是最有发言权的。

是的。小明，我还想跟你讲讲关于砾岩。你知道，砾石一般会在什么地方出现呀？

河边、湖边、海边！

对了。砾石一般都分布在水体的边缘。当水体进一步加深的时候，泥沙就会盖在砾石层的上面，对吗？如果是海洋的话，当水逐渐加深的时候，盖在砾石层上的首先是沙，水体继续加深时，水动能越来越弱，沉积物就会逐渐变成粉砂和泥，若水体再继续加深，就可能发生化学沉积了，出现石灰岩或者白云岩的沉积。

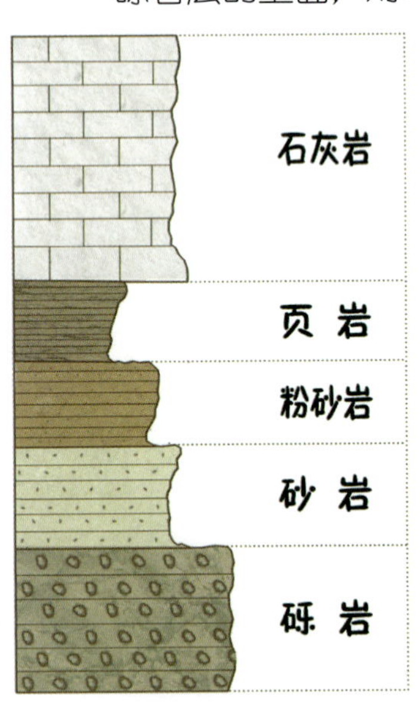

海进沉积序列

哦！好像水越深，沉积物颗粒就会变得越细，是吗？

是的。水越深，水的动能就越小，所能携带沉积物的颗粒自然就越细小。我们把这种随着水体加深、在地层中沉积物颗粒逐渐变细的沉积序列叫作“海进沉积序列”。

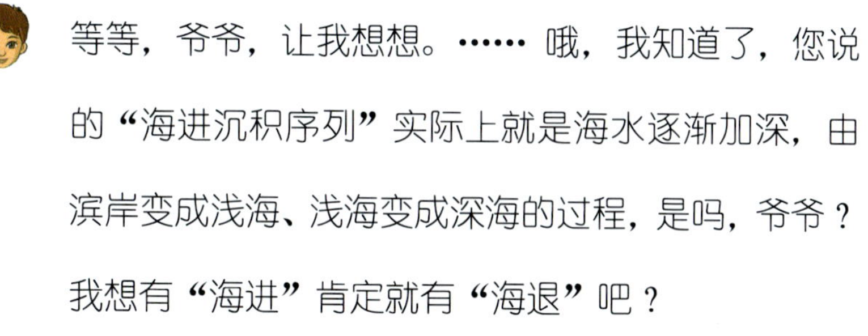

等等，爷爷，让我想想。…… 哦，我知道了，您说的“海进沉积序列”实际上就是海水逐渐加深，由滨岸变成浅海、浅海变成深海的过程，是吗，爷爷？我想有“海进”肯定就有“海退”吧？

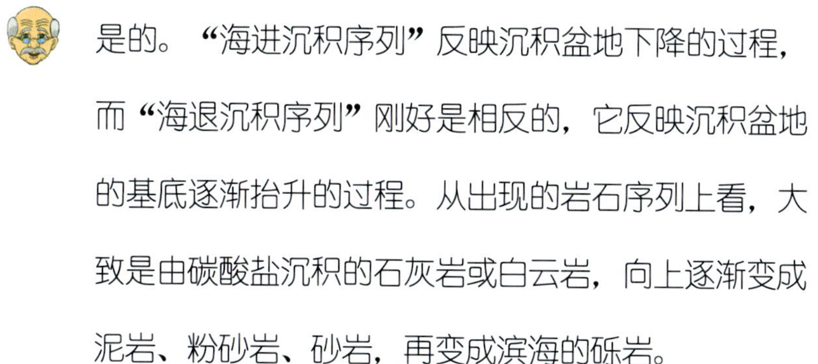

是的。“海进沉积序列”反映沉积盆地下降的过程，而“海退沉积序列”刚好是相反的，它反映沉积盆地的基底逐渐抬升的过程。从出现的岩石序列上看，大致是由碳酸盐沉积的石灰岩或白云岩，向上逐渐变成泥岩、粉砂岩、砂岩，再变成滨海的砾岩。

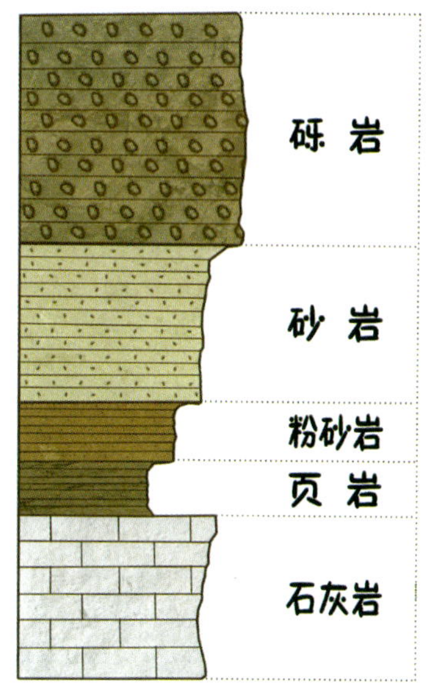

海退沉积序列

太有意思了！从岩层序列的变化就能分析出历史上地质环境的变迁。地质学家真了不起，我今后一定要当地质学家。

是的。从地层的发育情况，我们就能归纳出这个地方的地质发展历史，了解在这里曾经发生过一些什么样的事情。

地质学真是一门非常有趣的学问！

你再进一步想想，小明，如果我们按照时间的顺序，把全世界所有地方的地层都弄清楚了，然后进行一个横向的对比分析，互相补充，不就可以了解我们的地球在各个地质历史时期发生的各种事件了吗？

是呀！这确实是研究地球历史的最佳办法。

实际上，联合国教科文组织国际地层委员会和地质科学联合会（简称“地科联”）早就开始了这方面的研究。并且已经建立起了“国际地层表”和“国际地质年代表”。虽然还在继续完善的过程中，但“国际地层表”和“国际地质年代表”已经为分析全球的地质事件提供了基础。想想大陆漂移的生物化石证据和冰川证据，实际上不都是建立在这个基础之上的吗？

宙	代	纪	世	距今大约年代（百万年）	主要生物演化
显生宙	新生代	第四纪	全新世	现代	人类时代 现代植物
				0.01	
			更新世		
				2.4	
		第三纪（古近纪+新近纪）	上新世		哺乳动物 被子植物
				5.3	
			中新世		
				23	
			渐新世		
				36.5	
			始新世		
				53	
			古新世		
				65	
	中生代	白垩纪	晚 中 早		爬行动物 裸子植物
				135	
		侏罗纪	晚 中 早		
				205	
		三叠纪	晚 中 早		
				250	
	古生代	二叠纪	晚 中 早		两栖动物 蕨类
				290	
		石灰纪	晚 中 早		
				355	
		泥盆纪	晚 中 早		鱼
				410	
		志留纪	晚 中 早		蕨类
				438	
		奥陶纪	晚 中 早		
				510	
		寒武纪	晚 中 早		无脊椎动物
				570	
元古宙	元古代	震旦纪			古老的菌藻类
				800	
				2500	
太古宙	太古代			4000	

是的。真了不起！

为了在野外岩石地层中确定各个地质年代确切的分界点，地质学家创建了“金钉子”的概念。所谓“金钉子”是国际地层委员会和地科联，为了定义和区别全球不同时代所形成的地层之间的唯一标准或样板界线，在特定的地点和特定的岩层序列中将它标示出来，作为确定和识别全球两个时代地层之间界线的唯一标志。

爷爷，这是什么意思呀？我听不太懂。

宙	代	纪	世	期	年龄（Ma）	GSSP
					199.6±0.6	
显生宙	中生代	三叠纪	晚三叠世	瑞替阶	203.6±1.5	
显生宙	中生代	三叠纪	晚三叠世	诺利阶	216.5±2.0	
显生宙	中生代	三叠纪	晚三叠世	卡尼阶	228.0±2.0	
显生宙	中生代	三叠纪	中三叠世	拉丁阶	237.0±2.0	
显生宙	中生代	三叠纪	中三叠世	安尼阶	245.0±1.5	
显生宙	中生代	三叠纪	早三叠世	奥伦尼克阶	249.7±0.7	
显生宙	中生代	三叠纪	早三叠世	印度阶	251.0±0.4	⊤
显生宙	古生代	二叠纪	乐平期（世）	长兴阶	253.8±0.7	
显生宙	古生代	二叠纪	乐平期（世）	吴家坪阶	260.4±0.7	⊤
显生宙	古生代	二叠纪	瓜德鲁普期（世）	卡匹敦阶	265.8±0.7	⊤
显生宙	古生代	二叠纪	瓜德鲁普期（世）	沃德阶	268.0±0.7	⊤
显生宙	古生代	二叠纪	瓜德鲁普期（世）	罗德阶	270.6±0.7	⊤
显生宙	古生代	二叠纪	乌拉尔期（世）	空谷阶	275.6±0.7	
显生宙	古生代	二叠纪	乌拉尔期（世）	亚丁斯克阶	284.4±0.7	
显生宙	古生代	二叠纪	乌拉尔期（世）	萨克马尔阶	294.6±0.8	
显生宙	古生代	二叠纪	乌拉尔期（世）	阿瑟尔阶	299.0±0.8	⊤

举个例子吧，比如古生代的二叠纪与中生代的三叠纪是两个连续的地质时代。通过对野外剖面的详细勘察和研究，发现这两个地质时代的界线就出现在这个剖面中的某一点，而且这个点是世界上二叠系、三叠系最精确的分界点，这个点就代表了这两个时代地层之间的界线，就可以在这里定上一颗“金钉子”。国际上认可的二叠系、三叠系分界的“金钉子”就定在我们中国浙江省长兴县煤山。

哇，真了不起！总共有多少“金钉子”呀？

地质探秘神农架

全球地层年表中一共有“金钉子”110颗左右，目前已经确立的有近60颗，其中有10颗在中国。每颗“金钉子”的确立都意味着数十年大量艰苦繁杂的探索和研究。

中国的地质学家做了不少艰辛的科学研究，向他们致敬！

“神农架群”形成于10多亿年前的元古宙。这套地层在全世界并不多见。而且，由于年代久远，研究难度相当大，元古宙的10颗“金钉子”目前只确定了1颗。神农架很有希望为中国争取这颗古老地层的“金钉子”做出贡献。

200万：人类诞生
45.5亿：地球形成
2.3~0.65亿：恐龙
人类
哺乳动物
陆生动物
动物
多细胞生物
真核细胞
原核细胞
约3.8亿：陆生脊椎动物产生
45.27亿：月球形成
约5.3亿：寒武纪大爆发
7.5~6.35亿：二次雪球地球
约40亿：月球灾难结束；生命产生
约35亿：光合作用开始
约23亿：大气富含氧气；首次雪球地球
冥古宙 40亿
太古宙
元古宙
古生代
中生代
新生代
46亿
6500万
2.51亿
5.42亿
10亿
20亿
25亿
30亿
36亿

真的吗，爷爷？我一定要当地质学家，为我们国家多钉几颗“金钉子”！

好呀，小明，有志气。好好地学习吧，想当个优秀的地质学家要学的东西还多着呢。

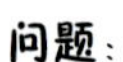

问题：

1. 什么是地层的“整合接触”“平行不整合”和“角度不整合”？
2. “底砾岩”是什么？辨认“底砾岩”有什么意义？
3. “海进沉积序列”和“海退沉积序列”是什么意思？它们反映什么样的环境变化？
4. 为什么要建立“国际地层表”和“国际地质年代表”？
5. 关于“金钉子”你知道些什么？

23

依山傍水的步游栈道把知识老人和小明带到了官门山景区的大门口。当那尊著名的“母爱”雕塑出现在眼前时，小明简直惊呆了！那是一对相拥的野人母子，人物形象栩栩如生：表面看来有些粗野的野人母亲，正蹲在地上，双臂拥抱着她的儿子，并深情地亲吻着他；小野人表现出对母亲的依赖和眷恋，神情单纯可爱。这对母子相拥的塑像把“爱”的普世价值和情感展现得如此淋漓尽致，令任何人都会为之动情。

小明注视着这尊雕像，慢慢地走出了官门山景区的大门。汪汪有点心神不定地跟在小明后面，静悄悄地，还多次不放心地回过头瞟望这对野人，有点胆怯地提防着它们。

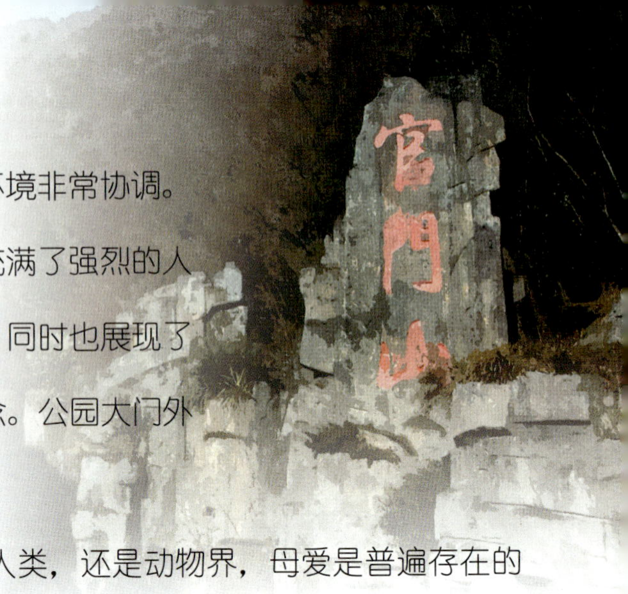

爷爷，这尊雕塑太完美了！它与这里的环境非常协调。

是的。这个公园大门的设计独具匠心，充满了强烈的人文精神，体现了普世皆存的高尚的母爱，同时也展现了人们尊重大自然，与自然和谐相处的理念。公园大门外的广场就叫作“母爱广场”。

“母爱广场”，这个名字真好。无论是人类，还是动物界，母爱是普遍存在的一种最珍贵的感情。

集地质地理、生物生态、科考科普，以及民俗展示于一体的官门山景区，以“母爱广场”作为参观游览的起点，具有特殊的意义。它引导人们不仅要珍视人与人之间的爱，而且还要培养人们对生物界、对大自然的热爱。如果每一个人都能以母亲对孩子的爱心，来珍惜大自然和地球环境，我们这颗绿色的星球必将变得更加美好。

爷爷，您讲得好极了！哎，那两个大石碑上刻的是什么字呀？

这边小点的石碑上刻的是“湖北神农架国家地质公园”，是 2005 年 9 月正式开园揭碑时竖立的。

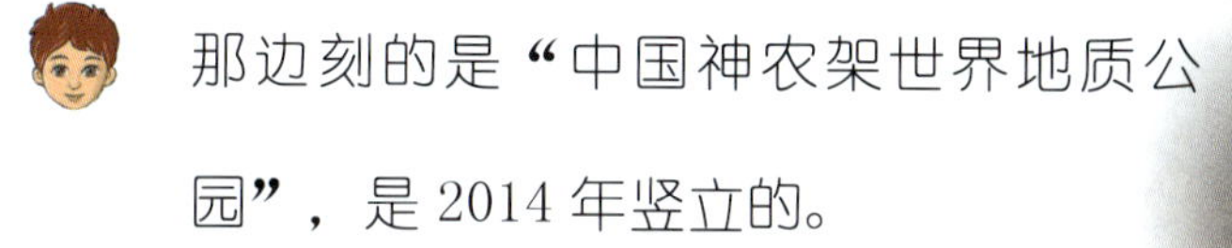

那边刻的是“中国神农架世界地质公园”，是2014年竖立的。

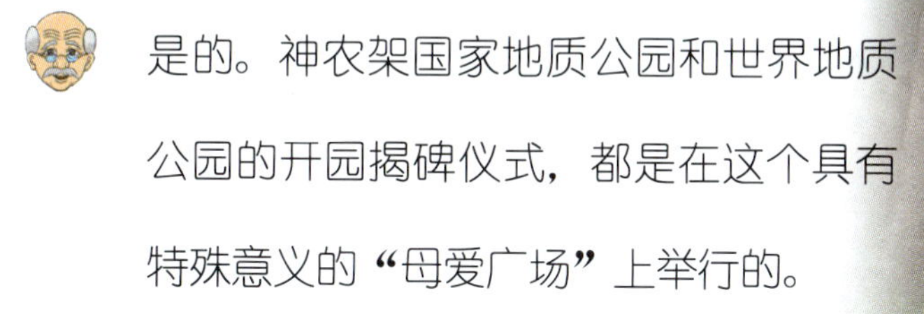

是的。神农架国家地质公园和世界地质公园的开园揭碑仪式，都是在这个具有特殊意义的“母爱广场”上举行的。

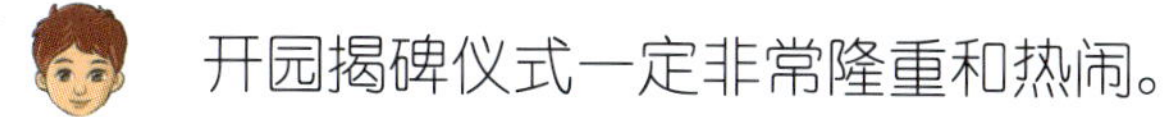

开园揭碑仪式一定非常隆重和热闹。

是的。小明，你过来看看，认不认得这广场周围是些什么岩石。

这个是……，哎呀，怎么又是像混凝土一样的，有点像神农顶上的隐爆火山角砾岩，但颜色不一样。

它确实是一种角砾岩，但是与火山作用无关。相反，它的形成与冰雪有密切的关系。

与冰雪有关？！

是的，确实与冰雪有关。它叫作“冰碛砾岩”。

哇，爷爷，大自然也太神奇了，火山能形成角砾岩，连冰雪都能造成角砾岩！快告诉我，这“冰碛砾岩”是怎样形成的！

这层“冰碛砾岩”被命名为“南沱组”。你看，这岩石以灰绿色泥砂质砾岩为主，砾石的成分复杂，为棱角状，大小不一，形状多样，磨圆度和分选性都很差。

什么叫磨圆度？什么是分选性呀？

磨圆度和分选性都是描述沉积岩石的常用术语。磨圆度是指岩石中的碎屑，比如石头碎块和矿物颗粒等，在流水的搬运过程中，经过长期反复的冲刷、滚动、撞击，它们的棱角被磨圆的程度。一般来讲，它们被搬运的距离越长，磨圆度就越好。当然与碎屑本身的硬度也有很大的关系。

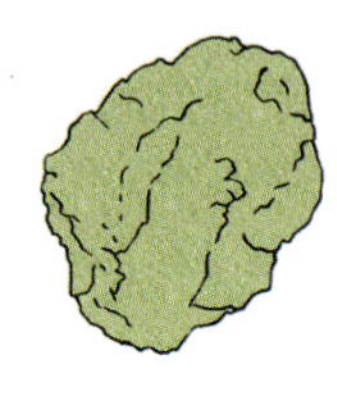

磨圆度示意图

石头碎屑经历了长时间、长距离的磕磕碰碰，肯定就变得又圆又光滑了，就像鹅卵石那样。

你说得很对。分选性则是指岩石中碎屑颗粒的粗细均匀程度。大小均匀的，就是分选性好的；大小混杂在一起的就是分选性差的。碎屑沉积物分选性的好与差，与沉积物运输载体的特征有关。

爷爷，什么叫运输载体呀？

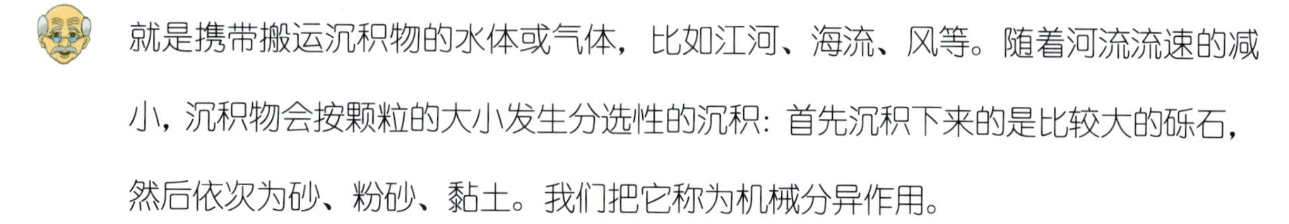

就是携带搬运沉积物的水体或气体，比如江河、海流、风等。随着河流流速的减小，沉积物会按颗粒的大小发生分选性的沉积：首先沉积下来的是比较大的砾石，然后依次为砂、粉砂、黏土。我们把它称为机械分异作用。

怪不得在山沟里看到很多巨大的石块，而在河边看到的常常是砂和泥。原来随着水流力量大小的变化，沉积物根据体积或重量就会分别沉积在不同的地方了。

是的！小明，你看看这“南沱组”的冰碛岩，它们有什么特征呢？

大大小小的砾石都混在一起，杂乱无章，好像个乱石堆。

所以它们显然不是河流带来的，而是冰川带来的。

冰川？

是呀，冰川也是一种运输载体呀。冰川在运移的过程中，会携带大量的泥砂石块等。当冰川融化的时候，它所携带的泥砂石块一下子就都沉积下来，堆在了一起。

哦，原来是这样呀！可是冰川不是在南极和北极，或者像喜马拉雅山那样的高山上才有吗？神农架这里过去也有冰川吗？

说来话就长了。这些冰碛砾岩形成于大约七八亿年前的新元古代，非常非常的古老。

什么是新元古代呀？

新元古代是地质年代表中的一个时代。地质学家根据绝对年龄测定和化石分析等科学手段，将地球46亿年以来的历史划分成了许多的时间单元。新元古代就是其中的一个阶段。它的时间延续大约是从10亿年到5.4亿年之间。

地质年代简表

时代划分		距今年限（Ma）
新生代		66.0
中生代		298.9
古生代	晚古生代	419.2
	早古生代	541.0
元古宙	新元古代	1000
	中元古代	1600
	古元古代	2500
太古宙		4600

哇。上10亿年呀，太古老了！但是我不懂，爷爷，干嘛要把46亿年的地球历史进行划分呢？

在地球漫长的历史中出现过许多重大的事件，这些地质历史事件在全球各个地方的表现都各不相同。我们现在研究地球历史的时候，一定要用一个同样的时间尺度去衡量。因此，我们必须要对地质年代有一个统一的划分，就形成了我们前面说的“国际地质年代表”。大家都根据这个时间划分来记录当地的地质事件。

哦。我明白了。但是在同一个时段，地球上各个地方发生的事情肯定是不一样的。

比如南北极是冰天雪地，但赤道地区则是烈日炎炎，说不定还会下大暴雨、山洪暴发呢。

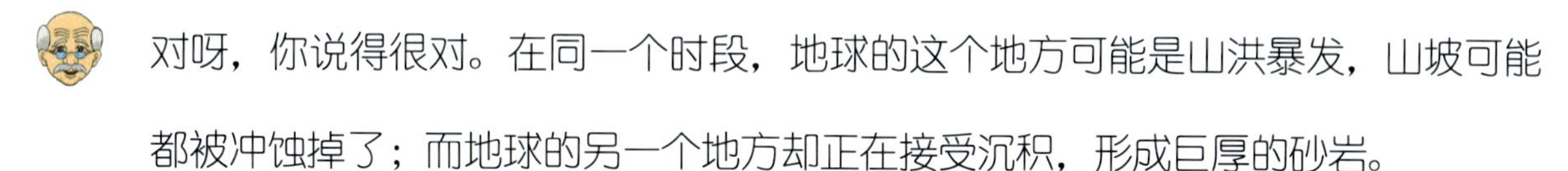

对呀，你说得很对。在同一个时段，地球的这个地方可能是山洪暴发，山坡可能都被冲蚀掉了；而地球的另一个地方却正在接受沉积，形成巨厚的砂岩。

可不是嘛。每个地方都有不同的发展历史。

是的。如果每一个地方都根据当地具体的地层和这个统一的“国际地质年代表”，建立起各自的地质历史，然后我们就能综合起来，对整个地球的发展过程进行归纳。

年代地层	系、组名称		
下古生界	志留系		
	奥陶系		
	寒武系		娄山关组—牛蹄塘组
			灯影组
新元古界	震旦系	上震旦统	
		下震旦统	陡山沱组
	南华系	上南华统	南沱组
			大塘坡组
			古城组
		下南华统	莲沱组
	青白口系		凉风垭组
中元古界	蓟县系	上蓟县统	神农架群
		下蓟县统	
	长城系	上长城系	
		下长城系	

哇，这可是个很复杂的事情，但很有意思！肯定是五花八门，乱七八糟，这里沉积、那里剥蚀；这里海进、那里海退；这里下大雨、那里火山爆发。

是的，确实很复杂。但是，在地球历史发展中，也曾经出现过一些具有共性的重大地质历史事件。

我觉得好像不太可能呀！爷爷，地球那么大，怎么会出现共同的现象呢？

那我就给你举一个就在我们眼前的实例：我们

面前“南沱组”的岩石，是在非常寒冷的环境下形成的。当时，也就是大约（7~8）亿年前，不光神农架这个地区被严寒统治着，世界上很多地方都被冰雪覆盖着。

是吗？您怎么知道的呢？

早在1964年，英国剑桥大学的哈兰德教授（Harland B W）发现大概在8亿年到5.5亿年前的这段时间内，在全世界各个大洲都出现了冰期沉积物。2000年，耶鲁大学地质学教授埃文斯（Evans D A D）等人，通过冰川沉积地层学、地质年代学和古地磁学的研究，指出（8~5.5）亿年这段时间内，地球上各个大陆冰期沉积的杂砾岩，大都集中出现在南北纬10°以内。小明，你想想，这意味着什么呢？

地球的赤道为0°，南北纬10°以内？天哪！那冰期沉积物不是出现在炎热的赤道附近吗？那怎么可能呀！

确实如此！在（8~5.5）亿年间，也就是新元古代，地球上的陆地基本上构成一个整体，叫作“联合古陆”，科学家们推测那时候地球曾经一度到处都被冰雪覆盖，赤道地区也不例外，整个地球变成了一个大雪球。当时冰盖有1km厚，地球的温度下降到摄氏零下50℃左右，并延续了上亿年的时间。这就是地球历

史上著名的“雪球地球事件”。

啊呀，太可怕了！幸亏当时我们不在。

哈哈哈哈！我们人类大约是在差不多6亿年之后才出现的，人类出现距今仅300万年左右。

爷爷，那后来呢？后来地球怎么变成现在这样的呢？

后来由于火山喷发，释放了大量的二氧化碳气体，经过长达1000多万年的积累，这些二氧化碳终于足够强大到造成“温室效应”，从而迅速融化了“雪球地球”，并使得全球海洋的温度上升到摄氏50℃以上。

啊呀，摄氏50℃以上！一下子又变得那么热，肯定会中暑呀！夏天气温上升到三十七八度就难受死了！

是的。后来“联合古陆”分解成许多陆块，海洋和陆地生物大量地繁衍，绿色开始统治了整个地球。终于，地球就逐渐形成了像现在这样适合于人类和各种生物生活的环境。

真是来之不易呀！爷爷，我觉得我们的地球好像是一个能够进行自我调节的、有生命的东西，它会努力把环境调节到最好。

是的，你的比喻很好。我们的地球确实具有自我调节的能力。可是，自从人类实现工业化之后，对大自然开始无休止地索求，采掘矿石、砍伐森林；同时，修建了大量的工厂，不断向大气排放二氧化碳和各种有毒气体，污染水源和土地，破坏环境。如今，地球已经快要支撑不住了，快要崩溃了。

那怎么办哪？爷爷，我们一定要帮助地球恢复良好的生态环境。

是的。目前地球面临的形势非常严峻：每年有将近 600 万公顷（1 公顷 =0.01km^2）的土地沦为沙漠、2000 万公顷的森林在消失；平均每一个小时就有一个物种绝灭！

哇，太可怕了！再这样发展下去，我们人类自己也将绝灭了！

所以，我们每个人都有责任和义务保护生态环境、保护地球资源！让地球处处都像神农架一样万年长青。好了，太阳快落山了，咱们赶快去“天生桥”走走吧。

问题：

1. 官门山景区大门前的“母爱广场”有什么象征性的意义？
2. 请讲讲“南沱组”冰碛岩的来历；为什么它的“磨圆度”和“分选性”都很差？
3. 什么是 8~5.5 亿年前的“雪球地球事件”？
4. 我们地球目前的生态环境状况面临什么样的挑战？
5. 应当怎样帮助地球恢复良好的生态环境？请列出日常生活中可实行的环保措施。

24

知识老人一边向小明进行讲解，一边走到“母爱广场”的中间，坐上了飞毯。太阳已经快要落山了，夕阳的余晖把周围的山顶映照得金碧辉煌。飞毯平稳地起飞，在夕阳的辉映下向群山环绕的“天生桥”景区飞去。

向下俯瞰，群山中一条平整的柏油路在绿茵茵的林中穿行，像带子一样缠绕在山腰。浓密的白云渐渐漫过山顶，像瀑布一样涌向山谷。当飞毯穿过云层，展现在他们眼前的，竟然是一个巨大的葫芦形的山洞。

这是一个非常漂亮的穿透型的溶洞，远看就像是一座桥横躺在山中。飞近一看，景色更加秀丽：岩洞旁边的石崖上，一股山泉从岩缝中喷涌而出，形成一道瀑布，如同一席水帘挂在岩壁上；山洞里激流轰鸣，湍急的溪流穿洞而下，形成高达10多米的瀑布，闪着耀眼白光注入到下面一连串的深潭之中；一条木质栈道依洞内悬崖攀升，几经曲折，穿过岩洞，向洞后碧绿幽深的山沟里面延伸过去。知识老人和小明乘坐的飞毯缓慢地向山洞口飞去。汹涌的激流穿过山洞奔腾而下，与一侧陡崖上的瀑布汇聚到一起，发出雷鸣般的轰响。

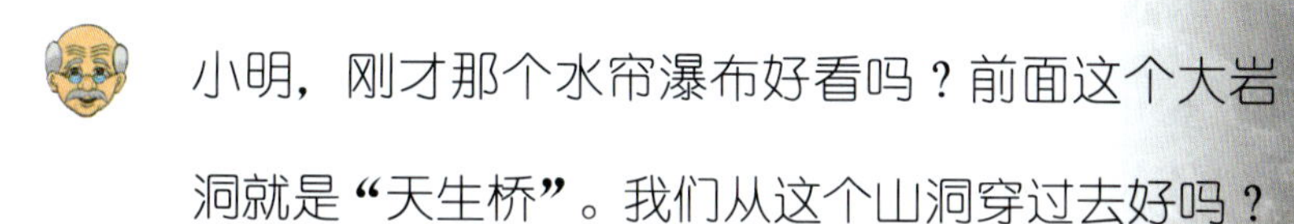

小明，刚才那个水帘瀑布好看吗？前面这个大岩洞就是“天生桥”。我们从这个山洞穿过去好吗？

那个水帘瀑布太美了！流水的轰鸣像千军万马在奔腾，好大的气势呀。您说“天生桥”，我怎么没看见桥呢？

远看这个山洞，不就像是一个巨大的桥洞吗。这种穿透型的溶洞在很多地方都以“天生桥”命名。

爷爷，这么大个溶洞是怎么形成的呢？

这个岩洞所在的山叫作“龙头山”。洞里的这条溪流叫作“黄岩河”。“黄岩河”起先是绕过“龙头山”向下游流去的。后来，由于“龙头山”的某个部位，可能出现了断层和许多的节理，使得岩石破碎，产生了大量的裂隙，加上这里的岩石又都是白云岩，非常容易……

我知道了！这里的石头非常容易被溶蚀。这些破碎的岩石在流水长期不断地冲击溶蚀之下，逐渐形成了这个溶洞。

是的。在开始的时候，可能只出现了许多小的裂隙，河水渗透进这些裂隙，逐渐使裂隙变大，越来越多的河水顺着裂隙向下流去，使河水发生了少量的分流。

哦，我明白了，刚才我们进洞前看到的那个水帘瀑布，应当就是“黄岩河”的一条分流，是吗？

是的。那个瀑布基本上是顺着岩层层理面形成的喀斯特溶孔流出来的。随着岩溶作用和垮塌的进一步发育，“黄岩河”逐渐把地下孔洞系统全部都打通了，慢慢地形成了一个巨大的喀斯特溶洞。最后，使“黄岩河”彻底地改道。以前绕过“龙头山”的老河道就被废弃了。“黄岩河”的流水一部分形成水帘瀑布，另一部分则穿过天生桥大溶洞，一路奔腾而泄，流向下游。

“黄岩河”真了不起，给神农架建造就了一座“天生桥”！

自然界的流水是一个伟大的工匠，现代山川地貌的形成，大多数都与自然界的流水有关。尤其是在碳酸盐岩分布的地区，类似天生桥这样的穿透型溶洞是非常普遍的。

我觉得像天生桥这样的穿透型溶洞喀斯特地貌特别好看，美极了。

是的，这就是我们所说的地质遗迹。它们不仅向我们揭示了一段非常有趣的地质发展历史，而且构成了地质公园不可再生的旅游资源，吸引着人们前来观赏、来研究、来欣赏大自然的创造。从而激发人们热爱大自然，保护生态环境的自觉性。

爷爷，正像您说的："自然界的流水是一个伟大的工匠。"这个"工匠"是那样地执着和坚持，真值得我们好好地学习。

你说得太好了！这就是大自然给人类的启示。会欣赏大自然的人，常常能从自然景观中得到许多启迪，变得更加聪慧。

是的。书中也说：人类的历史，就是探索大自然、向大自然学习的过程。

嗯，说得很对。小明你看，那边树荫下有一口井。这是我想特别向你介绍的一个地质遗迹景点。

是吗？我看见了，旁边的解说牌上写的是"天泉"。为什么叫天泉呀？它有什么特别之处呢？

这是一口很神奇的水井。不管春夏秋冬，这口井都会源源不断地有水涌流出来，从来不会干枯。

是吗，爷爷？怎么会这样呢？

从地质学上说，这个“天泉”是地下承压水的出水点。

地下水？我知道地下水，但是什么叫承压水呢？

你真的懂什么叫地下水吗？

地下水就是在地面以下的水呀，不是吗？

是的，但是不准确。根据地下埋藏条件的不同，水文地质学把地下水分为三大类，即上层滞水、潜水和承压水。

干嘛分那么细呀！有必要吗？地下水就是地面以下的水，多简单呀！

科学研究是很严谨的。每一个科学术语都具有明确的学术意义。“地下水就是地面以下的水”是一般人的理解。但是，小明，你不是想当科学家吗？

是的。那咱们严谨点吧。请您告诉我什么叫上层滞水？

上层滞水就是浸透在地表的土壤，或岩石裂隙里的雨水或融化的雪水。它们由于局部的隔水现象，比如地下的一层细密的黏土层，或者一层致密的石头，而停留在浅层的岩石裂缝或土层中。这些水会逐渐从地表的低洼处渗出来，汇聚到沟渠中，或者逐渐从土壤中蒸发消耗掉，或者向下渗透，汇聚到潜水中去。

那潜水是什么样的地下水呢？

潜水是埋藏在地表以下第一个稳定隔水层上面的地下水。

什么叫隔水层呀？

隔水层就是不透水的岩层。有很多岩石看起来很坚硬，但它们内部存在大量的细小孔隙，比如胶结得不太好的砂岩、砾岩等，水会充满孔隙，并顺着这些孔隙渗漏下去。但是，很多岩层基本上是不透水的，比如大多数的岩浆岩和泥岩就是很好的例子，水很难透过这些岩石漏到下面去，所以把它们叫作隔水层。

哦。我懂了。就是颗粒排列很紧密的石头。

隔水层上面储存的水就叫作潜水，也就是我们通常广义上所说的地下水。饮用或浇灌的井水，也大多是潜水。潜水的顶面叫潜水面，也就是我们常说的地下水位。每个地方根据地势和降雨量的多少，地下水位的深浅是各不相同的，而且还会经常发生变化。

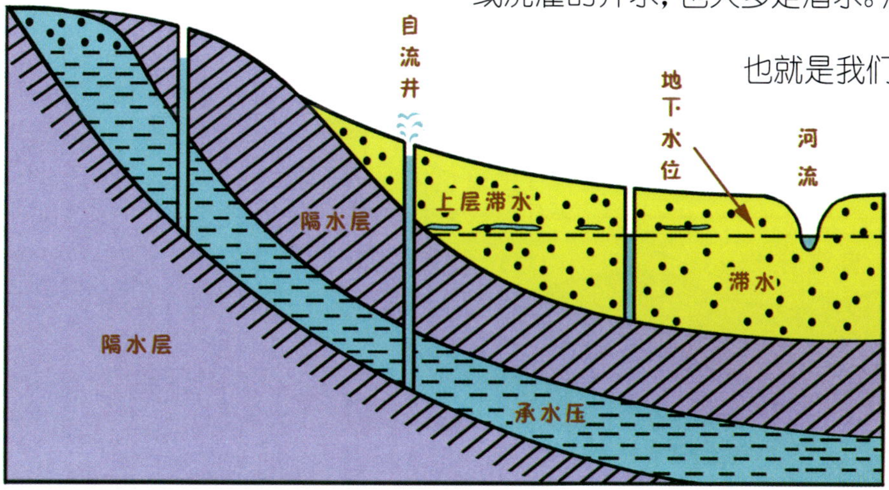

我知道，如果久旱不雨，地下水位就会降低，井里的水面就会变得很深。

当潜水从低洼处流出地面的时候，就形成了泉水。随着地形的变化，潜水还可能补充到小溪、河流、湖泊中去。

那承压水呢？承压水与潜水又有什么区别呢？

承压水是夹在两个隔水层之间的含水层中的水，含水层由比较松散的砂岩、砾岩等组成。形成承压水的条件是上下都有隔水层，而且水必须充满夹在隔水层之间的整个含水层。

为什么把它叫作“承压水”呢？它承受什么压力呢？

夹在两个隔水层之间的地下水常常受到地形的控制而被封闭在地下，它承受着静水压力，就像封闭在“U”形管里面的水，所以叫作“承压水”。由于顶部有隔水层，所以它的水源补给区要小于承压水的分布区。

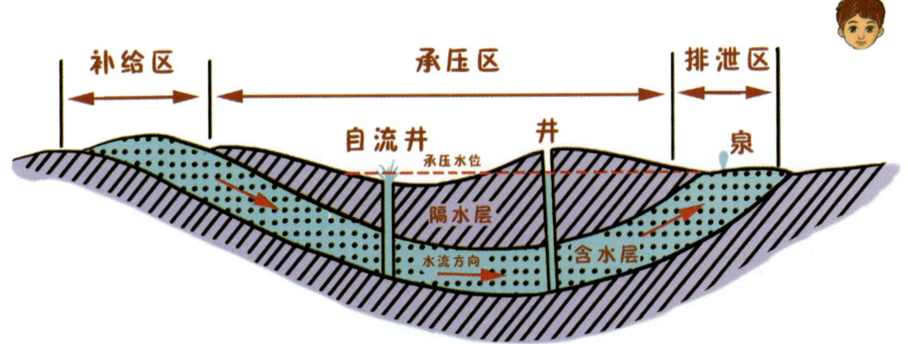

请您等一等，爷爷。我不懂，您说的“水源补给区要小于承压水的分布区”，这是什么意思呢？

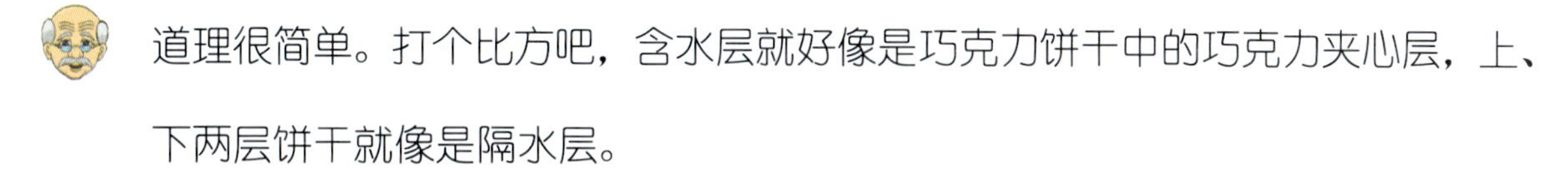

道理很简单。打个比方吧，含水层就好像是巧克力饼干中的巧克力夹心层，上、下两层饼干就像是隔水层。

嗯，还真有点像巧克力夹心饼干！

如果你还想再补充点“承压水”到含水层中去，那就只能从饼干的边缘往里加，对吗？

对呀，没错。……啊，我懂了，承压水的水源补给只能从隔水层的断缘处加进去！

你理解得很好，一般来说是这样的。由于受到隔水顶板的限制，因此承压水与大气圈和地表水圈的互动和联系比较微弱，不像潜水那样能经常参与地下的水循环。因此承压水不太容易遭受污染。绝大多数承压水来源于渗入水，所以，它们常常是比较清洁的、很适于饮用的淡水。

哦。那就是说“天泉”的水是很干净的，可以直接饮用，是吗？

是呀。神农架本来就地广人稀，没有什么污染。别说是“天泉”这样的承压水，就是很多山沟里的溪水，都是可以直接饮用的。

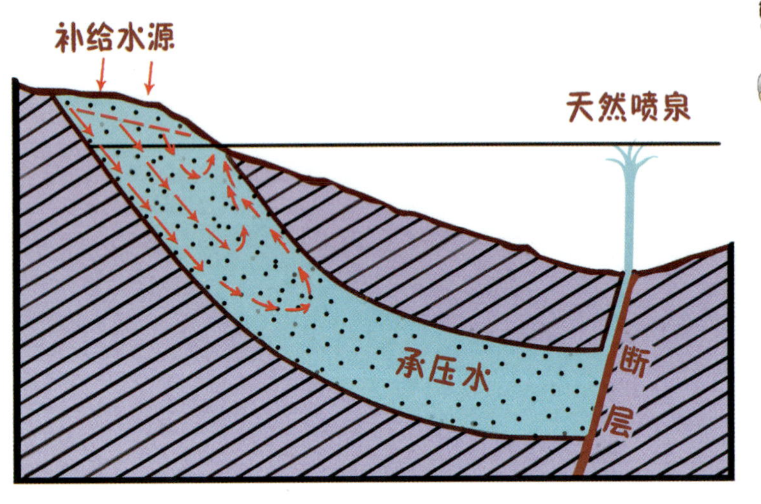

神农架真是个好地方！

承压水的形成与地质构造有密切的关系。形成承压水最有利的地质构造是向斜和单斜构造。它们分别构成自流盆地和自流斜地。

为什么叫“自流”盆地呢？水会自己流出来吗？

因为承压水在含水层中承受着静水压力，就像是一条灌满水的“U”形管，当你把这个“U”形管的一端抬高，水就会因压力从另一端喷流出来。对吗？

对呀，没错。

那么，在适当的地形和构造条件下，比如一条从地表通到地底含水层里的断层，就可能构成承压水的通道，使受到静水压力作用的承压水涌出地表，形成一个天然的喷泉。济南的趵突泉就是一个很好的实例。

哦！我懂了。神农架的“天泉”就是这样形成的，是吗？

对极了！世界上有很多大型的“自流盆地”，比如法国巴黎自流盆地、中国四川自流盆地、澳大利亚大盆地等都为人类提供非常优质的水源。

它们比神农架的“天泉”要大很多吗？

哈哈哈哈！小明，你还记得“地质思维”吗？我们以前介绍“地质思维”的时候，提到要从漫长的时间角度来考虑地质问题。现在，看来还得加上一条：那就是“地质思维”还必须考虑到研究对象的巨大规模。比如一个背斜构造可以横跨几十千米，一条大断层可以切过好几个省份呢。

哇，那么大的规模呀？！

这都还不算大呢！刚才我说到的“澳大利亚大盆地”就更加大了：它位于澳大利亚大陆中部偏东，介于东部大分水岭与西部沙漠高原之间。面积超过 175 万平方千米，将近占澳洲总面积的 1/4 呢。

真有那么大“承压水”的盆地？那简直是个地下海！

确实可以说是个地下海。这个自流盆地的东侧是大分水岭，由于受到来自太平洋东南风的影响，降水较多，丰富的雨水为自流盆地提供了充足的补给。

人们如何利用这些地下水呢？

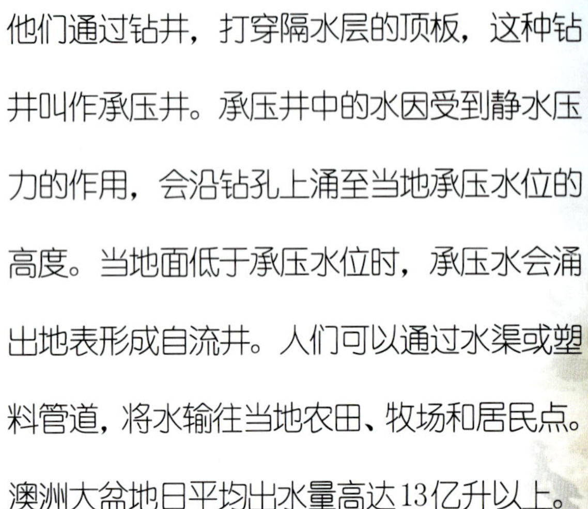

他们通过钻井，打穿隔水层的顶板，这种钻井叫作承压井。承压井中的水因受到静水压力的作用，会沿钻孔上涌至当地承压水位的高度。当地面低于承压水位时，承压水会涌出地表形成自流井。人们可以通过水渠或塑料管道，将水输往当地农田、牧场和居民点。澳洲大盆地日平均出水量高达13亿升以上。

真了不起！大自然为人们修建了一座好大的地下水库呀！

可不是嘛。大自然为人类创造了许多美好的东西。但是我们一定要小心地计划着使用，要爱护资源。这样才能保障人类社会可持续的发展。

是的。水是自然界的宝贵资源，是支撑人类生存的基本物质。

可是小明你知道吗，中国是一个水资源非常缺乏的国家。

不会吧。您看，咱们这里到处都是水呀！

是的，这里到处都是水，可是在我们国家，有许多地方到处都找不到水，连吃水都非常困难呢。

那是在沙漠里吧？那里当然找不到水。

不仅是在沙漠。根据全球的统计，中国是一个水资源极端缺乏的国家，在世界上排名第121位，是全球13个人均水资源最贫乏的国家之一。人均水资源只有2200m^3，仅为世界人均水资源的1/4。

啊！也就是说，全世界每个人平均能喝4杯水，咱们中国人只能喝1杯水。这是真的吗？问题真有那么严重呀？平日打开水龙头就来水啦，从来没想到我们竟是一个水资源非常匮乏的国家。

很多人都像你一样，不知道我们国家水资源形势的严重性。所以，我们有责任把这些情况告诉大家，让大家认识到这种危机，保护水资源，随时随地节约用水。

我一定要告诉我所有的朋友们，大家都来爱惜水资源。

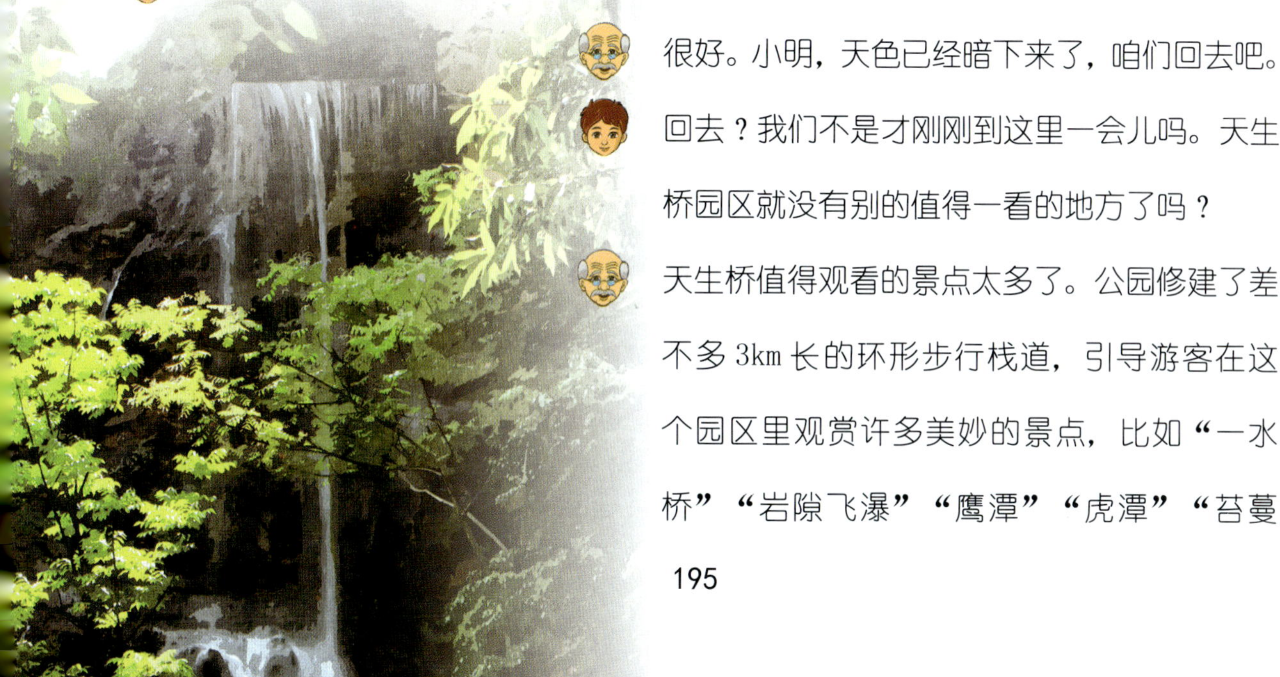

很好。小明，天色已经暗下来了，咱们回去吧。

回去？我们不是才刚刚到这里一会儿吗。天生桥园区就没有别的值得一看的地方了吗？

天生桥值得观看的景点太多了。公园修建了差不多3km长的环形步行栈道，引导游客在这个园区里观赏许多美妙的景点，比如“一水桥”“岩隙飞瀑”“鹰潭”“虎潭”“苔蔓

溪”“牵山桥”“老君驿站”“清风桥”“明月潭”等。

哇，这么多景点呀！

沿着栈道，人们还可以看到重塑远古人类生活的“巴人部落”，以及介绍当地的民间手工艺品和过去人们生产日常食品的各种作坊，比如“糖坊”“酒坊”“榨坊”“豆坊”“面坊”等。这些作坊再现了民间古老的生产方式，充分展现了神农架山区劳动人民巧妙利用自然资源的技巧和聪明智慧。

那咱们去看看呀，爷爷！

时间来不及了。你看，天都快黑下来了。要不，咱们很快地飞一圈，大致看看，下次来了再仔细地游览，好吗？

好吧。……哎，爷爷您看，那个圆形的建筑是干什么的呀？

啊，那是一个叫作“堂戏”的演出戏台。

什么是“堂戏”呀？

“堂戏”就是当地老百姓在家里堂屋表演的自娱自乐的文娱节目。神农架人爱唱歌，这里山歌的内容涉及劳动和生活的方方面面，包括田间耕种劳作、采药狩猎、祭祀庆典、婚丧嫁娶、爱情表达，甚至军旅征战等。

我听过神农架梆鼓，很好听，挺豪壮的！

天生桥景区特意在这里搭建了戏台，再现山区的古老风俗。

哇，太好了！下次来天生桥一定要听听堂戏，仔仔细细地好好游览一下。

好的。太阳落山了，咱们也该回武汉了。

好，走吧。汪汪，咱们要回家了。再见了，天生桥！再见了，神农架！

问题：

1. 你能分析一下“天生桥”穿透性大溶洞是如何形成的吗？
2. 有哪些重要因素控制溶洞的形成？
3. 请以“天泉”为实例，解释什么是承压水。
4. 阅读本书后，你是如何理解“地下水”的？
5. 我们应当怎样爱惜和保护我们的水资源？请举出日常生活中的例子。

尾声

飞毯载着知识老人、小明和汪汪，轻快地飞过龙头山。小明俯瞰着暮色中的群山，心中泛起无限的留念。他高声地呼喊：“神农架，我很快会带小伙伴们来看你的！”他的叫喊声像在平静的池塘中扔进了一块大石头，在山野中激荡起重重叠叠的回声。飞毯在夕阳映红的晚霞中翱翔，就像一叶小舟在群山翻动着的绿浪中前行，越飞越远，渐渐地变成了天边的一个小黑点。在小明的心中，一个美好的计划正在酝酿着：神农架，我一定要再回来好好看看你！

Shennongjia UNESCO Global Geopark Popular Science Series

The Geologic Discovery of Shennongjia

Li Xiaochi

Cataloguing-in-Publication Data
The Geologic Discovery of Shennongjia / by Li Xiaochi—Wuhan:China University of Geosciences Press, May, 2017.
(Shennongjia UNESCO Global Geopark, Pupular Science Series)

ISBN 978-7-5625-4029-8
Chinese version library CIP data Number(2017) No. 088829

Author: Li Xiaochi
English Translator: Feng Qinggao
English Editor in Charge: J.A. Grant-Mackie (NZ) Wang Min
Profreading: Lin Quan
Art Designer: Lai Liangxin
Art Editor: Splendid Color Creative Studio
Print Runs: 2000
Price: 128.00 Yuan
889mm×1194mm 1/20 Number of words: 590 000

Published by China University of Geosciences Press
388 Lumo Road
Wuhan 430074, Hubei,China

The Geologic Discovery of Shennongjia

Editorial Committee

Editor Affiliation for Book Series:

Administration of Shennongjia National Park

Preface

Dear readers, ***The Geologic Discovery of Shennongjia*** is a generous gift for you from the Shennongjia Geopark, China. Shennongjia is a scenic area of fantastic mystery. With the beautiful legends of Emperor Yan(Yandi), Shennong, the glorious epic “Darkness” of the Chinese nation, the lush virgin forests, the majestic mountains constituting the “Roof of Central China”, the Yinyuhe Valley comparable to the famous Grand Canyon in North America, the vast and spectacular Laojun Mountain and many supernatural stories of the “wild men” or “Man-Monkeys”. Shennongjia is a wonderful natural attraction awaiting your exploration.

As a member of the World Network of Biosphere Reserves, of UNESCO’s “Man and the Biosphere Reserve Program”, a site of world natural heritage, Shennongjia UNESCO Global Geopark carries the important responsibility of protecting the earth's natural resources and geological relics, advancing scientific research and popularization of geosciences, and promoting the economic development of the local community as well. Both earth environment and geological relics are valuable non-renewable resources. However, the rapid industrial development and humanity's ever increasing demands towards nature have unfortunately caused environmental damages and natural resource consumption to a dangerous degree. The fundamental key to resolving these problems is to educate people who should undertake their due responsibility and act together to protect the earth for ourselves and for the survival and happiness of our descendants as well.

This book is a popular science reading, not only for adolescents and children but also for adult tourists to Shennongjia. This book can serve as an interesting story, a book of scientific knowledge and a tourist brochure. It gives a comprehensive scientific introduction to the main attractions of Shennongjia Geopark and characteristics of various geological relics: how they came into being and the scientific principles of their formation. By combining popularization of scientific knowledge with interesting

reading, this book is sure to stimulate the reader's interest in learning and loving nature. It will also educate and instruct people how to respect nature and conscientiously fulfill their duties of protecting the earth's environment.

This book will bring the reader into the beautiful fairy-tale world of Shennongjia through a story of a professor and a little boy Xiaoming who are visiting Shennongjia Geopark on a flying carpet. Their dialogues will inspire the readers to explore. The 120 questions raised in this book are broad in coverage of scientific knowledge. Some are simple questions for primary and secondary students; some are interesting questions for adults' deeper consideration and discussion. There are also open-ended questions that have no direct answers or conclusions from general reading of this book. These questions are designed to inspire the readers' extensive reading or online enquiry by using modern information tools so as to expand their knowledge. In nature, geology and earth environment are always inseparable just like the relationship between "skin" and "hair". Specific geological conditions would inevitably produce specific natural environment. Everything is based on geology in nature. Accordingly, it is really hard to distinguish between the geological and geographical environment. Thus, though entitled *The Geologic Discovery of Shennongjia*, this book actually goes far beyond geology itself.

Dear readers, I sincerely hope you will like this book and take it as a key to unlock the gate to the mysterious world of Shennongjia. Welcome to Shennongjia!

Dr. Li Xiaochi
December 22, 2016

01

In the early morning, a senior professor and Xiaoming are starting their long-awaited journey to Shennongjia on a flying carpet with the little dog Wangwang from Wuhan. High buildings of the city are becoming clearer and clearer in the dim light. Down below is the sky-reach TV tower standing aloft on Guishan Hill, the resplendent Yellow Crane Tower shining above the Yangtze River and beetle-like cars bustling on streets and bridges. A new busy day has begun.

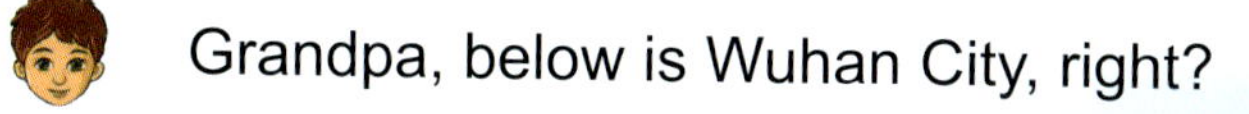

Grandpa, below is Wuhan City, right?

Right. Look at this beautiful city! Wuhan is the largest and the most popular city of Hubei Province and also the provincial capital.

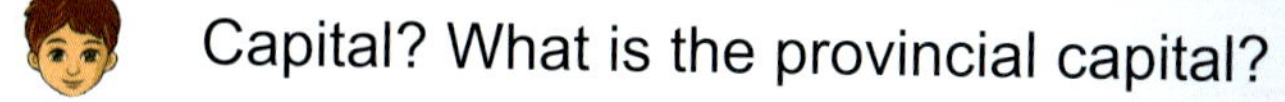

Capital? What is the provincial capital?

A provincial capital is the site of the provincial government, for example, Changsha is the provincial capital of Hunan, and Nanchang is the provincial capital of Jiangxi. Wuhan is the provincial capital of Hubei Province.

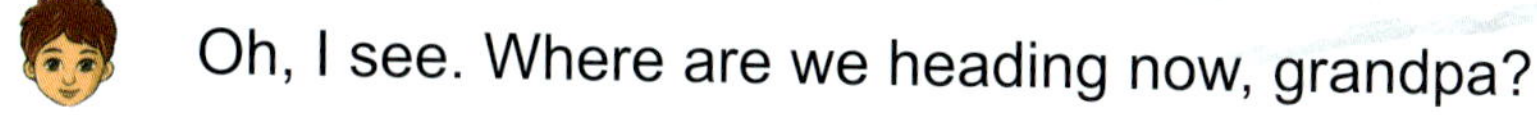

Oh, I see. Where are we heading now, grandpa?

The morning sun is right behind us. So then, which direction are we facing now?

Well, we are now flying toward the west, since the rising sun is behind us.

That's right. Shennongjia is to the north-west of Wuhan City. It is close to Chongqing City.

Grandpa, is Shennongjia far away?

No. It is in the northwest of Hubei Province, only 480km away from Wuhan.

Wow! Look, Grandpa. How wide the river is on the left!

Oh, yes. It is the Yangtze River. It's to our left, or to the south. So, can you decide which direction the river flows?

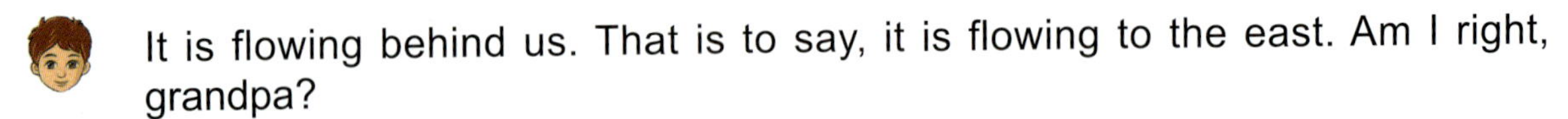

It is flowing behind us. That is to say, it is flowing to the east. Am I right, grandpa?

Absolutely right! As the song goes, the Yangtze River is rolling eastward. The two longest rivers in our country, the Yangtze River and Yellow River, both flow from west to east.

Look there, grandpa! There's another river over there in our right.

Well, that is the Han River, the longest branch of the Yangtze River. The area between the Yangtze River and Han River and that surrounding the confluence make up Jianghan Plain, the granary of Hubei Province. Xiaoming, can you tell me why the Yangtze River, Han River and Yellow River are all flowing eastwards?

Let me think. Oh, I see. The terrain of our country is higher in the West and lower in the East. Water seeks its own level. Therefore, large Chinese rivers all flow from west to east.

That's right. Do you see those distant mountains over there? They are Exi Mountain Regions. And Shennongjia is located there.

Exi? Where is Exi?

Exi refers to the western region of Hubei Province. "E" is the abbreviation of Hubei Province. In China, each province is abbreviated to a single character. For example, "Gan" stands for Jiangxi Province, "Xiang" for Hunan, "Yu" for Henan, and so on.

Oh, I see. Our Chinese characters are really amazing. They are concise and accurate. Wow! Grandpa, look there. How huge the Exi mountains are! The clouds are wafting around the mountainside.

Exactly. There are 6 peaks higher than 3000 m. Among them Shennongding is the highest. Its altitude is 3106.2 m.

What is the meanning of altitude?

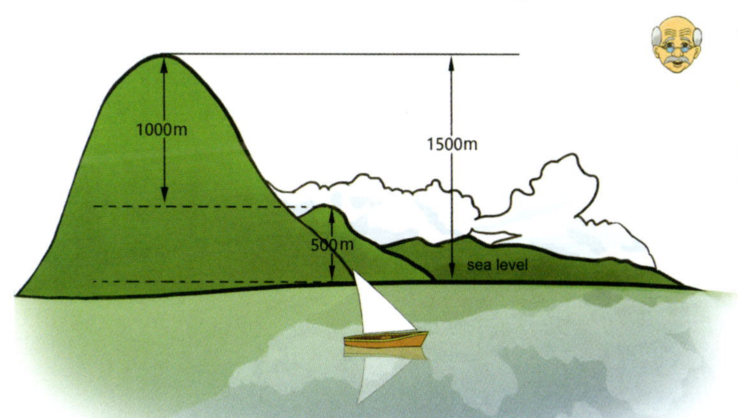

Well, altitude means the height calculated above sea level. Shennongjia rises so abruptly like a skyreach pillar that it has got the name of "Roof of Central China".

Roof? Do you mean Shennongjia is the highest place in Central China like the roof of the world Himalayas?

Sure it is.

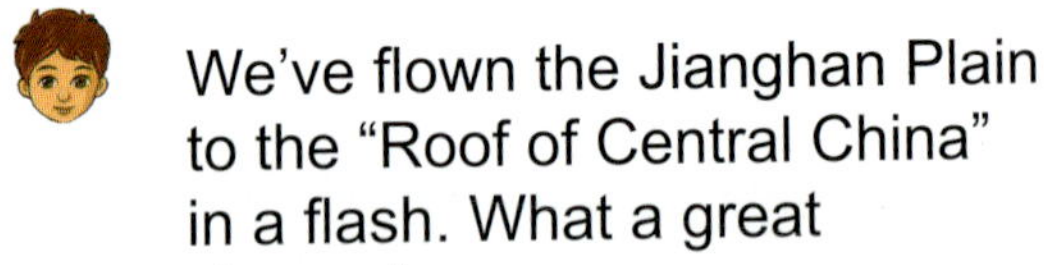

We've flown the Jianghan Plain to the "Roof of Central China" in a flash. What a great change!

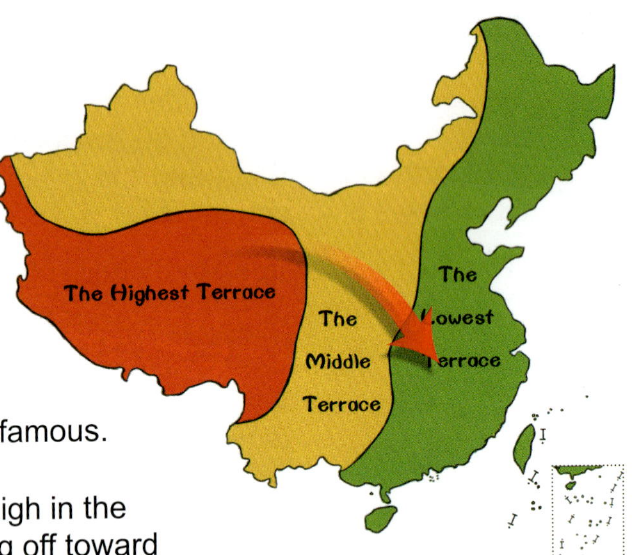

Yes. China is a big country with a complex and varied terrain. There're high mountains, hills, plains and lowlands below sea level.

Yeah, the Tibet Plateau is world famous.

The general terrain of China is high in the West and low in the East, sloping off toward the sea. It can be divided into three geomorphological ladders.

Three ladders? How are they divided?

The first ladder refers to the Tibet Plateau. Its average altitude is more than 4500 m.

Then, how about the second ladder?

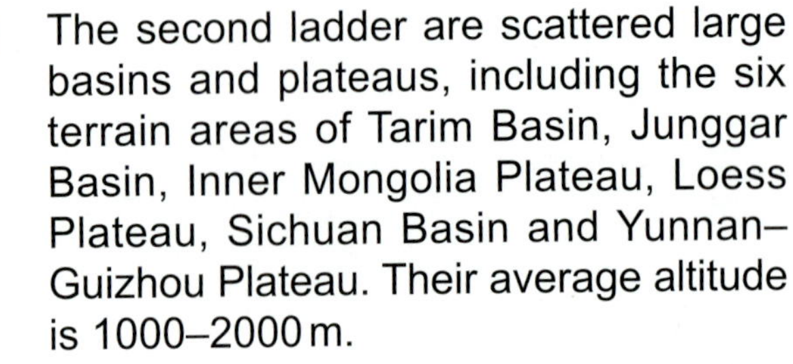

The second ladder are scattered large basins and plateaus, including the six terrain areas of Tarim Basin, Junggar Basin, Inner Mongolia Plateau, Loess Plateau, Sichuan Basin and Yunnan–Guizhou Plateau. Their average altitude is 1000–2000 m.

Then, must the third ladder refer to the coastal plains?

Yes. Scattered on this ladder are vast plains, as well as hills and low mountains at an altitude of less than 500 m above sea level, such as the Northeast Plain, the North China Plain and the middle-lower Yangtze Plain. They make the three areas of natural terrain.

Wow, just like a three-storey building. We are living on the first floor, or on the first ladder, aren't we?

Well, you're very good at imagination. We are on the first floor while Shennongjia is up on the second floor and even at the top of the stairs.

Is that so? What do you mean?

Shennongjia is located in the northwest of Hubei Province, where the second and third ladders meet. In terrain it extends from the southwest to the northeast starting from the eastern Qinling Mountains. And in landforms, it is characterized by mountains merged with rivers, showing conspicuous contrast of high and low levels.

Or, Hubei province seems to be an advantaged place. It occupies two floors.

OK, Let's "go upstairs" right now.

Well, "upstairs"!

Shall we go and visit Shengnongding first? Shennongjia forestry district is very vast with an area of 5000 km^2. The Geopark alone covers an area of more than 1000 km^2.

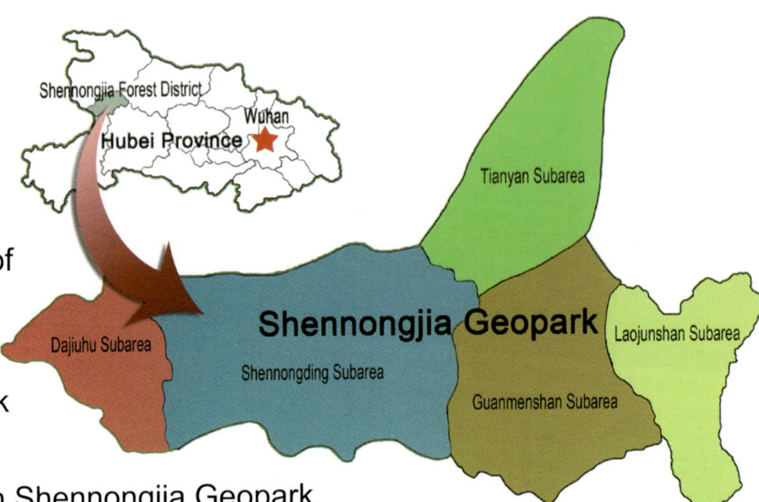

Wow, what a big park! How can we have a look at it?

There are five zones in Shennongjia Geopark, namely Shennongding, Guanmenshan, Tianyan, Dajiu Lake and Laojun Mountain. Our main task this time is to visit Shennongding and Guanmenshan, for many scenic spots are concentrated in these two zones. We may have a general impression of Shennongjia Geopark after our visit to them. Shennongding is the highest of the five zones. The temperature will drop a lot because of the great height of the peak. Now put your coat on.

Grandpa, why is it colder on higher mountains?

A very interesting question. Maybe you can find the answer by reading references.

Questions:

1. How many provinces are there in China? Do you know the abbreviated name and the capital city of each province?
2. How do you judge directions without a compass?
3. Where is Shennongjia located? Why is it called the “Roof of Central China”?
4. Can you systematically describe the main features of Chinese terrain?
5. Why the higher the mountains, the lower the temperature?

02

With the senior professor, Xiaoming and his Wangwang on it, the carpet is flying high up over the fields and mountains toward Shennongjia, leaving the white clouds behind. It goes up and up gradually with the terrain. As the sun rises, the clouds are gradually scattering and the visibility is getting better and better. The mountains are covered by vast forests. It seems a wild profusion of vegetation, seething and teeming with life.

Wow! The woods are so dense in Shennongjia! Look down there! The boundless forests are just like a sea of green gentle waves with the ups and downs of the terrain.

That's true. Forest covers 96 percent of Shennongjia. It gives Shennongjia the only well-preserved subtropical forest ecosystem in today's global mid-latitudes.

What is mid-latitude, then?

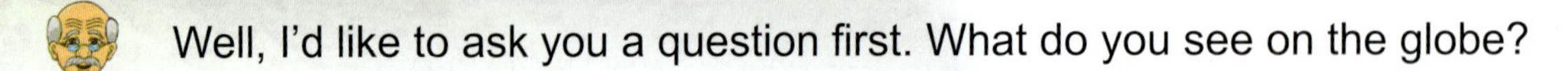

Well, I'd like to ask you a question first. What do you see on the globe?

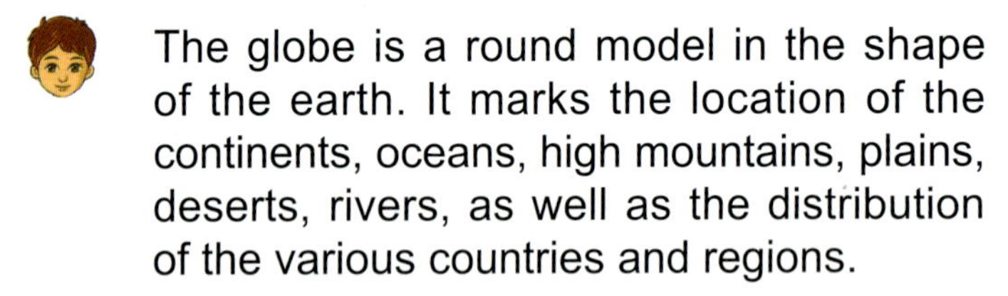

The globe is a round model in the shape of the earth. It marks the location of the continents, oceans, high mountains, plains, deserts, rivers, as well as the distribution of the various countries and regions.

Anything else?

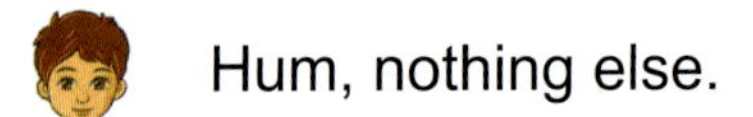

Hum, nothing else.

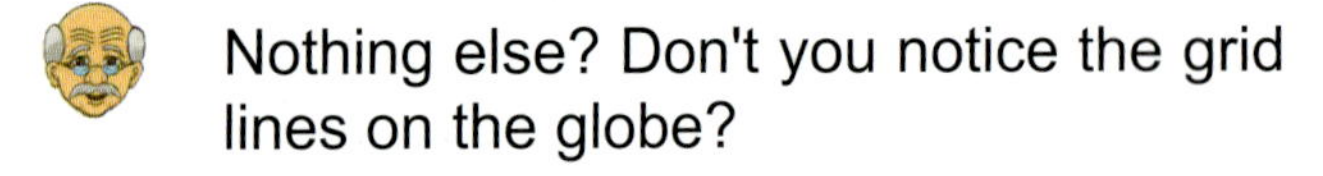

Nothing else? Don't you notice the grid lines on the globe?

Oh,yes. But what's the use of these lines? I believe they do nothing but divide the earth into a lot of small pieces.

Don't dismiss these lines. They are very important. The vertical lines connecting the north and south poles are called "longitudes". Any one of the longitude circles can divide the earth equally into two hemispheres. The horizontal circles are "latitudes". They intersect the longitudes at right angles. The longest "latitude" in the middle is what we call the "equator".

Oh, I see, I see. It is the equator that divides the earth into the southern hemisphere and the northern hemisphere.

That's right. The latitude and longitude of the earth are marked in "degrees". The total degrees for global longitude are 360°. But for latitude, it is divided into 90° from 0° at the equator to North Pole and South Pole respectively. The latitude and longitude form a complete coordinate system to divide the earth into different parts according to certain rules. Every specific location in the world can be accurately identified by its "longitude" and "latitude".

Just a minute. I think I've got to learn why radio and television tell us where in degrees east longitude and degrees north latitude the epicentre is located in reporting an earthquake. That's the function of the longitude and latitude. Right?

Yes. But they do more than that. For example, the division of global time is also associated with longitude, while the temperature of different parts of the earth is closely related to the latitude, too.

Well, Grandpa, how about the mid-latitude you've just mentioned?

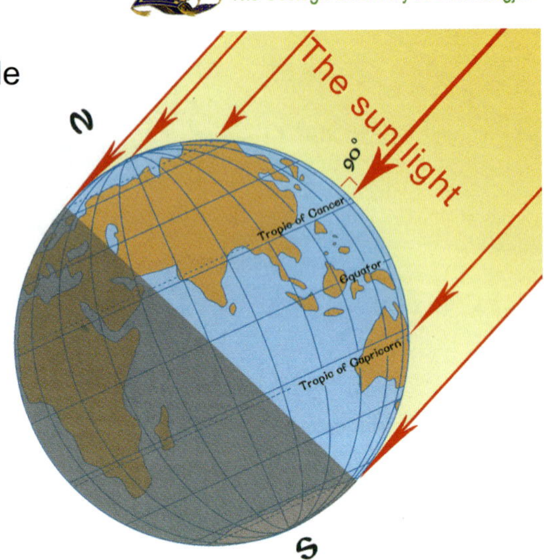

As you know, the earth is divided into two hemispheres by the equator. Near the South Pole and the North Pole, there are two circles of dotted lines (located 23°26' south and north latitudes respectively), named as "Tropic of Capricorn" and "Tropic of Cancer".

What strange names! It would be too much trouble to draw so many lines. Who did it?

This is determined by the observations and studies of scientists. It seems a little bit complicated but it provides a lot of convenience for the study of geography. For instance, the Tropic of Capricorn and Tropic of Cancer respectively stand for the southernmost and northernmost positions where the direct sunshine can reach on the earth within a year. Direct sunlight will not move further towards the poles but return toward the equator from here, hence the name of tropics, or return lines.

OK. I've got to know the original meaning of tropics.

According to the observation of scientists, direct sunlight repeats its movement between the equator and the Tropics of Capricorn and Cancer. When direct sunlight is shining between the equator and the Tropic of Cancer, we have summer in the northern hemisphere while it is winter in the southern.

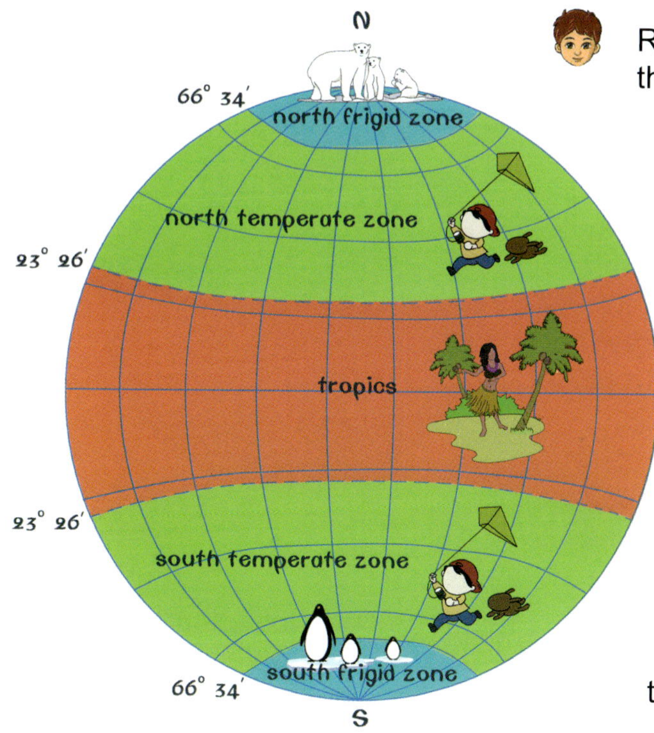

Really? So seasons are opposite in the two hemispheres?

No doubt. When the sun shines directly over the equator and moves into the southern hemisphere, winter begins in the northern hemisphere.

Wow, what a wonderful earth we've got. The direct sunlight always moves to and fro with the equator as the centre, sometimes to the South, sometimes to the North.I think that's why the equator is so hot all year round.

Yes. The equator is the tropical zone of the earth. The zone between the Tropic of Cancer and the Arctic Circle is the North Temperate Zone. The corresponding region in the southern hemisphere is the South Temperate Zone. Scientists have called the regions in the Arctic Circle and Antarctic Circle “high latitude regions”.

What's the difference between “high latitudes” and other regions? Wait a minute, grandpa. Don't tell me. I think “High latitudes” refers to the regions of the Antarctic Circle and the Arctic Circle. It is especially cold throughout the year as the sun never shines straight down there. So, the equatorial region must be called “low latitudes”, right?

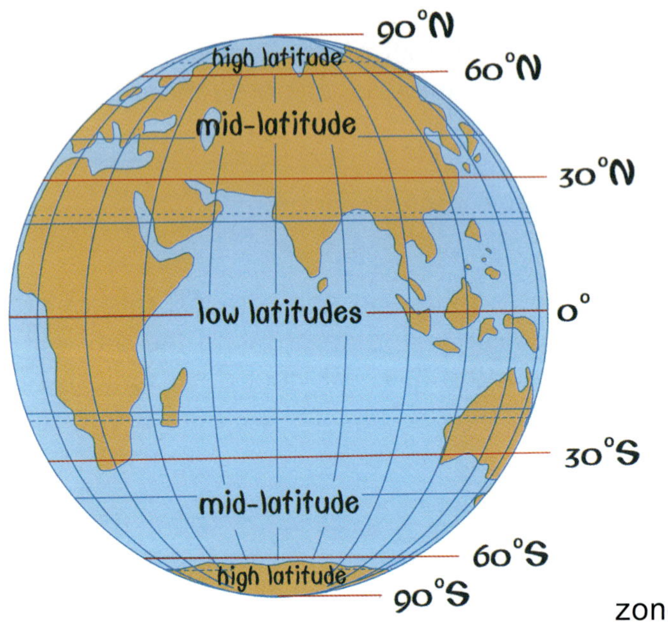

You're so clever, Xiaoming. So you understand what is called "mid-latitudes", don't you?

Yes. I guess the mid-latitude region must refer to the "north temperate zone" and "the southern temperate zone".

Roughly right, but not so exact. The "middle latitude region" is indeed in the south or north temperate zones, but the scientific term of "mid-latitudes" specifically refers to the zone between southern latitude 30° and northern latitude 30°.

Wow, it's so complicated. Why should we set out a middle latitude region?

Because it is a special region. The cold air mass of high latitude and the hot and humid air mass of low latitude overlap here. This results in frequent cyclone activities and complex climate changes especially in the Northern Hemisphere because a lot of land is concentrated here to make things more complicated.

Well, that's why scientists investigate this region as a special area, right?

Yes. The non-periodic variation of the weather and the change of precipitation

in the middle latitudes are very complicated. Besides, the continental area is vast in the northern hemisphere. The effect of sea-land thermodynamic contrasts and complex terrains make the climate of this area more perplexing and changeable.

But then, what is sea-land thermodynamic contrast, grandpa?

There is a great difference between the storage and release rate of solar energy in the land and the sea. The solar heat stored during the day will be released slowly at night in the sea, but the land will cool down quickly. The thermal differences between land and sea will become greater as one of the important causes of monsoon.

Then, what is monsoon?

Oh boy! You are so inquisitive. Now we are approaching Shennongding. As for what is monsoon, you may leave it as a question and search for the answer afterwards. It’s really important to understand what the monsoon is.

Questions:

1. What are the globe and its function? Can you name all the lines on the globe?
2. How are the earth's longitude and latitude divided? And what are their functions?
3. What is the “mid-latitude region”? And what are its features as compared with other parts of the earth?
4. How do you understand the professor’s remark that the division of global time is associated with longitude, while the local temperature of different parts of the earth is closely related to the latitude?
5. Why do scientists classify the “mid-latitude region” as a special area? And how is the climate of this region influenced by the monsoon?

03

Through the white clouds the carpet is climbing very quickly toward Shennong Peak, the highest peak of Shennongjia area. Below the carpet are lofty mountains and green rolling waves of forests in the sunshine and gentle breeze. What a lush, vibrant natural beauty. Now the carpet is getting closer and closer to Shennong Peak.

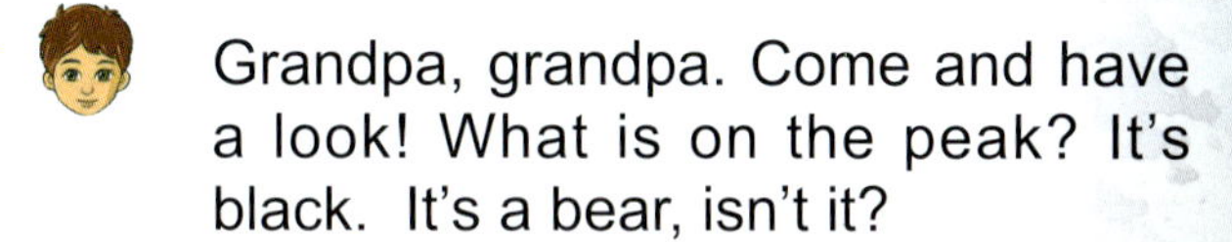

Grandpa, grandpa. Come and have a look! What is on the peak? It's black. It's a bear, isn't it?

You may see it more clearly as we get closer. It's a big bronze tripod, called "Shennong Tripod".

A bronze tripod? Why is it put there at the mountain top?

"Tripod" and "peak" are homophones in Chinese. Both are pronounced as "Ding". People put the tripod on the top of Shennongjia to express their worship and respect Emperor Yan in commemoration of his noble spirit of defying difficulties for the benefit of the people. Visitors to Shennongjia should first of all know who Emperor Yan is and what he did for us. That's why I asked you to find materials about Emperor Yan.

I did it. Emperor Yan is the legendary founder of Chinese farming civilization. And he...

Look out! We are going to land right now.

Shall we land on the platform beside Shennong Tripod, grandpa?

OK. Let's go round the tripod twice and take a closer look at it, shall we?

Good. Wow! What a huge tripod! It must be five or six metres high, I think.

You got it accurately. Shennong tripod is 5.9 m in height.

Look, grandpa. Do you see the words inscribed on the tripod?

Yes. The central part of the tripod is in the shape of 羊 (sheep) to mean "Three sheep bring bliss". The tripod ears are modeled after the hieroglyphic writing of the Chinese word "cloud" to match Shennongding's landscape of wind and clouds. The tripod body is carved with some Chinese hieroglyphical characters like 东 (east), 西 (west), 南 (south), 北 (north), 天 (heaven), 地 (earth), 山 (mountain), 川 (river), 金 (gold), 木 (wood), 水 (water), 火 (fire), and 土 (soil). They are all important elements of traditional Chinese culture.

Wow! It's really sky-high. Really awesome!

Yes. It sure is. The local people have created some free verses to express their wish. One of them goes, Shennongding(Shennong Peak), Shennongding(Shennong Tripod), there Shennong Tripod is standing on Shennong Peak. You can view the sun and the moon on Shennong Peak, and make a wish by the Shennong Tripod.Mountains are rolling like horses, white clouds floating in the sky. Not fearing the floating dust will obscure my vision, because I'm here on Shennong Peak.

Oh, what a big tripod. Who put it up there? And how ?

Well, you see there's a road down the mountain. To put the big tripod on Shennongding, people had to pull it apart first, and then trudge the separated parts up the mountains bit by bit along that steep road until the whole tripod was assembled at last.

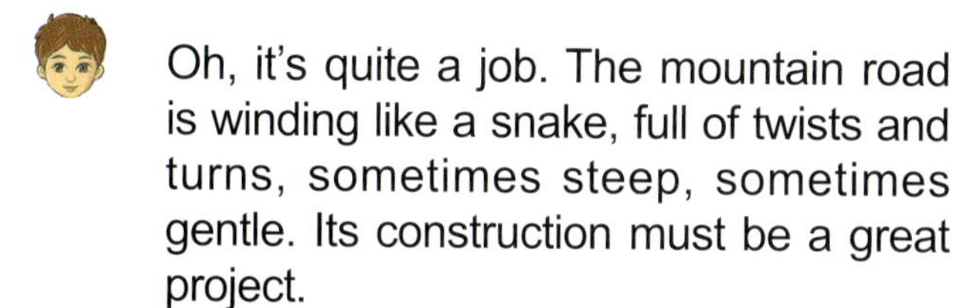

Oh, it's quite a job. The mountain road is winding like a snake, full of twists and turns, sometimes steep, sometimes gentle. Its construction must be a great project.

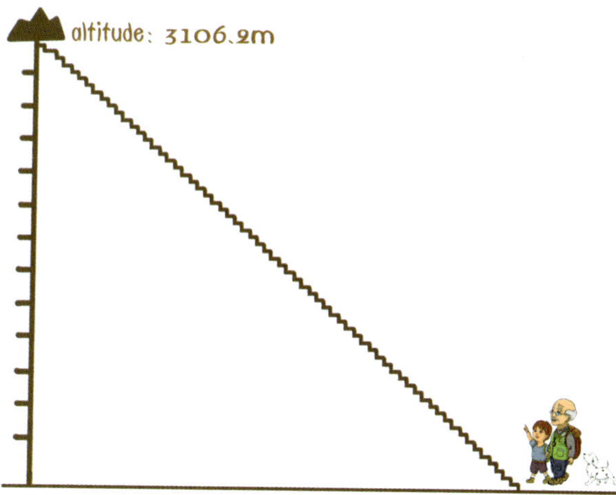

I think so. They named this road "Qingyun Ladder" to mean a ladder of ongoing success in one's promotion to higher positions. The road is 1600m long with 2999 steps altogether.

Oh, Shennongjia people are great indeed.

Sure. To build the road, they spent 310 days transporting almost 30 thousand pieces of black stone, 3000m^3 of sand and 625 tons of cement.

Wow! How could they get so many things up the mountain?

Most of them were carried up the mountain on shoulders and backs. Mules were once employed to move sand and cement. Four mules were said to have been "sacrificed" by overwork.

Wow, that's really great! It's good to have such a mountain road built by Shennongjia people at an immense cost. They have done a great thing for tourists.

Visitors to Shennongjia always take it as an honor to stand on Shennongding, the highest top of Shennongjia. At the starting point

of the “Qingyun Ladder”, a fully-equipped mountaineering base, the “Shennong Camp” was specially built by Shennongjia Geopark to serve those visitors who want to climb the “Roof of Central China”.

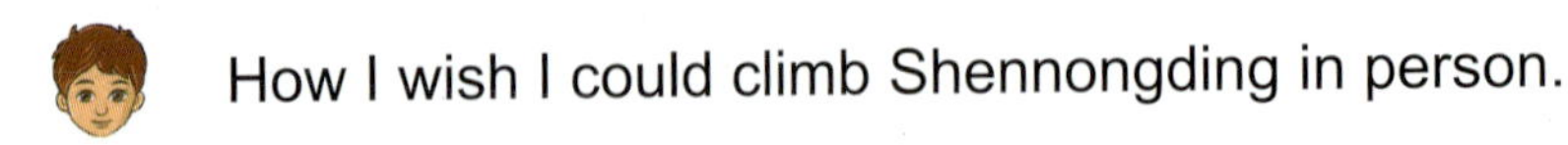

How I wish I could climb Shennongding in person.

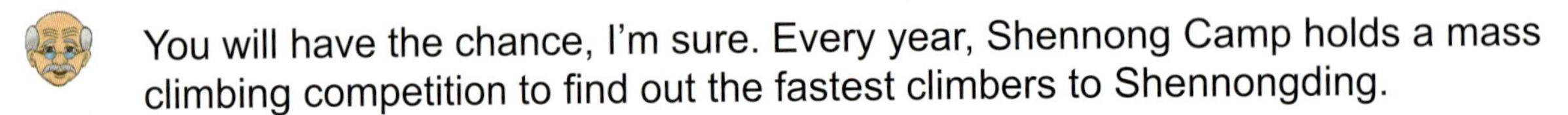

You will have the chance, I’m sure. Every year, Shennong Camp holds a mass climbing competition to find out the fastest climbers to Shennongding.

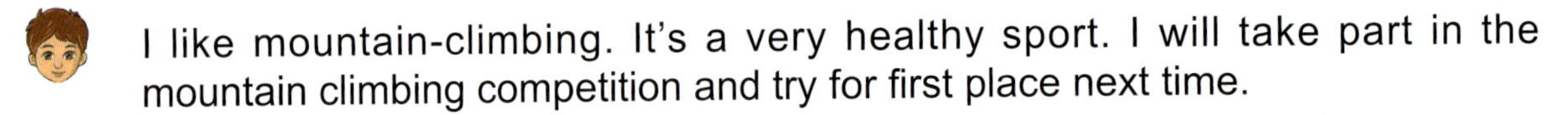

I like mountain-climbing. It’s a very healthy sport. I will take part in the mountain climbing competition and try for first place next time.

That’s good. But then you should know the summit championship record is 20 minutes.

Wow! It's a big challenge for me. I think I should do more exercises.

Shennongjia is the best destination for recreation, tourism and the promotion of one’s health. Visitors come here to enjoy the original natural environment, the clean air and the fresh food. They get a rest and improve their health immersed in nature.

The environment here is very fine.

Yes. In some places of good natural environment, Shennongjia Geopark has built a lot of walking paths to help people exercise while enjoying the beautiful scenery.

That's really very thoughtful!

Shennongjia enjoys a beautiful environment and vast territory. Every year the geopark organizes many kinds of mountain climbing and off-road activities. Everybody will be fascinated by Shennongjia, especially school students of the summer camp. Teenage pupils can enjoy nature and commune with it in camping and trekking. They have the chance to exercise while learning a lot about mountains and forests, and experience field survival training.

Okay then, I'm going to join the summer camp. Shennongjia has prepared such good conditions for tourists. Let's salute the people of Shennongjia.

Yes, we should. We should also express our heartfelt thanks to the builders, managers and organizers who have contributed so much to tourism industry.

Questions:

1. Who is Emperor Yan? What did he do for people? And what are some of his historical contributions?
2. What is a "tripod" and its use? Why was it put on Shennong Peak?
3. Will you describe to your friends your personal feeling about "Qingyun Ladder" if you have climbed it?
4. Why did Shennongjia people build "Shennong Camp", "Qingyun Ladder" and the walking paths?
5. Do you know the preparations you should make for mountaineering?

04

The senior professor flew the carpet slowly and did two aerial loops around the tripod before it landed gently on the viewing platform beside the tripod. Wangwang jumped off the carpet and ran back and forth excitedly along the boardwalk of the viewing platform. Xiaoming opened his arms to take a breath of the fresh air on the mountain top. Then he brought out a telescope from his pack and looked around the wooden railing.

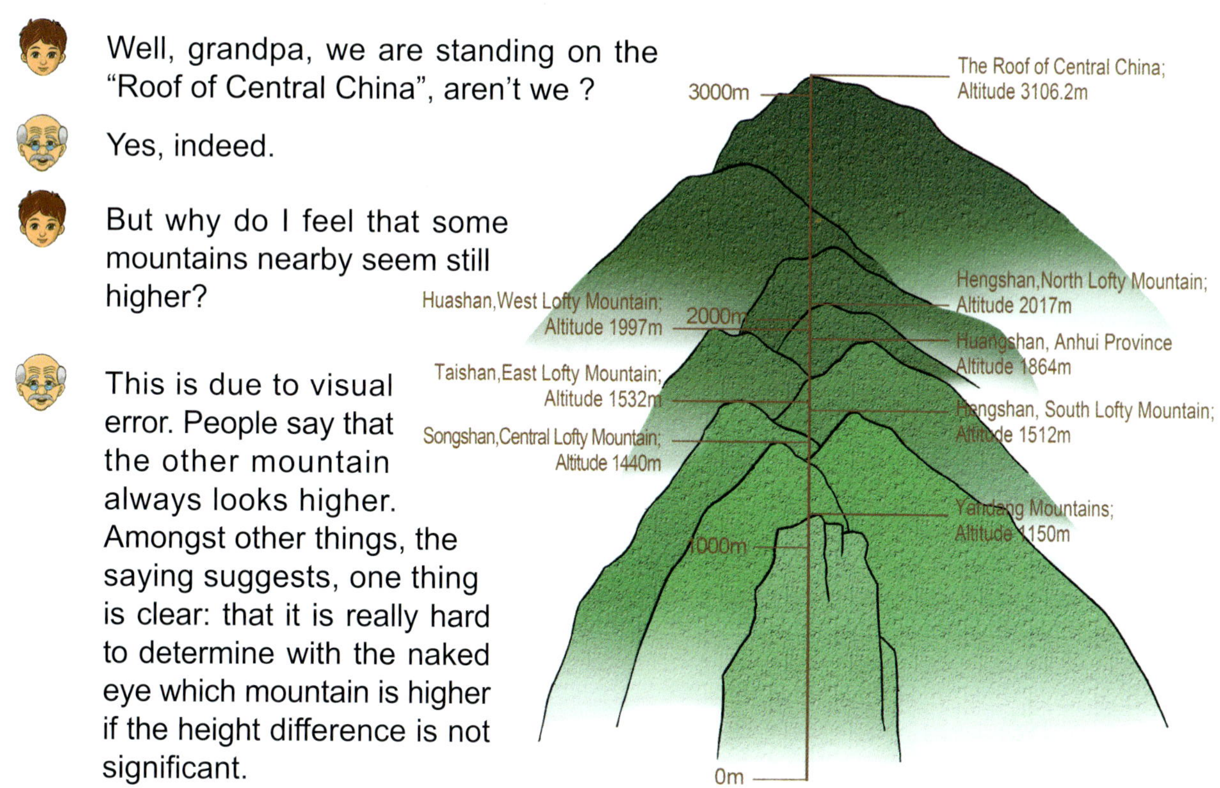

Well, grandpa, we are standing on the "Roof of Central China", aren't we ?

Yes, indeed.

But why do I feel that some mountains nearby seem still higher?

This is due to visual error. People say that the other mountain always looks higher. Amongst other things, the saying suggests, one thing is clear: that it is really hard to determine with the naked eye which mountain is higher if the height difference is not significant.

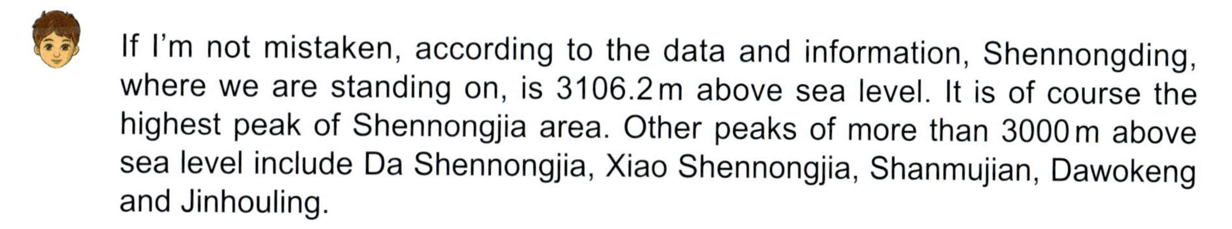

If I'm not mistaken, according to the data and information, Shennongding, where we are standing on, is 3106.2 m above sea level. It is of course the highest peak of Shennongjia area. Other peaks of more than 3000 m above sea level include Da Shennongjia, Xiao Shennongjia, Shanmujian, Dawokeng and Jinhouling.

Ah, your memory is really excellent. In the central region of China, the six peaks that rise more than 3000 m above sea level are all situated in Shennongjia. So Shennongjia is nicknamed the "Roof of Central China".

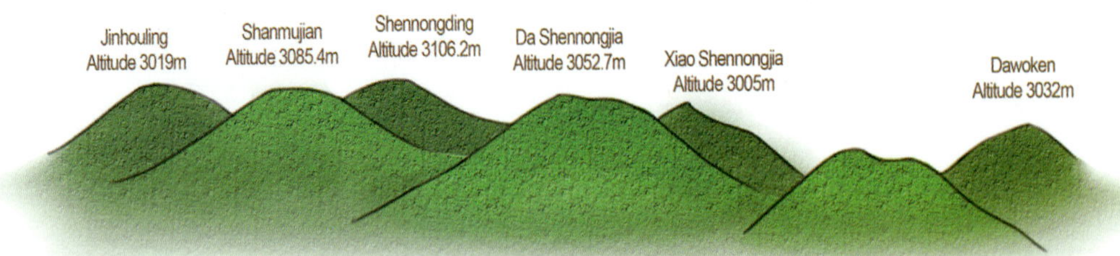

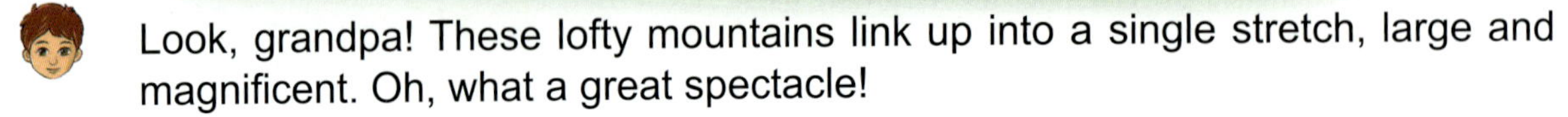

Look, grandpa! These lofty mountains link up into a single stretch, large and magnificent. Oh, what a great spectacle!

Yes. These mountains in Shennongjia constitute a natural geographical barrier and the watershed of the Yangtze River and Han River, too.

Watershed? Does it mean the water on one side of the mountain flows into the Yangtze River, and that on the other side flows into the Han River?

That's right. High as it is, Shennongjia actually feeds four river systems. They are Nan River and Du River flowing into Han River, and Xiangxi River and Yandu River into the Yangtze River.

What is a river system then?

A river system consists of the river and all its tributaries.

The bigger and longer the river, the more complex the river system. Am I right?

Yes, that's true. We may take the Nan River water system of Shennongjia for example. Nan River is a tributary of Han River, but it is made up of quite a few branches of its own, such as the Guanmen River, Gushui River, Luoxi River, etc. Within the Shennongjia, the Nan River basin area amounts to 500 km^2 or so.

Grandpa, what do you mean by basin area?

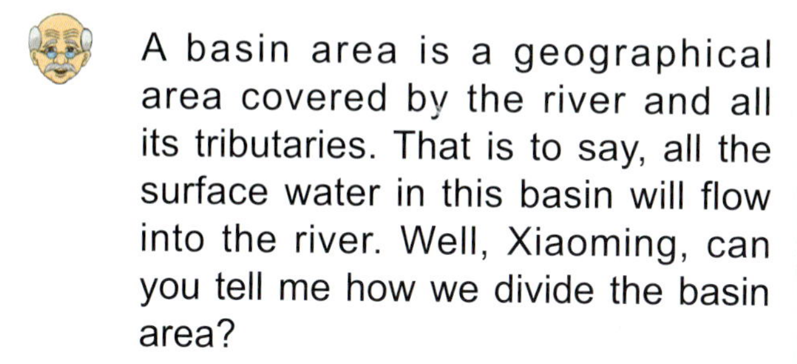

A basin area is a geographical area covered by the river and all its tributaries. That is to say, all the surface water in this basin will flow into the river. Well, Xiaoming, can you tell me how we divide the basin area?

Oh. Let me see. All the tributary water will collect in the main river. As water flows downwards, the dividing line should be on the top of the mountain. Oh, yeah, it is the mountain top that divides the basin area. And that's what you just mentioned about “watershed”.

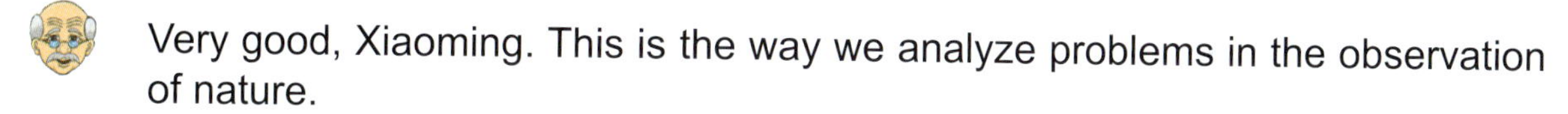

Very good, Xiaoming. This is the way we analyze problems in the observation of nature.

Grandpa, you just mentioned the four water systems of Shennongjia. Besides the Nan River, the Du River is another tributary of the Han River, isn’t it?

Yes. But the Xiangxi River and the Yandu River are tributaries of the Yangtze River.

I think the basin area of Yangtze River tributaries should be bigger?

Maybe, maybe not. The area of a river basin is determined by the specific terrain, for example, the basin area of Xiangxi River is the largest in Shennongjia region which covers more than 3000km^2. The Yandu River, also called the Shennong Stream, is another tributary of the Yangtze River. Its

basin area is only 200 km^2, less than half of the Nan River system that flows into the Han River.

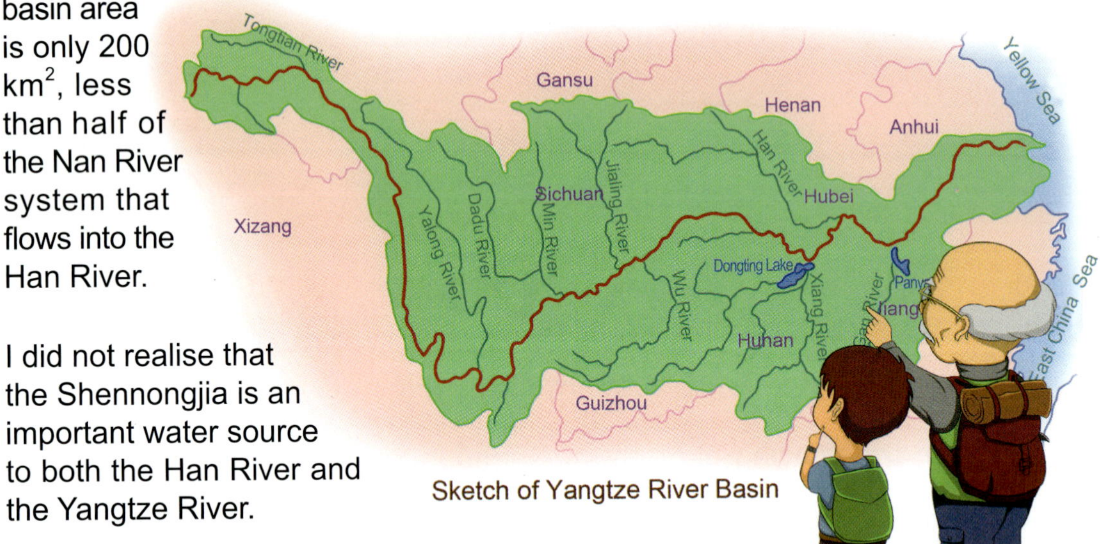

Sketch of Yangtze River Basin

I did not realise that the Shennongjia is an important water source to both the Han River and the Yangtze River.

But you should know that the basin area I've mentioned now only refers to the calculation of Shennongjia. The whole basin area a river covers may be much larger.

Questions:

1.Why is Shennongjia named the "Roof of Central China"? Which mountains in Shennongjia are more than 3000 m above sea level?

2.Can you name the water systems of the Shennongjia area?

3.What is a watershed? Can you draw the watershed between rivers on a topographic map?

4.What is water system? Check what tributaries are included in the water systems of the Yangtze River and the Yellow River respectively.

5. What is basin area? Can you check and find out which of the two rivers, the Yangtze River or the Yellow River has got a larger basin area?

05

Walking on the viewing platform of Shennongding with Xiaoming, the professor kept looking about while Xiaoming, leaning on the railing, observed carefully with his telescope. Now and then he took pictures of the beautiful scenery. The little dog ran back and forth excitedly, watching all around, too. The view is very open here. Looking out, they could clearly see the top of Shennongding, which dominates all other mountains around. Sitting on the viewing platform ladder, the weary boy seemed completely absorbed by the majestic mountains in front of him.

Grandpa, why is Shennongjia so high?

Well, it's hard to explain in a few words. Let's begin with the structure of the earth.

The surface of the earth is stone, and the interior is magma, isn't it?

But things are not that simple. Just like a hard-boiled egg, the earth is roughly made up of three layers: crust, mantle and core. Its crust is the solid circle like the egg shell and its mantle is equal to the white protein portion of the egg.

And the core is equivalent to the egg yolk, right?

Yes, exactly. The average thickness of the earth's crust is about 17 km and that of the continental crust is 33 km. The crust of the plateau or mountain area is much thicker. The thickest place can even exceed 70 km. On the other hand, the ocean crust is much thinner than the continental crust, only a few kilometers thick.

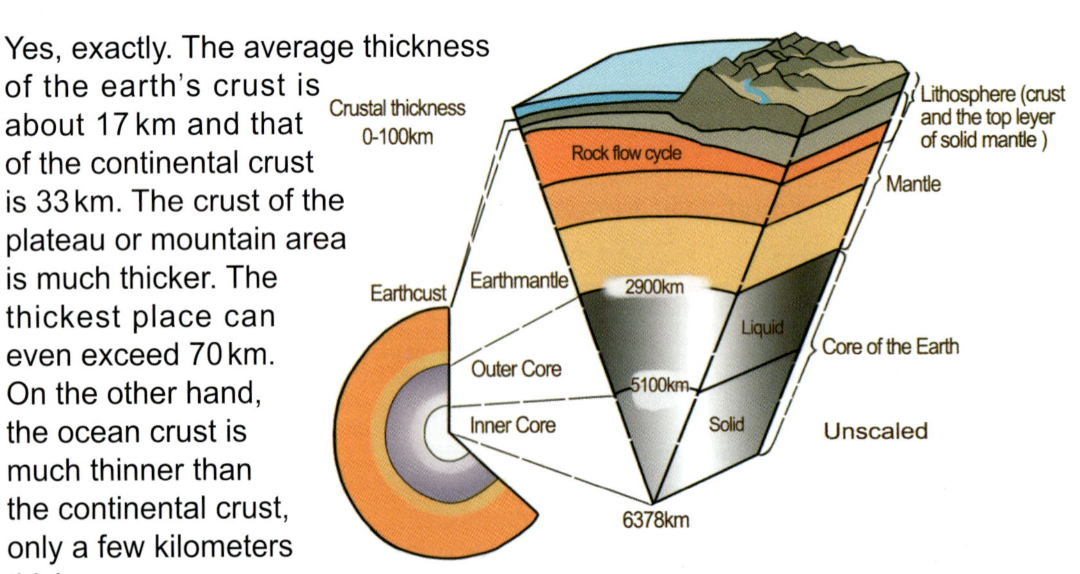

The egg shell is not so thick. Now what about the mantle and the core?

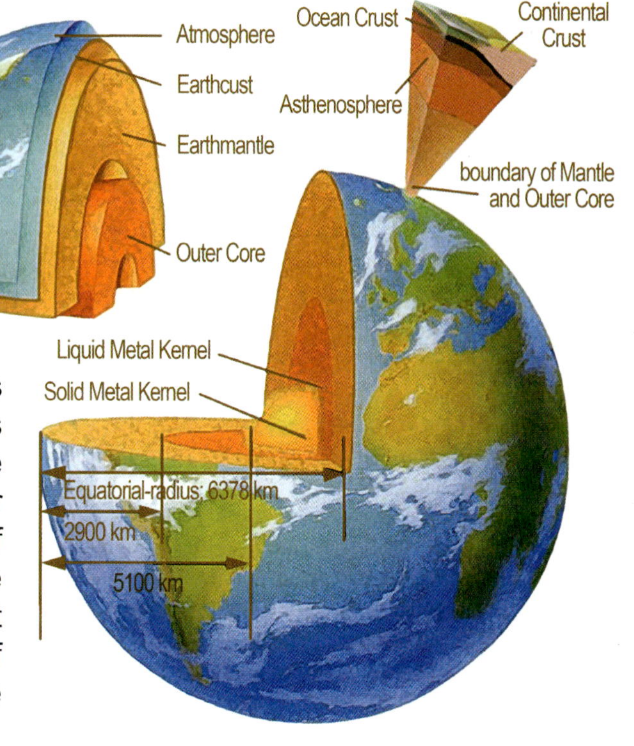

The egg can never be as big as the earth. Relative to the size of the earth, such thickness of a few kilometers or more than a dozen kilometers is indeed insignificant for the earth's crust. The mantle beneath the earth's crust is the middle layer of the earth mainly composed of dense rock forming materials. The earth's core is the central part of the earth, mainly composed of iron and nickel elements. It is the equivalent of the egg yolk.

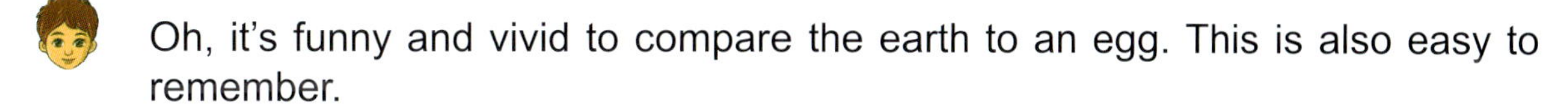

Oh, it's funny and vivid to compare the earth to an egg. This is also easy to remember.

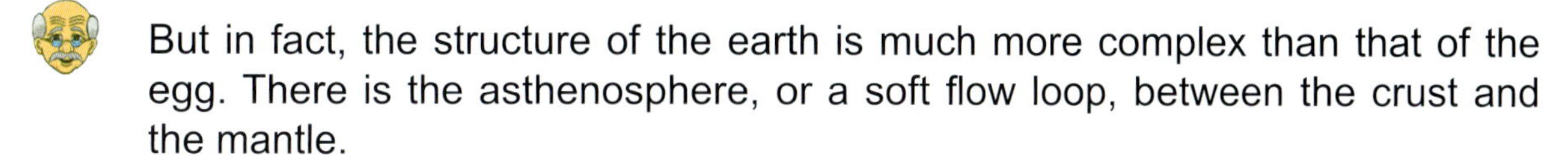

But in fact, the structure of the earth is much more complex than that of the egg. There is the asthenosphere, or a soft flow loop, between the crust and the mantle.

A soft flow loop? Is it really soft?

Yes, it's soft indeed. The soft flow loop is continuously distributed around the world. That is to say, it exists beneath the continental and oceanic crust alike.

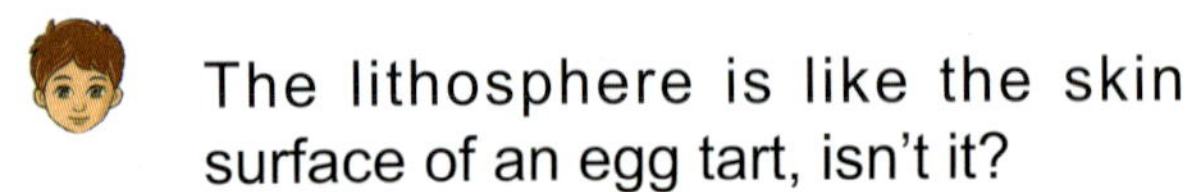
It presents a gradual transition between the lithosphere and asthenosphere, with no obvious boundary.

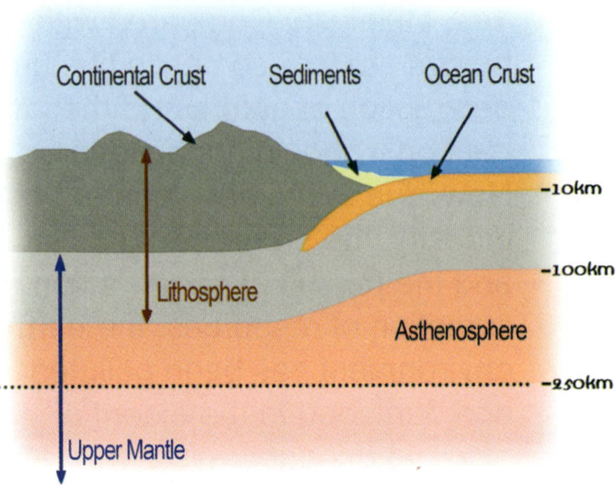

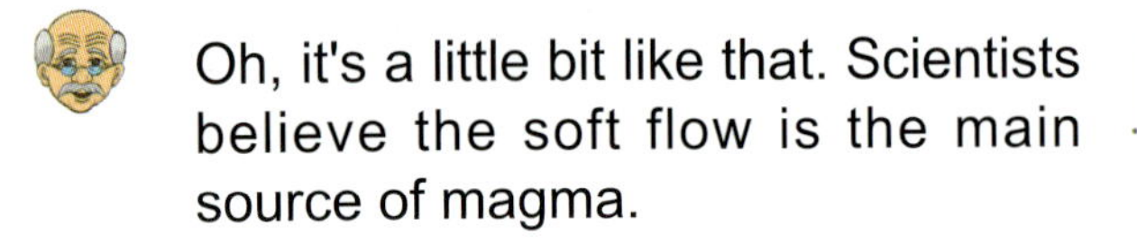
The lithosphere is like the skin surface of an egg tart, isn't it?

Oh, it's a little bit like that. Scientists believe the soft flow is the main source of magma.

Wow, the earth is much more complex than the egg indeed.

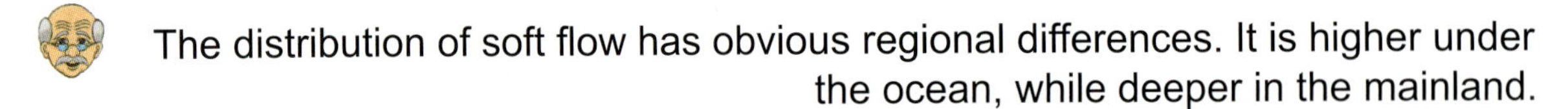
The distribution of soft flow has obvious regional differences. It is higher under the ocean, while deeper in the mainland.

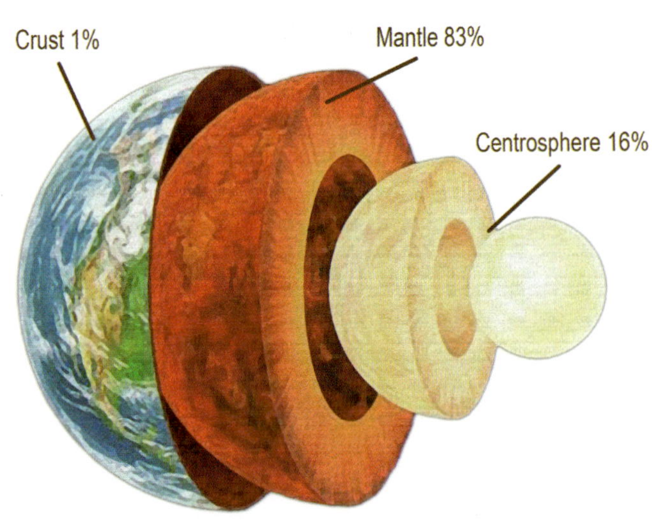

Well, the earth is so amazing!

In the total mass of the earth, the core accounts for 16%, mantle for 83%, and the crust only for 1% though it is the part most closely related to human beings.

Compared with the earth, human beings are really so small.

Yes, I think so. Compared with the universe and nature, human beings are indeed negligible, but our destruction of nature and the impact on it should not be ignored. Now, due to the development of industry, the progress of society and the pursuit of people's consumption of resources, the earth's environment has been polluted and damaged to the extent of extreme danger.

That's true. In many places now there are no fish in the rivers, no birds in the sky; the trees are cut down, and many animals are dying out. We must immediately stop the pollution and destruction of water, air, forests, and land.

If humans do not protect the earth that sustains them, the consequences will be disastrous. In this sense, Shennongjia is considered the last remaining piece of land of idyllic beauty in the relatively developed regions of Central China. It is a green pearl, and “ecological lungs” in the central region of China to “get rid of the stale air and take in the fresh” to ensure a healthy human living environment.

Questions:

1.Can you draw the interior structure of the earth and illustrate the three layers?

2.What are continental crust and ocean crust? What are their respective characteristics?

3.What is the relationship between the soft flow loop and the continental crust and the oceanic crust?

4.Can you cite one or two examples of human destruction of the earth's environment?

5.What does an “ecological lungs” mean? Why does the professor call Shennongjia the “ecological lungs” in the regions of Central China?

06

The professor's words plunged Xiaoming into thought. The little dog was crouching at his feet quietly. After a passionate talk about environmental protection, the professor suddenly thought of the question by Xiaoming. He turned around and began to explain why Shennongjia is so high.

OK. Let's go on with the structure of the earth. Have you ever heard of "plate tectonics"?

Yes. But I can hardly understand it.

Well, plate tectonics is an important geological concept based on the corollary of "continental drift" and "seafloor spreading".

Continental drift? Drifting like a boat?

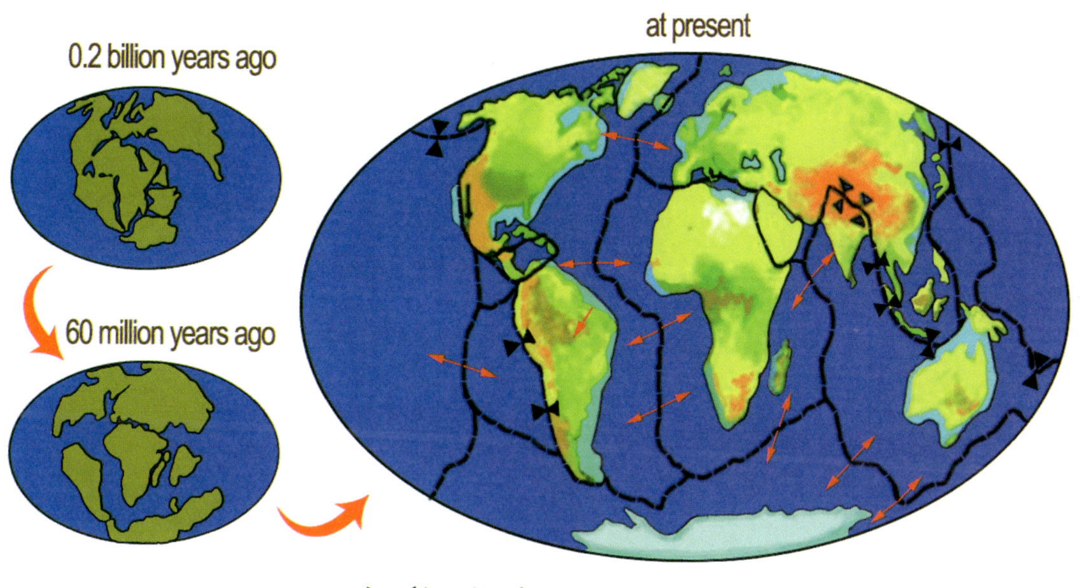

Yes. German scientist Wegener found that both sides of the Atlantic, especially the contours of the coast of Africa and South America seemed like they could be spliced together perfectly. Then Wegener boldly speculated that they had been originally linked together until they drifted apart later.

Yeah, on the globe Africa and South America do seem to be a perfect match.

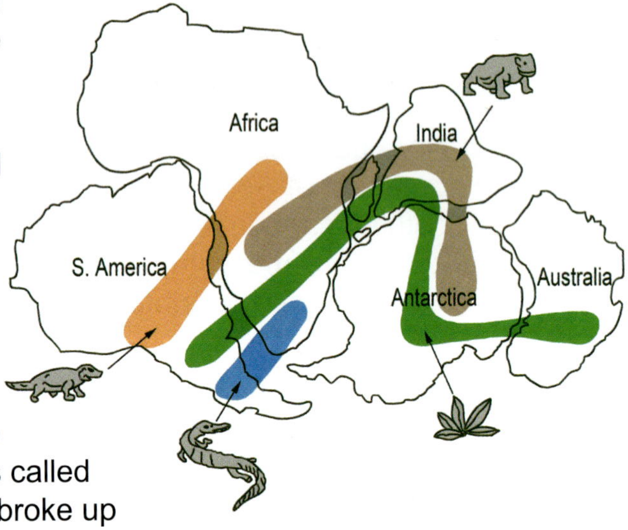

After a lot of investigations and research Wegener first proposed the theory of continental drift in his work *The Formation of the Mainland and the Sea* in 1921. According to him, all the present continents once comprised one large landmass called "Pangea". This "Pangea" later broke up and drifted apart, and gradually became the present geographical arrangement of continental fragments.

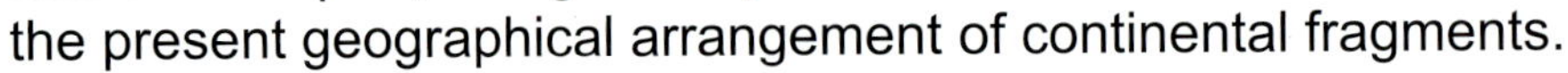

Wow. That's funny. It's a bit like this. It sounds reasonable, I think.

Wegener also collected some hard evidence, for example, fossils of the same era and the same kind were identified in some lands now separated by the sea. Of course, these creatures (including plants) were not capable of traveling overseas themselves. The only possibility is that at the time these creatures lived these continental blocks were connected to each other, after which they migrated and spread. Later, they were separated by the ocean as the continents had drifted apart.

Yes, this is very persuasive indeed.

In addition, glacial remnants also provide good evidence.

But how can the glaciers prove the continental drift?

Some glacial relics of the same period have been discovered in the southeast of South America, South Africa, South India and South Australia, where glaciers are quite impossible today. However, if these continents were put together, you would find these sites of glacial relics could also come together and the direction of the glacier flow seems to indicate the ancient earth ends just like today's North and South Poles.

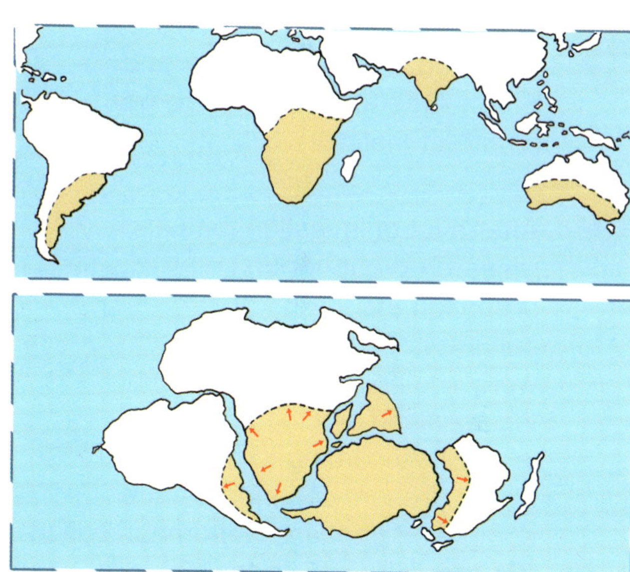

Sketch map of ancient glacier

Wow, it's really magic! It seems the continental drifting really did happen in the history of the earth evolution.

Sure. But can you tell us what forces caused the continents to drift?

Oh yeah. What is the force to drive the continental drift? It must be powerful enough to do so anyway.

No doubt. With no definite solution to this problem, Wegener's theory of continental drift has always been a defective one, and was hardly accepted by any scientists.

Of course not. A scientist is always serious about conclusive evidence. He is never slipshod in anything.

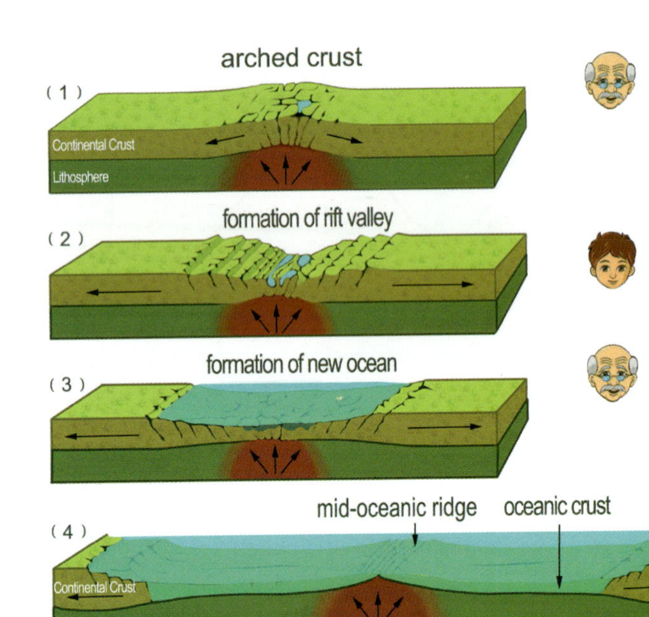

Marine geologists Hess H H and Dietz S R respectively proposed the theory of seafloor spreading almost at the same time in 1960.

Seafloor spreading? That sounds quite awful.

Yes. Seafloor spreading is indeed a terrible geological phenomenon. From field surveying, Hess found that due to the thermal convection inside the earth, the hot rising lava flow out of the ground would break through the oceanic crust and keep pouring out to form mid-ocean ridges and new oceanic crust, gradually driving the earlier formed oceanic crust apart on each side of the ridge and constantly forming new oceanic crust between.

It seems like pushing a folding door, doesn't it?

Right. You're a very smart boy full of imagination. This is certainly a continuous process. The lava keeps pouring out to form new oceanic crust, continuously pushing the previously formed oceanic crust apart on (to) both sides. Thus the continents on both sides would drift farther and farther away like floating ice blocks on the water.

Oh, oh! This is the way continental drift happened!

The sea floor spreading theory has solved the insurmountable problem of power in Wegener's continental drift theory proposed forty years before. It

has also laid a foundation for the theory of plate tectonics. Since then, the continental drift theory has become the mainstream view and is widely acknowledged.

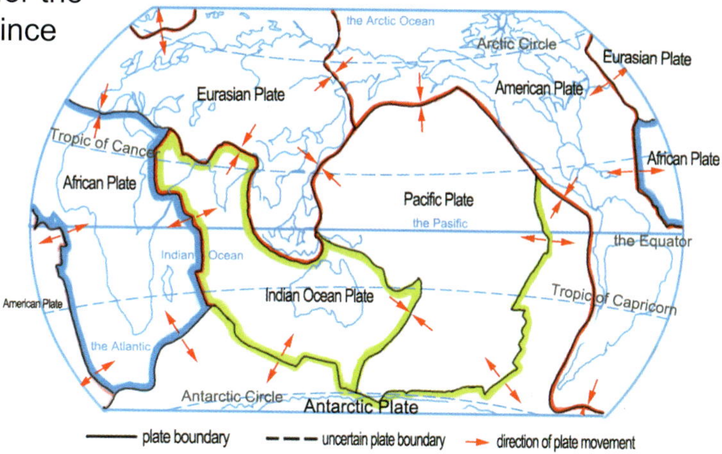

Global distribution of 6-plates

Wegener's proposal got widely recognized more than forty years later. The scientific development is really difficult.

Yes. Scientific research and discovery is full of hardships and setbacks.

Grandpa, you have gone a long way, but you don't tell me why mountains in Shennongjia are so high.

Well, I think the theory of plate tectonics can answer your question.

Really? I've heard of "plate tectonics". It is said that the earth's five continents and four oceans are made of plates just like a jigsaw puzzle.

That's true. The theory of plate tectonics was proposed by French scientist Robison in 1968 on the basis of the "theory of continental drift" and "seafloor spreading".Robison believed that the earth's lithosphere is not a single piece, but is divided into many units called "plates". They are floating above the "asthenosphere" and are in constant motion because of the thermal convection inside the earth. In some places they drift apart; in others they come together.

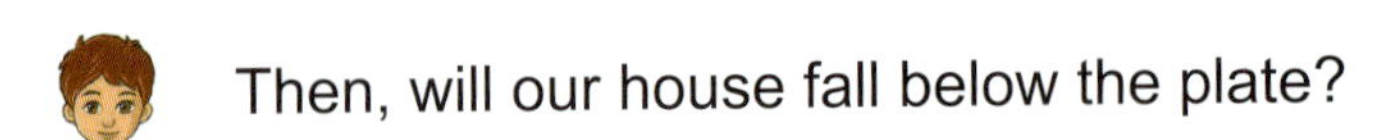

Then, will our house fall below the plate?

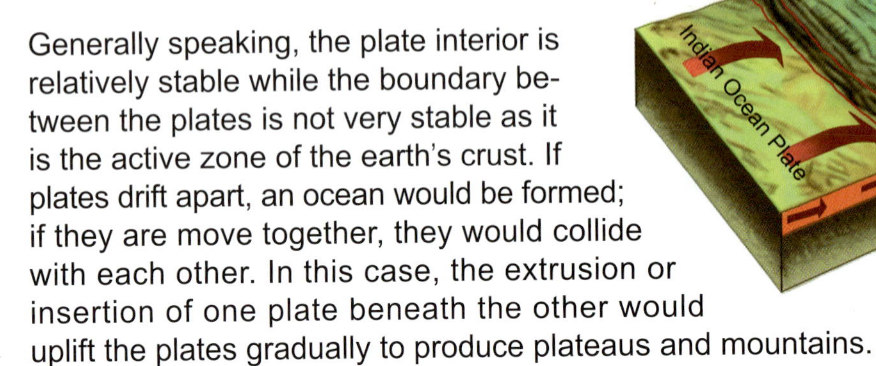

Generally speaking, the plate interior is relatively stable while the boundary between the plates is not very stable as it is the active zone of the earth's crust. If plates drift apart, an ocean would be formed; if they are move together, they would collide with each other. In this case, the extrusion or insertion of one plate beneath the other would uplift the plates gradually to produce plateaus and mountains.

Oh, I got it, grandpa. It is said that the Tibet Plateau and the Himalayas were uplifted and became the "roof of the world" because the Indian Ocean Plate inserted beneath the Eurasian Plate.

Yes. Because of the plate collision, some plates would dive into the mantle under the opposite plate and get cut and melted; some plates would rise to make mountains and plateaus. The plate edges, as a result of compressive fold, would often form mountains extending in a certain direction.

Well, grandpa, I understand Shennongjia was formed by compressive uplift, right?

You're right, Xiaoming. You see, in our country and even all over the earth, most of the large mountains have a certain extension direction. Why? Because they basically represent the oldest welded plate boundary.

Wow? It's incredible. Scientists are really great. I'm going to be a geologist in the future.

The formation of Shennongjia mountains as the "Roof of Central China" is actually associated with plate tectonics movement. Some scientists believe that Shennongjia itself belongs to a large plate edge, which is uplifted as a result of plate convergence, extrusion and compressive folding. But some other scientists think Shennongjia is a "micro plate" between two large plates. When the large plates drifted together, this area was compressed and uplifted to become the "Roof of Central China".

Oh, amazing! But how did all this happen?

Nobody clearly understands what had happened in the end, and how Shennongjia became a mountain area step by step. This waits for your exploration, you future geologist.

OK. As I grow up, I will devote myself to the study of how Shennongjia became the "Roof of Central China".

Questions:

1.Can you briefly narrate Wegener's theory of continental drift? Why was it recognized by scientists over forty years later?

2.What is the theory of seafloor spreading and its significance?

3. Can you explain the theory of plate tectonics? Why do large mountains on the earth show some direction of extension?

4.What happens as plates collide?

5.Please refer to the geological references of the Great Rift Valley and take it as an example to explain to your friends the plate tectonic movement.

07

Strolling on the viewing platform with his dog, Xiaoming kept looking around, completely intoxicated with the sun and wind. Suddenly, the dog Wangwang started barking toward the boardwalk of the viewing platform. Down below, a hare is seen running quickly into the rubble nearby. Xiaoming wanted to find out where the hare was hiding when it disappeared in the rock cracks without trace.

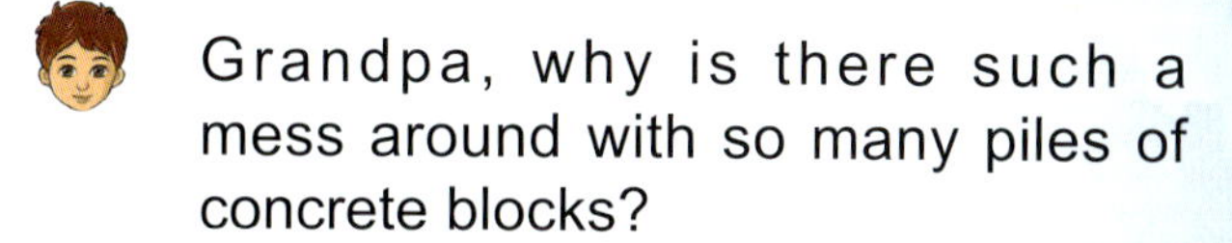

Grandpa, why is there such a mess around with so many piles of concrete blocks?

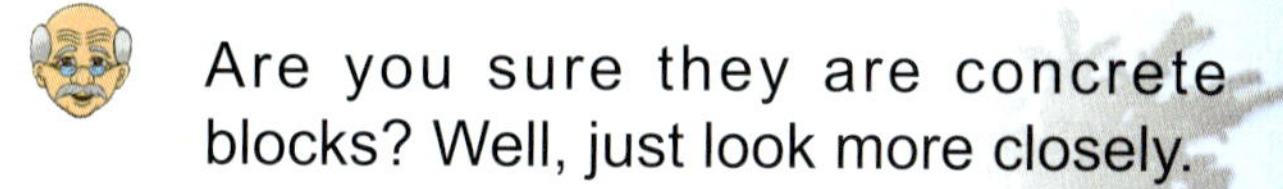

Are you sure they are concrete blocks? Well, just look more closely.

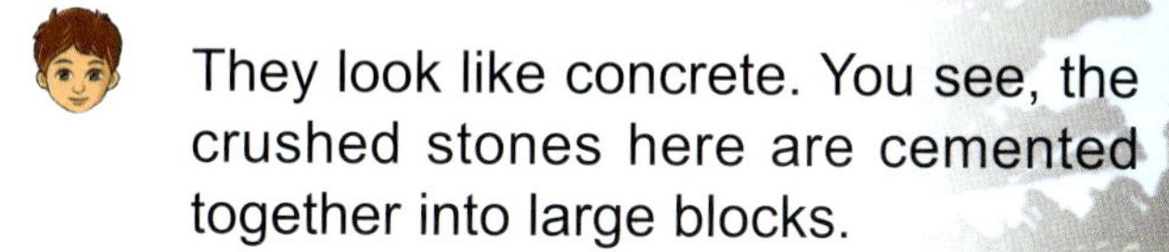

They look like concrete. You see, the crushed stones here are cemented together into large blocks.

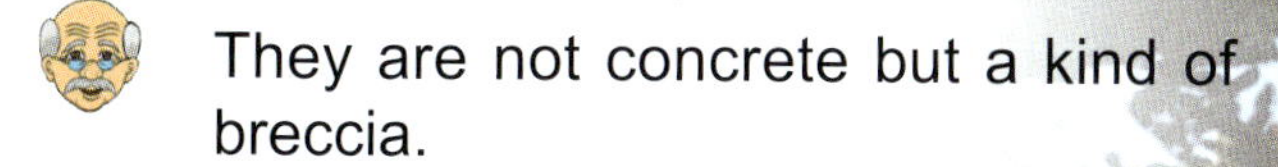

They are not concrete but a kind of breccia.

Breccia?

Yes. There are many different kinds of breccias in the natural environment. And this is a very unique volcanic rock.

Volcanic rock! Then there must be a volcano here?

Don’t fear. There’s no volcano here. In geology, this type of rock is named “cryptoexplosive breccia” means “hidden explosive volcanic rock”.

Oh, why such a strange name!

It is so named due to a special background. As the magma beneath the earth is crushed, it rises close to the land surface along fissures and becomes gasified at once due to the rapid pressure decrease and superheating of the surrounding groundwater by magma, resulting in a powerful explosion breaking the surrounding rocks and causing the lava to become solidified. Then the broken rocks are cemented by the lava that keeps coming up to form this kind of breccia like concrete.

Wow, this process is too terrible.

Yes. Geological processes are often accompanied by terrible and destructive phenomena. Although there is a strong explosion to form concealed explosive volcanic rocks, the explosion doesn't make smoke and fire as a general volcanic eruption does. Hidden under the ground, they are therefore named "hidden explosive volcanic rocks".

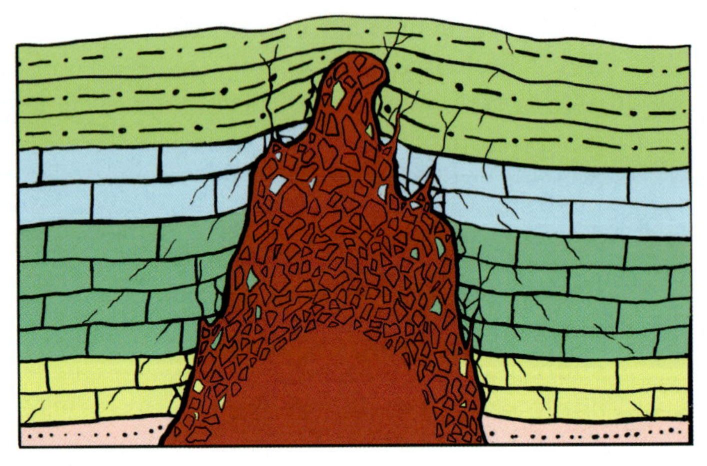

Diagram of cryptoexplosive breccia

Haha, the name sounds really appropriate for those rocks. As they are formed under the ground, how are they now piled here on this mountain? Who dug them out?

Nobody did it. The perennial weathering and denudation of nature removed the stones and soil layers above, just letting them "see the sun". This kind of volcanic rocks is much harder than other rocks around and, therefore, more difficult to be weathered. And that's why they are often found as the tops of huge mountains.

These "concealed explosive volcanic rocks" should be a single piece, how do they become such rubble piles? Who broke them?

They were not broken, but a masterpiece of nature.

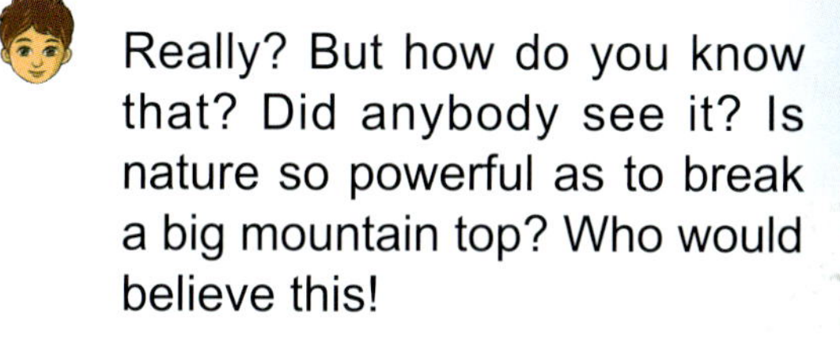

Really? But how do you know that? Did anybody see it? Is nature so powerful as to break a big mountain top? Who would believe this!

Don't argue like this, Xiaoming. The strength of nature is always unparalleled. Don't you know such geological events as volcanic eruptions, earthquakes, and landslides can release tremendous energy to cause sudden human disasters.

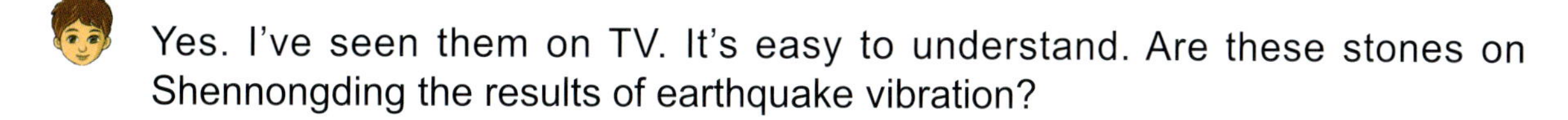

Yes. I've seen them on TV. It's easy to understand. Are these stones on Shennongding the results of earthquake vibration?

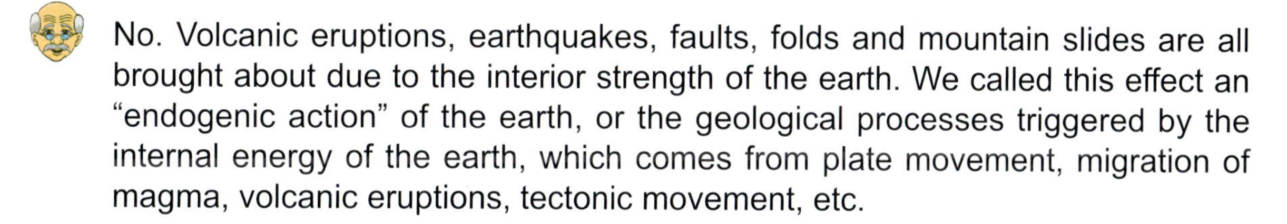

No. Volcanic eruptions, earthquakes, faults, folds and mountain slides are all brought about due to the interior strength of the earth. We called this effect an "endogenic action" of the earth, or the geological processes triggered by the internal energy of the earth, which comes from plate movement, migration of magma, volcanic eruptions, tectonic movement, etc.

To correspond with the "endogenic action", there must be an "exogenic action". Is that so?

Yes, you're right. The "exogenic action" refers to the effect by the external factors of air, water, solar energy, and organisms upon the earth rocks such as wind and rain, rivers, waves, tides, glaciers, wind erosion, erosion of plant roots, and so on.

I believe the "exogenic action" cannot be so powerful as the "endogenic action".

Yes, that's true. But don't dismiss the "exogenic action" of the earth. Do you understand the old saying: "Dripping water wears through a stone "?

Yes. It dictates the power of a drop of water. Very small as it is, a drop will be able to wear a hole in the stone if it drops on the stone constantly through months and years. This proverb indicates that we should have patience and perseverance in doing everything. If we persist, we will be able to succeed in the end.

Very good. The earth's "exogenic actions" such as sunlight, wind, water, etc. are the most patient and persistent. For millions of years, they have always been repairing and reforming our planet.

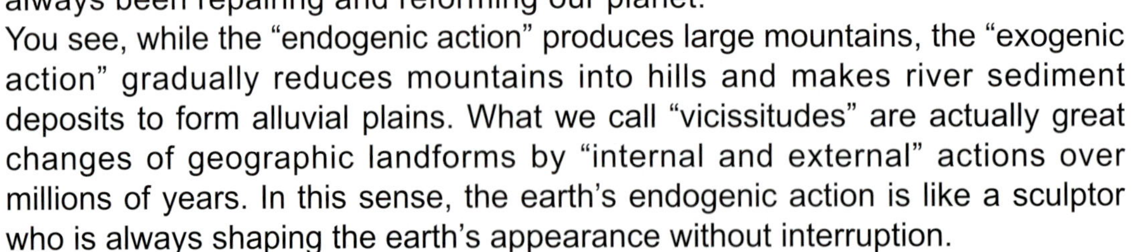

You see, while the "endogenic action" produces large mountains, the "exogenic action" gradually reduces mountains into hills and makes river sediment deposits to form alluvial plains. What we call "vicissitudes" are actually great changes of geographic landforms by "internal and external" actions over millions of years. In this sense, the earth's endogenic action is like a sculptor who is always shaping the earth's appearance without interruption.

Grandpa, you have not yet told me how rocks on this mountain were broken by the "exogenic action". I can't imagine how the sun, wind, and water are able to break rocks of a whole mountain.

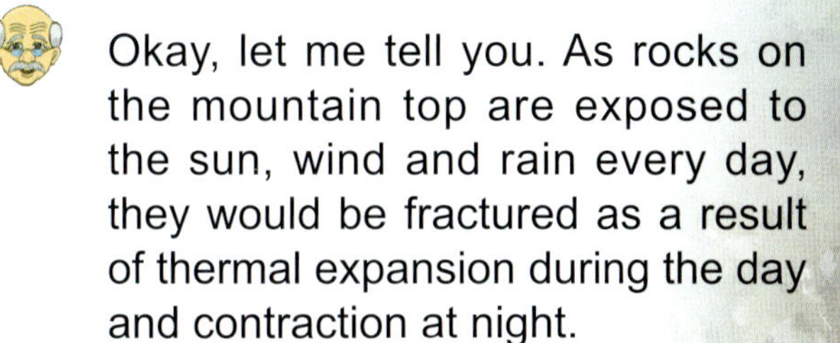

Okay, let me tell you. As rocks on the mountain top are exposed to the sun, wind and rain every day, they would be fractured as a result of thermal expansion during the day and contraction at night.

But how can the small cracks break rocks of the whole mountain?

A good question. The small cracks are not fatal at all. But if water penetrates into them, these cracks would cause trouble.

What kind of trouble?

Mountains in Shennongjia are very high. When temperature drops at night, the water in the cracks will become ice. And during the freezing process, its volume will...

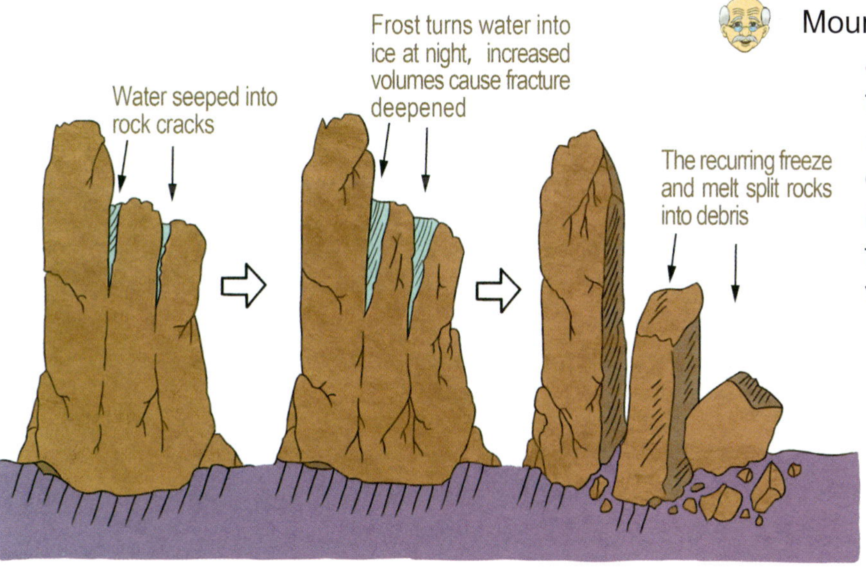

Oh, I see, I see. When water freezes, its volume will consequently increase to push the cracks open.

Pretty good, Xiaoming. It's quite right of you to apply knowledge to problems you meet. We learn in order to practice. Now can you tell me how water breaks rocks?

Let me try. It's very simple indeed. First of all, the expansion and contraction makes the rock fractured and brings in some rain water. When the temperature drops, the water freezes and its volume increases to make the crack wider. As more water penetrates and freezes, the crack expands further to allow still more water in. This is an infinite loop, which will gradually split and break the hard rock in the end.

Oh, you've given a wonderful explanation. For any interesting geological phenomenon, you must think carefully first, and then try to explain the geological phenomenon that you have learned before.

Thanks, grandpa. I'll do so.

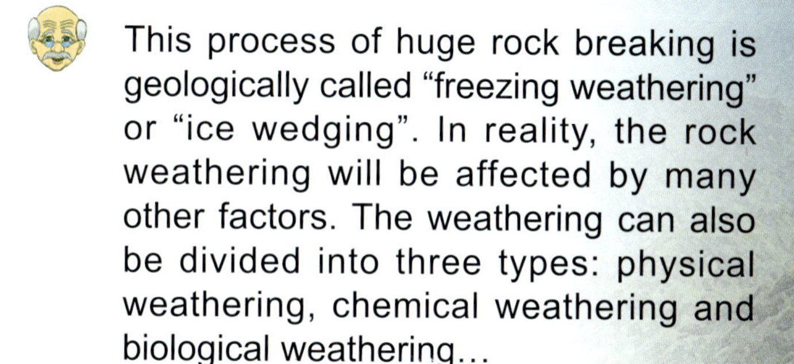

This process of huge rock breaking is geologically called “freezing weathering” or “ice wedging”. In reality, the rock weathering will be affected by many other factors. The weathering can also be divided into three types: physical weathering, chemical weathering and biological weathering…

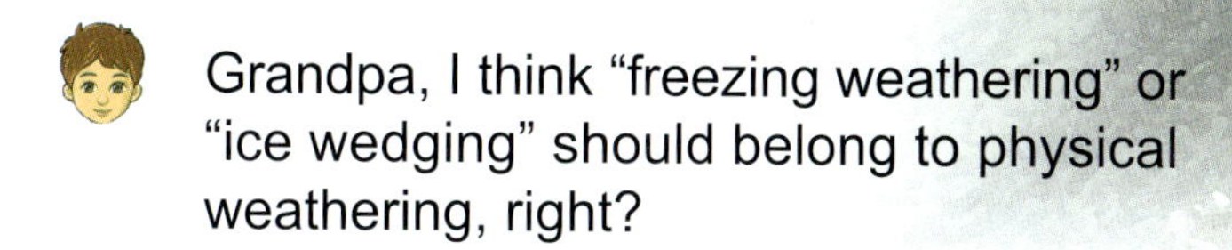

Grandpa, I think “freezing weathering” or “ice wedging” should belong to physical weathering, right?

Yes. “Freezing weathering” is a typical physical weathering, which will only lead to changes in the volume and appearance of rocks.

Then what does chemical weathering and biological weathering mean respectively?

We’ll discuss this question in our visit to Wall Rocks (Banbiyan) and Shennong Valley (Shennonggu). Now it’s getting late. Let’s hurry up to Jinhouling Subarea.

Ok! Follow me, Wangwang!

Questions:

1. What are “concealed explosive volcanic rocks”? What are other types of breccias?
2. Why are such “concealed explosive volcanic rocks” exposed on the top of mountains in Shennongjia?
3. Can you give examples to show the effect of “endogenic action” on the global environment?
4. What is the “exogenic action”? Does it have any concrete manifestations in nature?
5. Can you describe the specific process of “freezing weathering” or “ice wedging”?

08

The professor unrolled the carpet and seated himself on it, followed by Xiaoming and his dog. The carpet was slowly rising steadily, heading for the dark green mountains. As the carpet shaved the treetops, Xiaoming observed the thick woods closely. The dog was jumping up and down around him as if to find something interesting to bark at in the woods below.

Grandpa, are we flying towards Jinhouling? There must be many golden monkeys there.

Sure. In Shennongjia area, Jinhouling is one of the six mountains higher than 3000m above sea level. Characterized by the thick primitive forest, Jinhouling got its name because the endangered Golden Snub-nosed Monkeys reportedly inhabiting this area.

Grandpa, I've seen a lot of photos and videos of Shennongjia Golden Snub-nosed Monkeys. They are really very beautiful!

The famous Golden Snub-nosed Monkey is a primate native to China, a first-class national protected animal like the giant panda. There are three classes of golden monkeys in China, namely, Sichuan Golden Monkey, Guizhou Golden Monkey, and Yunnan Golden Monkey. Shennongjia Golden Snub-nosed Monkey was originally defined as a subspecies of Sichuan Golden Snub-nosed Monkey. Now experts are quite sure that they are a unique population of Shennongjia, where the unique geographical location and good ecological environment offer an ideal habitat for Shennongjia Golden Snub-nosed monkeys.

Grandpa, I've seen a lot of photos and videos of different golden monkeys and I think Shennongjia Golden Snub-nosed Monkeys are the most beautiful and charming in the world for their strong body, long blonde hair and bright eyes.

Yes, they are very beautiful and believed to be Shennongjia wild elves. A research center of Shennongjia Golden Snub-nosed Monkey is found in Dalongtan. Chinese first research project of Golden Snub-nosed Monkey fixed point feeding at field sites has been successful here.

What does "fixed point feeding" mean?

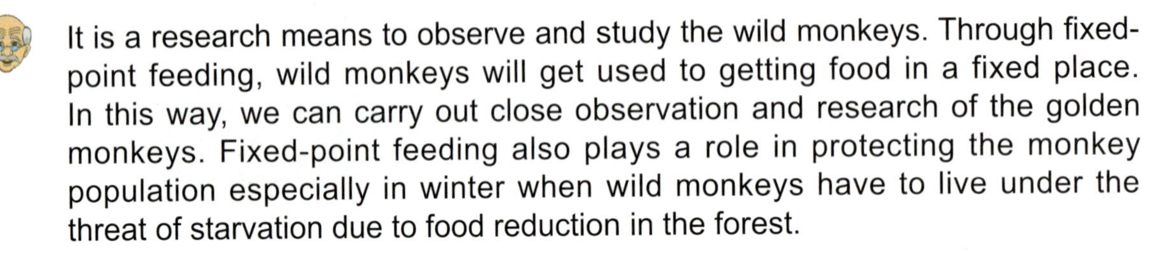

It is a research means to observe and study the wild monkeys. Through fixed-point feeding, wild monkeys will get used to getting food in a fixed place. In this way, we can carry out close observation and research of the golden monkeys. Fixed-point feeding also plays a role in protecting the monkey population especially in winter when wild monkeys have to live under the threat of starvation due to food reduction in the forest.

Yeah. In the forest wild animals can be very distressed in winter. Some would certainly starve to death with no food.

Yes, they would. But it is very difficult for the wild monkeys to develop a habit of feeding themselves in certain fixed places.

I think so, too. They don't understand what we say and they don't realize what we are doing is for their own good.

The Golden Snub-nosed Monkey is an animal with a very high vigilance. Poachers' hunting and killing in the past have made them so afraid of and vigilant against humans that they won't allow anybody to get close to them. This has brought about great difficulties for scientific research and protection of the golden monkeys.

Then, how did the researchers of Shennongjia set up the fixed points? How could the monkeys get so obedient?

Well, it's a long story. Researchers in Shennongjia had spent three years finding out the golden monkey's living habits. They had suffered a lot, especially in the cold winter to track the monkeys in the vast woodlands and put foodstuffs for them. However, the golden snub-nosed monkeys were so watchful that they would rather go hungry than have the apples, carrots, and peanuts we provided.

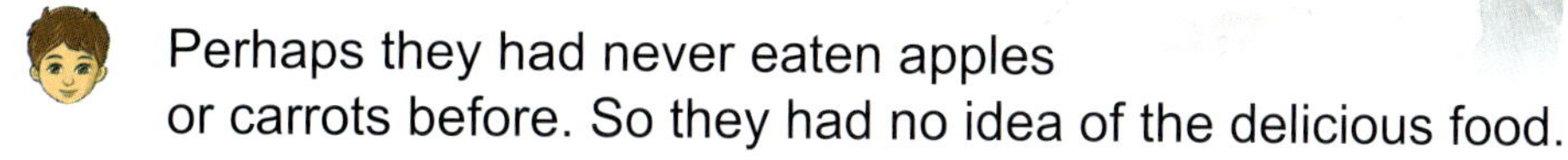

Perhaps they had never eaten apples or carrots before. So they had no idea of the delicious food.

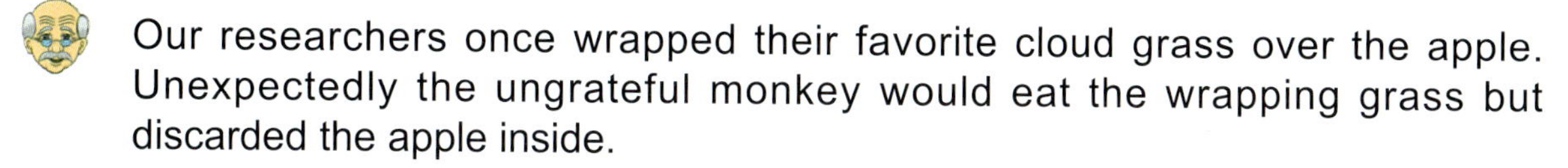

Our researchers once wrapped their favorite cloud grass over the apple. Unexpectedly the ungrateful monkey would eat the wrapping grass but discarded the apple inside.

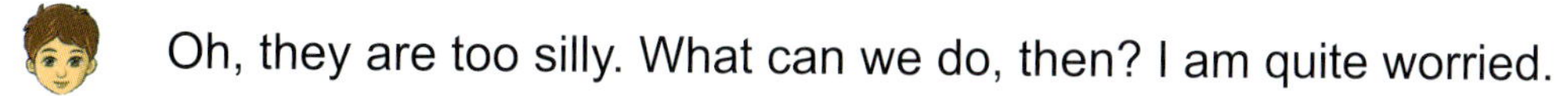

Oh, they are too silly. What can we do, then? I am quite worried.

The researchers were ready to give up, when a golden monkey made its first bold attempt. It was on December 28, 2005, a very momentous day as the researchers put in some food as usual. A golden monkey picked up an apple and then looked closely at it with hesitation.

Don't hesitate any more. Just eat! Quickly! I'm really anxious!

The hungry monkey examined the apple for a long time. Then it sampled the first bite hesitantly, but wolfed down the whole apple with relish. This is a great demonstration!

Oh, it is really a brave monkey like the first man to eat crab. I think the monkey would tell its friends very soon that "Apple is really delicious! Come on and try it. It's Ok!"

I think so. Later the researchers nicknamed this monkey "Bold" as a good example. Since then, golden monkeys began to trust our researchers and gradually developed the habit of feeding in fixed places.

Wow, how wonderful! They've finally come to understand our good intentions.

The fixed-point feeding has protected the healthy growth of the Golden Snub-nosed Monkey population, the number of which has more than doubled consequently. In addition, the successfully established mutual trust relationship between the researchers and the golden monkeys has made close observation possible. In this respect, Shennongjia Golden Snub-nosed Monkey Research Center has achieved a number of world-class scientific research results.

Those scientific research workers are really great. I must salute them!

Questions:

1. How much do you know about the golden monkey? Can you describe Shennongjia Golden Snub-nosed Monkey to your friends?

2.Why did Shennongjia researchers set the Golden Snub-nosed Monkey fixed feeding siutes in the field?

3.How was it established in Shennongjia?

4.Why do scientists study golden monkeys and other animals?

5.What kind of relationship should be formed between humans and wild animals?

09

As the carpet was slowly hovering over Jinhouling, the senior professor and Xiaoming looked down to enjoy the high mountains, green forests and crystal clear streams and springs. Suddenly, the dog was jumping up and down on the carpet, beginning to bark towards the thick forest as if it had found something unusual below.

 Xiaoming, watch your dog. There may be golden monkeys below.

 Really? Come here Wangwang. Don't make a fuss. We're going to have a good look at the golden monkey.

 Look! Dozens of golden monkeys are just walking through the woods.

 Wow! How agile they are. They can swing from tree to tree.

 Golden monkeys are accustomed to living at 1600m above sea level just as every species has its own zone of activity.

 What do they eat there?

  They have wild fruits, twigs, leaves and grassroots. Their favorite food is usnea on tree branches and trunks of primitive forests.

Come on, grandpa. There's another group of monkeys behind! Wow, even some baby monkeys.

 That's normal. The group in front is an "all male unit"and the group that follows is a "monkey family".

 "All male unit"? What does it mean?

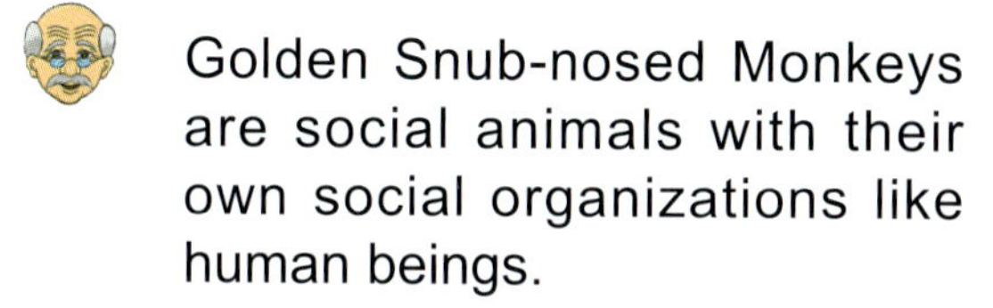

Golden Snub-nosed Monkeys are social animals with their own social organizations like human beings.

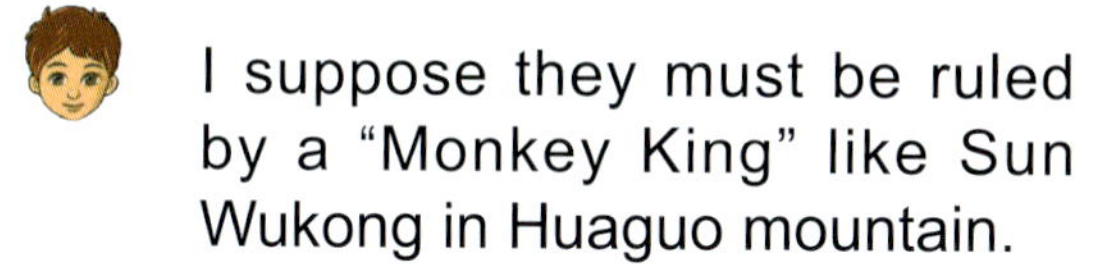

I suppose they must be ruled by a "Monkey King" like Sun Wukong in Huaguo mountain.

Yes, many people think so. But in fact, each golden monkey group lives in an organization of "family"and "all male unit".

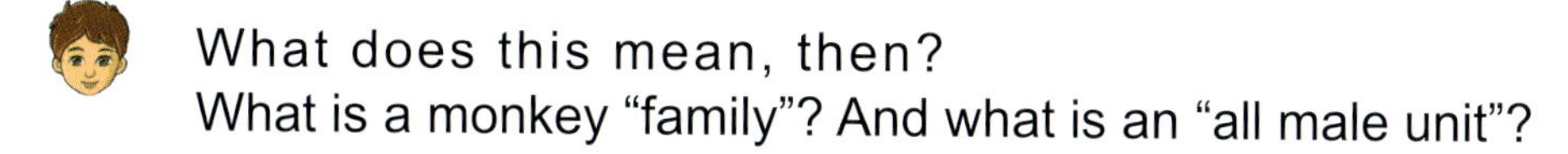

What does this mean, then?
What is a monkey "family"? And what is an "all male unit"?

A golden monkey group has 50 to 200 members, including an "all male unit" and several "families".

Then, what is the "all male unit"?

Well, you can probably guess the meaning from its name. It means members of the unit are all male monkeys.

Wow, there seems to be sexual segregation among monkeys, too.

Yes, something like that. An "all male unit" is the equal of the golden monkey "army". In addition the task to look for food, this "army" is also responsible for the safety and security of the whole group.

Haha, it's interesting indeed. It's just like our Liberation Army, isn't it?

Yes. When the group is foraging and feeding during migration, members of the "all male unit" will work in front and back of the whole group. Members in front are responsible for path finding while those in the back will urge the fellows to follow closely and make sure that nobody drops out.

Oh, they seem to have a strong sense of organizational discipline. They are really something.

Golden monkeys take a nap at noon every day. At that time, the male unit would keep watch and safeguard the whole group.

Wow! They are quite responsible. Now then, what is a golden monkey "family"?

There are several "families" in each monkey population led by an adult male monkey, who is referred to as the "parent". As the most authoritative in the "family", he has the "right" to "marry" more than one female monkey. His "wives" are supposed to keep the "family" in peace and raise infants under the leadership of the "parent" monkey.

Grandpa, come on. There are several baby monkeys over there. And one is still at the breast. How lovely they are with beautiful golden brown hair!

Yes. Golden monkey cubs are the loveliest animals. They are innocent and full of curiosity.

I really want to hug them. You see, even Wangwang stops barking now. He wants to play with them.

In a golden monkey community the cub will get the care of the whole family. Young females will stay in the family, until they "get married". And once grown up, a young male monkey will be rushed by its "father" to the "all male unit" as a man "joins the army".

Wow! How cruel the "father" is.

This is the law of nature. In the golden monkey society, they have their own "social codes", which are essential to guarantee the continuation and prosperity of the population.

Nature seems so cruel, I think.

That's true. Natural organisms are constrained by the natural environment, and their survival is not easy. Therefore, we should do our best to protect them.

Yeah. If everyone takes it as their responsibility to protect wildlife and natural environment, our nature will be more harmonious and beautiful.

Questions:

1.Can you describe to your friends the golden monkeys you saw in Shennongjia?

2.What's the relationship of the "all male unit" and "family" in a golden monkey group?

3.Why is it possible for the "social codes" to ensure the continuation and prosperity of the golden monkey population?

4.Do you know what other animals have similar strict "social structure" as the Golden Snub-nosed Monkeys do?

5.Why should everyone take it as their own responsibility to protect wildlife and natural environment?

10

Observing from the flying carpet, the senior professor and Xiaoming were overlooking the lovely golden monkeys. The monkeys were so cheerful that they made happy cries now and then during their eating and playing. They kept jumping up and down among the trees, causing branches to sway constantly with leaves glittering in the sun. Then as they disappeared quickly behind the ridge, the joyful noises reduced gradually and all was quiet again. As soon as the carpet landed on a small square, the little dog jumped down and rushed towards the mountain woods.

Come here, Wangwang. Stop running about.

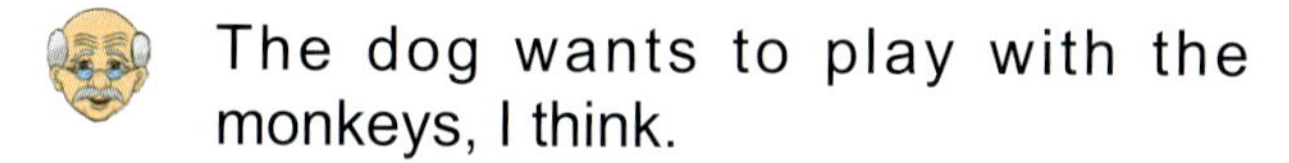

The dog wants to play with the monkeys, I think.

Come back, you silly dog. The golden monkeys are leaving.

Yes, they have all left. Don't disturb them. We may walk in the forest.

Ah! There is spring water here. This must be the Jinhou Stream (Jinhouxi)! How beautiful it is.

The most attractive aspect of the Jinhouling Subarea is the thick forest and Jinhou Stream, a stream that runs all the year round through the forest.

Grandpa, trees on this mountain are very tall and dense!

Yes. That's characteristic of a typical primitive forest.

Then, what are the characteristics of primitive forest? Just some old and rough trees?

It is not that simple. The primitive forest has lots of unique phenomena and characteristics. Let's walk up along the path to the mountain and look for some of these characteristics, shall we?

OK. Hey, grandpa, look, there are a few flowers, pink and white. They are very beautiful.

Yes. You can find more flowers here. They are orchids. Xiaoming, you have got the first important feature of the primitive forest.

Have I? What are the characteristic of the primitive forest? What is the relationship between the orchid and the primitive forest?

Orchid and fern are relatively ancient species on the earth. Ferns appeared about 300 million years ago. A primitive forest must have wild orchids and ferns to prove its primitive and age-old nature. So this is the first characteristic of the primitive forest.

Oh. What are other characteristics of the primitive forest?

Do you see a few trunks fallen on the ground there and over there? There is another one over there.

Yep, but they are all rotted and covered over with moss. And I find a few small mushrooms, too.

These are the other two characteristics of the primitive forest. There should be naturally fallen trees in the primitive forest.The dead trees would naturally fall to the ground without any human intervention until they are completely rotten. Everything is in a natural state.

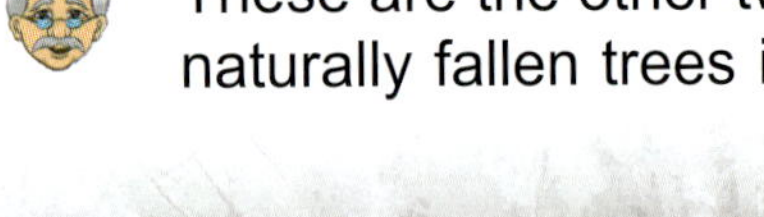

That's what it is. Since it is a primitive forest, there should be no human disturbance. But do you think mushrooms are necessary for a primitive forest?

Not absolutely necessary. The primitive forest is thick with a lot of tall trees. Because of the limited

light in the lower part of the forest, there are lots of fungi, usneas and mosses, which prefer a dark and humid environment. Of course, mushrooms are one of these plants. You see, some green filaments are attached or hanging on many branches. They are usneas, the favorite food of golden monkeys.

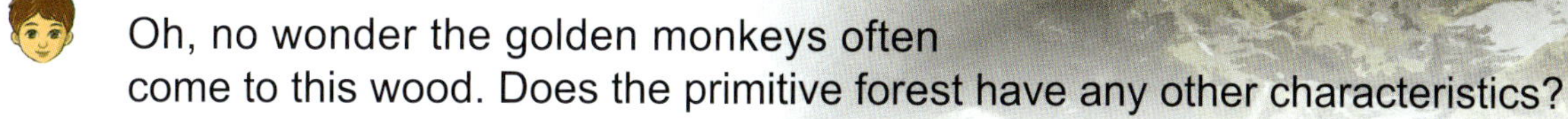

Oh, no wonder the golden monkeys often come to this wood. Does the primitive forest have any other characteristics?

Of course. The primitive forest will show various plant growth levels: moss, grass, shrubs and tall trees from the bottom upward.

And?

The primitive forest generally has a certain thickness of humus soil. The thick layers of humus are formed from rotting branches and leaves, which reveals the age of the forest, indicating it has been there for many years.

This is quite interesting and reasonable.

You can judge whether or not Jinhouling is a primitive forest according to these characteristics I just spoke of.

We have got them all here. So Jinhouling certainly is a typical primeval forest. As soon as I entered the woods, I felt the fresh air invigorating.

But do you know why? The primitive forest not only conserves but also generates oxygen. It is actually a huge generator of negative oxygen ions.

Negative oxygen ion?

Yes. A negative oxygen ion is the negatively charged ion of oxygen. Colourless and tasteless, oxygen ions are regarded as "vitamin in the air". It's essential for promoting human blood circulation, increasing blood oxygen content and cheering up the spirit. So it has the sedative, hypnotic and antitussive effect, and is useful to increase appetite and lower blood pressure.

Wow! Negative oxygen ions are so important!

Yes. Usually, the negative ions will increase in the air after a thunderstorm to make people feel particularly energised. The international index of good air environment is that which contains more than 2000 negative oxygen ions per cubic centimeter of air. In the Jinhouling primitive forest, the average content of negative oxygen ions reaches up to 160 thousand per-cubic centimeter. Hence, the forest here is praised as a "Natural Oxygen Bar".

No wonder I'm feeling very energetic. The primitive forest in Jinhouling is precious indeed.

Yes. Forests are the most valuable resources and wealth on earth. Every tree here is priceless.

Priceless? How can trees be thought priceless? They are only used as building materials.

In the past, when people cut down forests for timber, they would only calculate the volume of timbers produced by the forest. According to the calculations of scientists, a 50 year old tree is worth 300 US dollars or so in the market. However, if calculated

according to its ecological benefits, a tree of the same age can produce oxygen worth 31,250 US dollars each year. In addition, forest trees can reduce air pollution, conserve water and prevent soil loss. Rotting branches and leaves are good for soil fertility.

Moreover, forests are also the home to birds and animals.

That's true. Lots of biological species are raised in forests. Altogether, the ecological value of a tree can be higher than 200,000 US dollars.

Wow! These trees are priceless indeed. We must protect the forest.

Now, Xiaoming, have you noticed the disappearance of the running stream?

Oh yeah. We don't see the stream any longer. But I can hear it. Where has it gone?

Just listen by the stone.

Wow! I can hear the rumble of running water under the stone. There seems to be a river underground.

Yes. This is the mysterious underground river of Jinhouling. We can hear it although we cannot see it on the ground as if we're standing in front of a waterfall

long away ahead. So this place is named as “Jinhoutingpu”, where golden monkeys listen to the waterfall.

It’s really incredible. How can there be an underground river here? And how did it form?

This is a good question. I leave it for you to check the information later. Now let’s go down. Look carefully. Walking down along the road you can find clear mountain springs gushing out of rock gaps and tree roots to form mountain brooks, which will collect somewhere as the source of Jinhou Stream eventually.

Okay. Come back, Wangwang. Let’s go.

Questions:

1.Why should there be fungi, usnea, moss and other plants that do not require a lot of light in the primitive forest?

2.Why is it possible to infer the age of a primitive forest from the thickness of humus?

3.Can you sum up the characteristics of primitive forests?

4.Why does the professor say that every tree in Shennongjia is priceless? How should we understand the effect of forest on the environment?

5.What is an underground river? How is it formed?

11

The carpet was flying to the lower Jinhou Stream. The little stream has now developed into a mountain river, flowing down through the forest like a wild horse running away. Suddenly, a steep cliff came into view. Here the stream water is dashing down to form a spectacular waterfall.

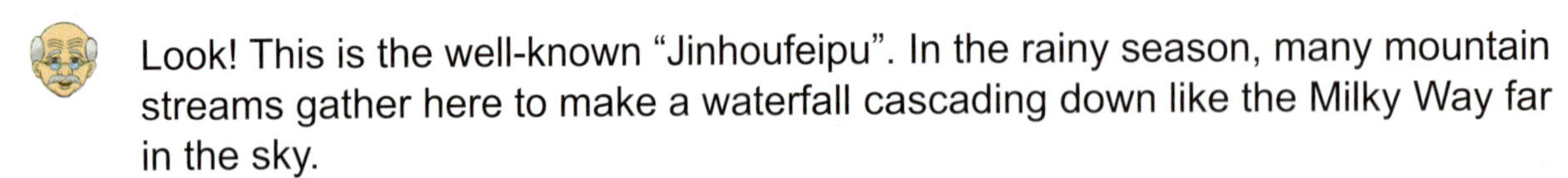

Look! This is the well-known “Jinhoufeipu”. In the rainy season, many mountain streams gather here to make a waterfall cascading down like the Milky Way far in the sky.

It’s so beautiful that I don’t want to leave Jinhouling, the primitive forest and the cliffside waterfalls.

There are many more beautiful scenes in the Shennongjia. We’re getting to the famous Shennong Valley(Shennonggu) soon.

Shennong Valley? It must be the largest valley in Shennongjia, I guess.

It is not the largest but it enjoys the central position of the Shennong Peak Subarea. Its formation has much to do with the geological structure of the "Shennongjia anticline”.

Shennongjia anticline? What does that mean?

Anticline is a geological term. The rock stratum of Shennongjia is a sedimentaryrock formed more than one billion years ago in the sea. Do you remember the differences between sedimentary,magmatic and metamorphic rocks? I told you to check the information before we came here.

Oh, yeah. Magmatic rocks are formed by magma, while sedimentary rocks are mainly layered rocks formed in water. Formation has much to do with the geological structure of the “Shennongjia anticline”.

Yes. They ought to be in a horizontal state, right?

Yeah. But few of them on the mountain are horizontal. They are all tilted and twisted, instead. How could this be?

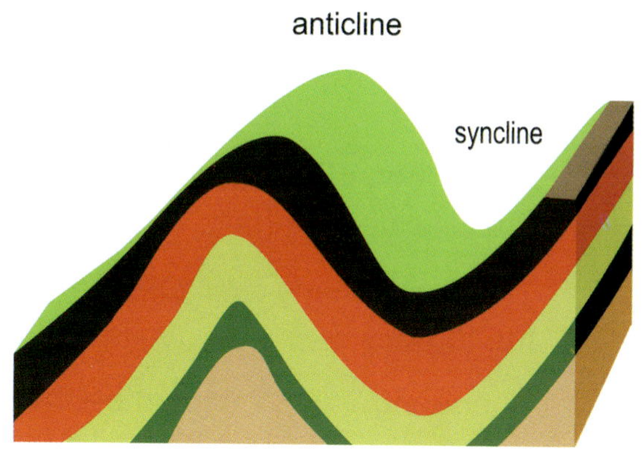

This is because those rock strata became squeezed, folded, twisted and broken during the geological history of billions of years. When a rock folds, there might be two important forms: either anticline by upward arching or syncline by downward concaving.

Oh, I understand anticline and syncline.

The formation of Shennong Valley is closely related to the Shennongjia anticline.

Really? But you just said downward concaving forms syncline. Accordingly, the valley should be syncline. How could anticline make the valley Shennong Valley?

Xiaoming, here is another question for your consideration. Which do you think is easier to form a mountain, an anticline or a syncline?

Anticline, of course because it is arching upward.

Don't hurry to answer, Xiaoming. Just think again.

Well, but it beats me. Do you mean it's easier for a syncline to form a mountain?

Yes. In nature, a syncline is actually more likely to form a mountain.

Why? A syncline is concave downwards! How can it form a mountain?

This should be considered from the long-term nature of geological effects. The general public often find it hard to consider problems from a "geological point of view".

Geological point of view? What is it, then?

All the geological phenomena we observe today are actually the historical results of actions over tens of millions of years or even hundreds of millions of years. In investigating geological phenomena, we should think about them on a large time scale of millions of years. This is the way of geological thinking. Do you remember the proverb that constant dripping wears away a stone? And do you think it possible for water drops to wear down a stone without constant dropping of tens of millions of years?

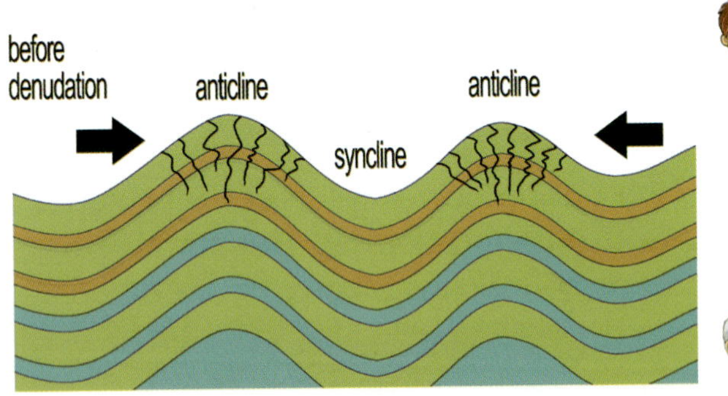

Certainly not. Nevertheless, the syncline formed by downward bending can surely never be higher than the anticline formed by upward arching even after tens of millions of years, then?

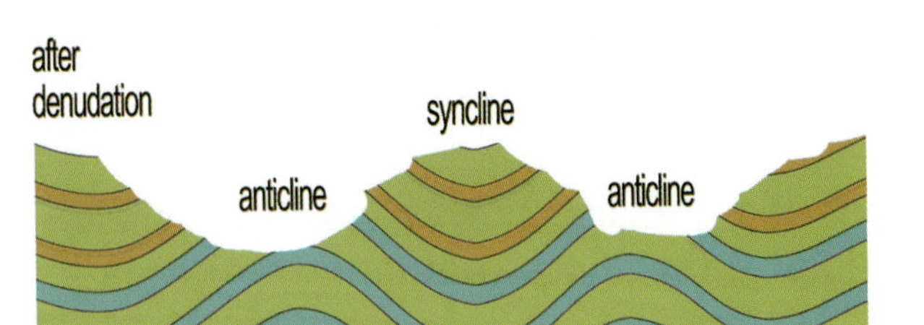

Not absolutely impossible. Let's think about this problem from a different angle. When the strata fold, they may form anticlines and synclines as well. Forces on them are completely different in direction. Consequently, the rock deformation should be completely different in nature.

 Really? Would you tell us how different the forces are upon anticline and syncline?

 Okay, let's suppose you bend a chopstick to the maximum extent. It will break at last, won't it?

 Yeah, that's certain.

 But where will it break?

 In the middle, of course.

 The top of the anticline formed during folding of the strata is equivalent to the middle part of the bent chopstick.

 Yeah. The rock strata will break at the crest of the anticline.

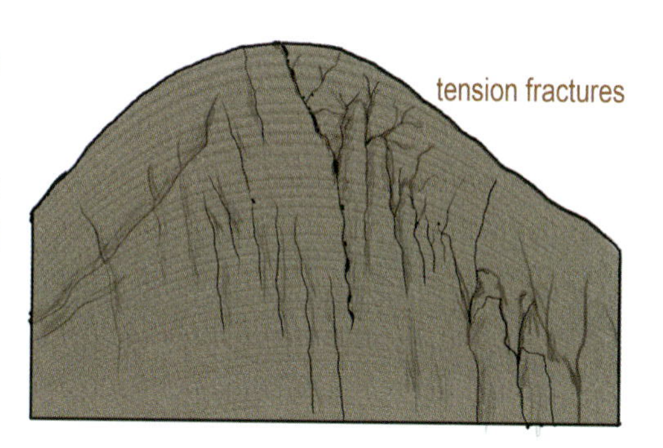

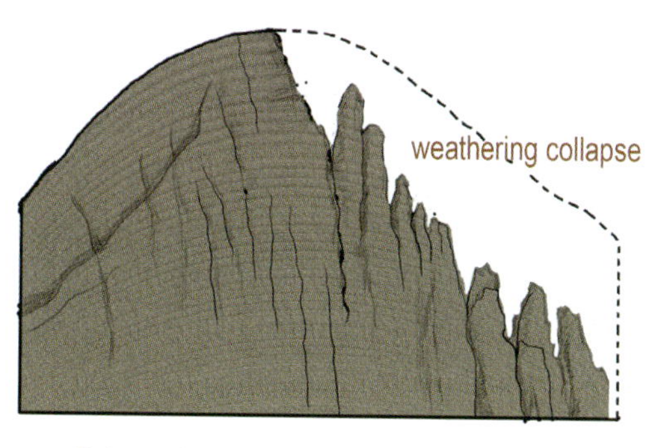

Schematic diagram of Shennongjia anticline

 You're right. As the rock strata break, the pulling power will result in a series of tensile cracks inside. These cracks are called "joints". Because of the numerous joints, some anticline rocks become much weakened in weathering resistance capability.

Oh, I see. It is the joints that make the rock more susceptible to weathering and erosion. As running water washes away the soil and stones produced by weathering and erosion, a valley is gradually formed there. So that's the way it is. Then how about synclines?

 When the syncline part folds, it suffers the pressing power, which will make rocks more compact and more capable of resisting weathering. And as a result…

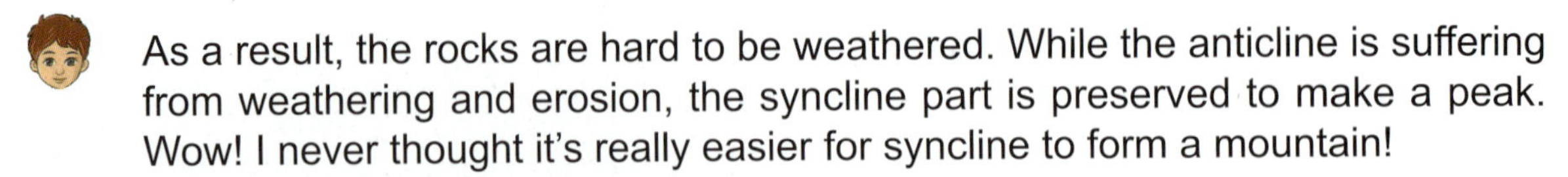

As a result, the rocks are hard to be weathered. While the anticline is suffering from weathering and erosion, the syncline part is preserved to make a peak. Wow! I never thought it's really easier for syncline to form a mountain!

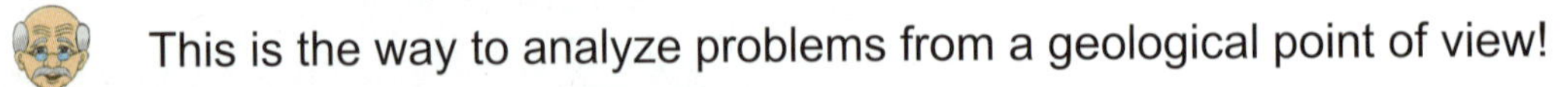

This is the way to analyze problems from a geological point of view!

Wow! It's great I've learned the geological thinking. There are so many things in nature that are easy to be misunderstood by ordinary people.

All the mountains and rivers around us are products of several million years of geological processes. For any magical geological landscape or phenomenon we should do an analysis from the geological point of view.

Yes, grandpa. I believe geological thinking is a very interesting method for analyzing geological phenomena.

That's true. Everything in nature will change a lot over a long period of time. To view the natural landscapes from the geological viewpoint is really a good way to improve our ability of logical thinking and reasoning. It is also an imaginative way in our observation of natural phenomena.

Well, geological thinking is interesting indeed from a geological point of view!

In observing natural phenomena, geological thinking will give us more fun in our sightseeing tour. Thinking has enlightening significance to help understand nature. Nature teaches us lots of useful knowledge. Beside synclines and anticlines, there exists another common geological structure which often forms valleys.

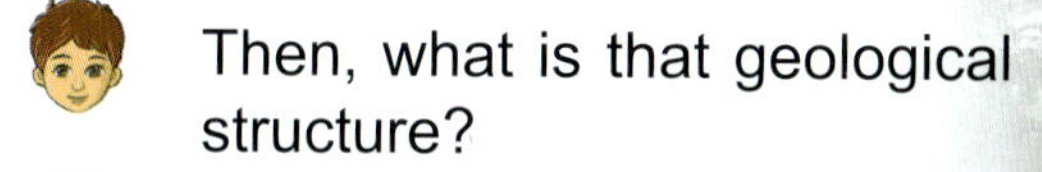

Then, what is that geological structure?

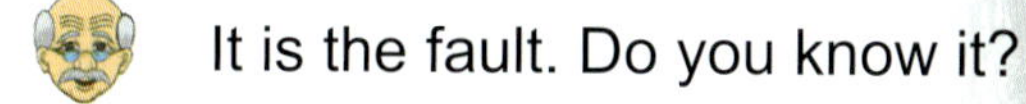

It is the fault. Do you know it?

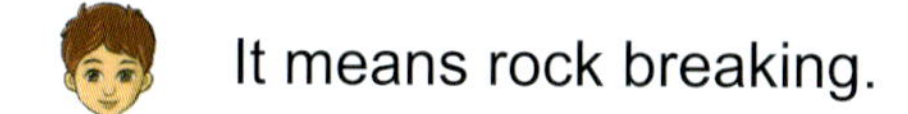

It means rock breaking.

That's right, but not exactly scientific. In geology a fault is defined as a geological feature in the crust, consisting of breaks due to stress with clear relative movement along the fracture plane.

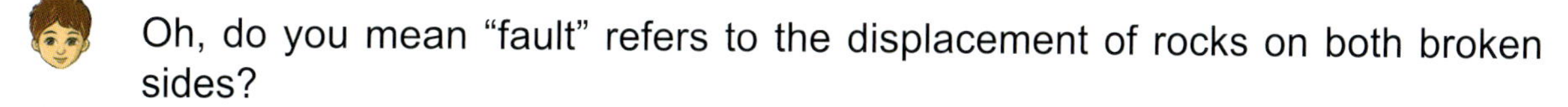

Oh, do you mean "fault" refers to the displacement of rocks on both broken sides?

Yes. If a rock only splits we just call it joint; when displacement occurs we call it a fault.

Now I understand why you say a fault is easy to form a valley. As the stratum breaks and dislocates, the crushed rocks would become very prone to weathering and water erosion. When this goes on, a valley would be naturally generated in the end.

You are quite right. According to the change of landform, geologists working in the field can judge if there is a fault and can even infer the properties of the fault.

Properties of the fault? What are they?

Based on the dislocation direction of rock strata on both sides of the fault, geologists divide the fault into three categories, namely, strike-slip fault, normal fault and reverse fault.

Oh, faults are so complicated. Then how are they divided?

We should first of all understand some related terms. The surface or face of the fault is called the fault plane, which is often tilted. Rock strata above the inclined plane are the hanging wall of the fault, while those below are the footwall.

That's easy to understand.

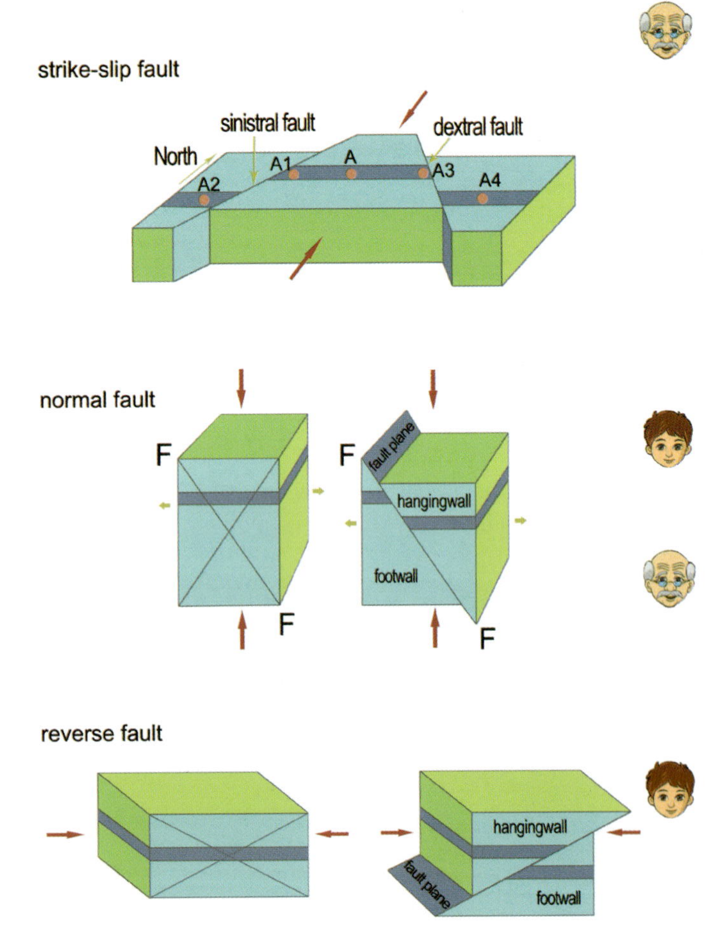

Okay. A fault sliding down along a fault plane is called a normal fault which is formed under the forces of tension. Influenced by gravity, the hanging wall will naturally slide down to form a fault which is often tilted. Rock strata above the inclined plane are the hanging wall of the fault, while those below form the footwall.

I guess the reverse fault must be formed under the squeezing pressure.

Xiaoming, you are so smart. A reverse fault is actually formed in this way. But can you imagine the relative motion of the upper and lower walls during movement?

Let me see. Yeah. During movement, the hanging wall of the fault will press upwards along the fault plane. Is that right, grandpa?

That's right. You've got a logical and clear mind.

Then what about the strike-slip fault? Does it mean its two walls move in the horizontal direction rather than moving up and down?

Yes, it's a very good deduction, Xiaoming. Learning requires both good thinking and contextual intelligence. Well, here we are in Shennonggu.

Questions:

1.Can you explain from the ecological perspective why there are so many springs and waterfalls on Jinhouling?

2.What is the "anticline" and "syncline" structure? What other geological structures can be observed in the fields?

3.Why is it often easier for synclines to form mountains?

4.What is geological thinking? How can you use it to explain the formation of river deltas?

5.What are the main features of the "strike-slip fault", "normal fault" and "reverse fault"?

12

Parking the carpet outside the entrance to Shennong Valley, the professor got off with Xiaoming and Wangwang, and headed for the stone staircase to the entrance. Walking to the top of the staircase, they saw the ink-and-wash-painting-like Shennong Valley in the rising mists and fog. Viewed from afar, the overlapping mountains became blurred, disappearing and reappearing in the mist. On one side of the valley stand several giant stone pillars, around which white clouds are drifting in the blue sky.

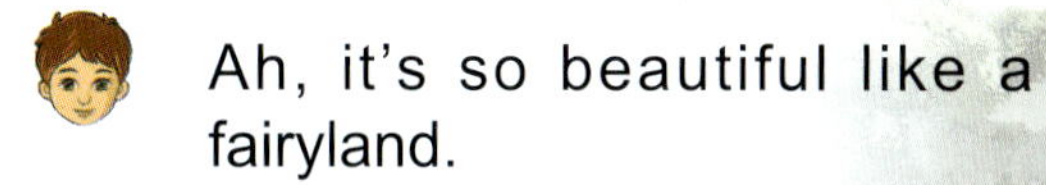

Ah, it's so beautiful like a fairyland.

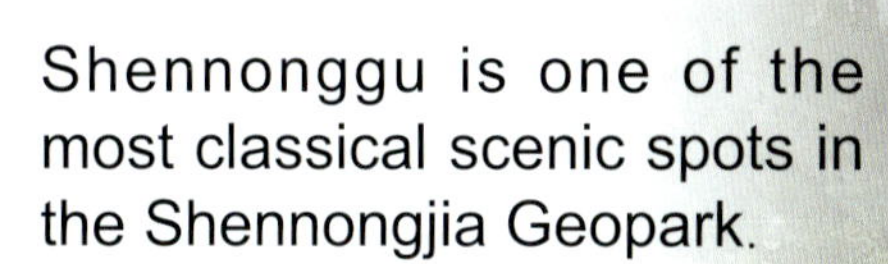

Shennonggu is one of the most classical scenic spots in the Shennongjia Geopark.

Grandpa, is it the valley Shennong Valley?

Yes. Just take a look at the geological condition first, and then try to analyze how the valley is formed. OK?

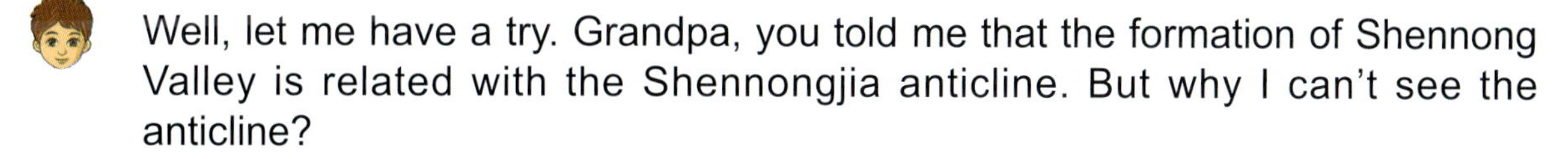

Well, let me have a try. Grandpa, you told me that the formation of Shennong Valley is related with the Shennongjia anticline. But why I can't see the anticline?

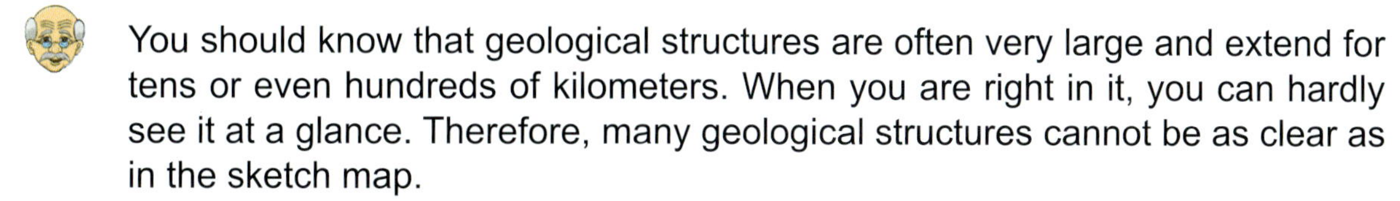

You should know that geological structures are often very large and extend for tens or even hundreds of kilometers. When you are right in it, you can hardly see it at a glance. Therefore, many geological structures cannot be as clear as in the sketch map.

Then, grandpa, shall we fly the carpet to observe it?

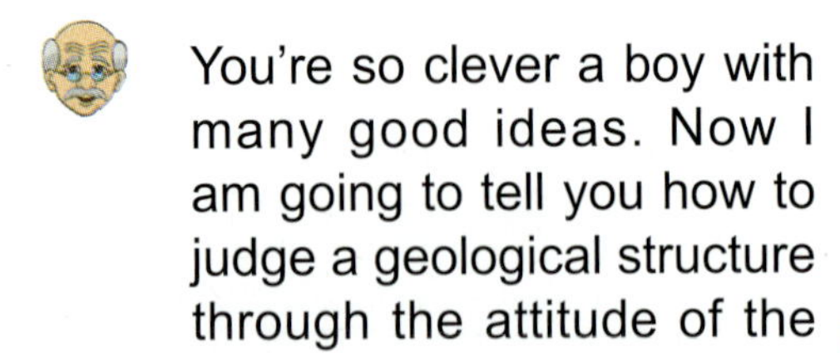

You're so clever a boy with many good ideas. Now I am going to tell you how to judge a geological structure through the attitude of the strata.

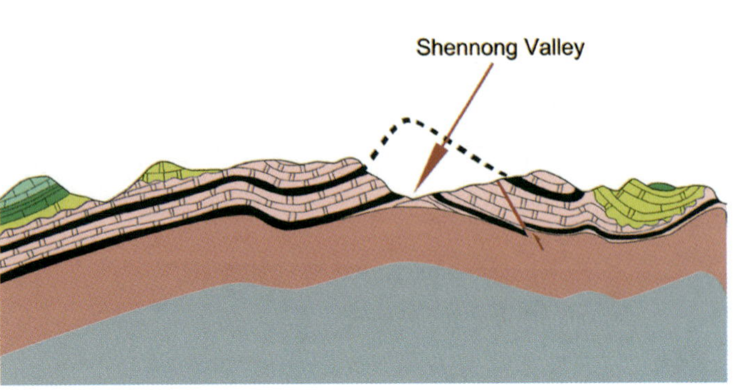

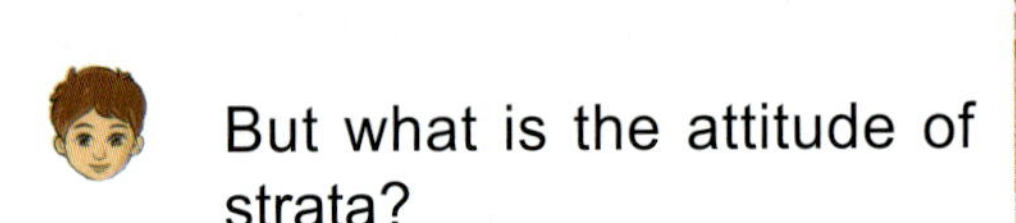

But what is the attitude of strata?

It is a geological term. In short, the so-called attitude of strata describes, for example, the level of the rock, horizontal or tilted, the direction and angle of the tilt, and its relationship with other geological bodies nearby, etc..

Oh, it sounds simple. Let me see, grandpa. The rocks on my left are tilted to the left; those to my right are tilted to the right. Well, grandpa, this is an anticline. Yes, an anticline indeed.

That's right. If the rocks on both sides are tilted to you, they are of the syncline structure.

It seems true that a valley is often made by anticline. Shennong Valley is really produced by it.

Our deduction using the geological approach is verified again. Now the next question: Why are there so many stone pillars here? Can you explain?

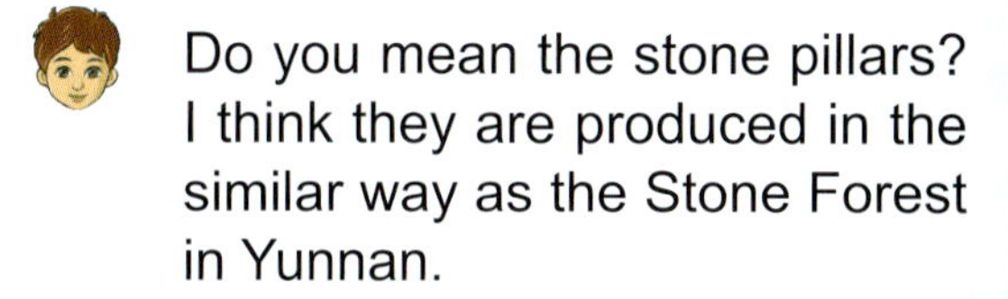

Do you mean the stone pillars? I think they are produced in the similar way as the Stone Forest in Yunnan.

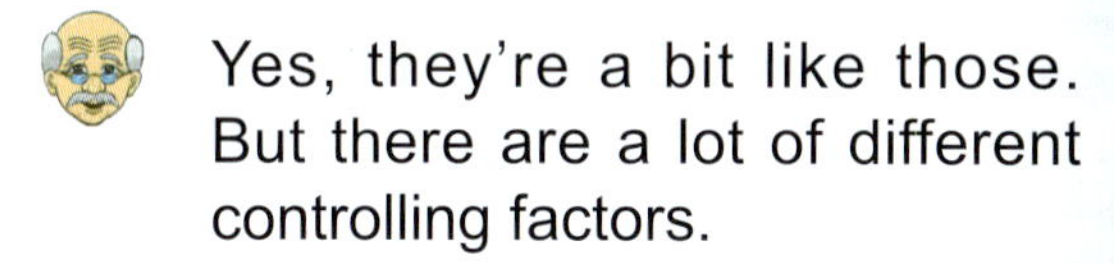

Yes, they're a bit like those. But there are a lot of different controlling factors.

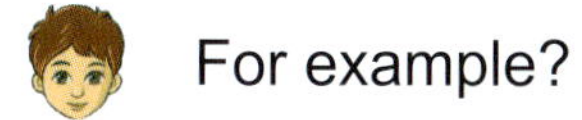

For example?

For example, the stone forest is often developed in a large area of limestone while the distribution of stone pillars in Shennong Valley is rather localized.

Yes. But then what causes it?

As you know the formation of Shennong Valley is related to the Shennongjia anticline. Therefore, stone pillars in Shennong Valley are confined to the area of the Shennongjia anticline.

That's true. But why are there stone pillars only here? Rocks nearby also belong to the Shennongjia anticline. They don't form stone pillars. Why?

First of all, the formation of stone pillars depends on the number of cracks in the rock, as well as the density and the morphological development of the cracks in the stratum's folding and bending.

I think so. The more cracks, the easier it is to form stone pillars.

Yes. In the course of anticline formation, a large number of vertical cracks will be generated somewhere in the rock stratum as a result of the pulling effect. In geology, the naturally formed cracks are called joints.

Yeah. You told me that before.

A large number of vertical joints can cut a rock into a lot of vertical columns. The subsequent weathering and erosion will happen along these vertical joints. Don't forget to apply the geological thinking to your consideration of weathering and erosion process.

Well, I know. After many years of weathering, erosion and collapse, the weathered rocks around the vertical column had been reduced to crushed stones and sand, which would be carried away by the flowing water. Thus, the remaining square pillars stand erect at last.

What a good understanding. After long-term exposure to the sun, wind and rain, as well as erosion and gravitational collapse, the jointed rocks finally turned into such a magnificent stone forest. This is nature's miracle creation.

Yeah. Nature is like a great sculptor, who has created so many beautiful sceneries.

Questions:

1.How do you observe synclines and anticlines in the field?

2.What is a joint? Do you believe you can identify one in the field?

3.What role does the joint play in weathering, erosion and landscape formation?

4.Can you specifically explain to your friends how Shennong Valley is gradually formed by an anticline?

5.Illustrate the formation of Shennonggu stone pillars, and explain their differences from the Guilin Stone Forest.

13

Much attracted by the picturesque valley Shennong Valley, Xiaoming could not help taking pictures one after another while Wangwang was jumping back and forth around him on the viewing platform. A bit frightened by the rising clouds and mist around the viewing platform, the little dog barked at them repeatedly as if facing a formidable enemy. Then as the clouds and mist faded away it began to look down at the valley, shaking its tail proudly like a winner.

Xiaoming, have you noticed that quite a few stone pillars here are relatively smooth? Do you know the reason?

Yes, they should have edges and corners since they have been weathered and denuded along the vertical joints. Why are many of them so round unlike the stone pillars I found in Zhangjiajie?

Well, let me tell you. This has something to do with the nature of the rock itself. Rocks of Shennongjia are quite different from those in Zhangjiajie. The latter are sandstones,while the former are much older sedimentary rocks, or dolomite, which were formed in the ocean over a billion years ago.

"Dolomite". What a nice name!

Like limestone, dolomite is a kind of carbonate rock, mainly composed of calcium and magnesium carbonate. They are easily dissolved by water and result in peaks, caves, sinkholes and so on.

Oh, do you mean their formation is so much like that of Guilin mountains?

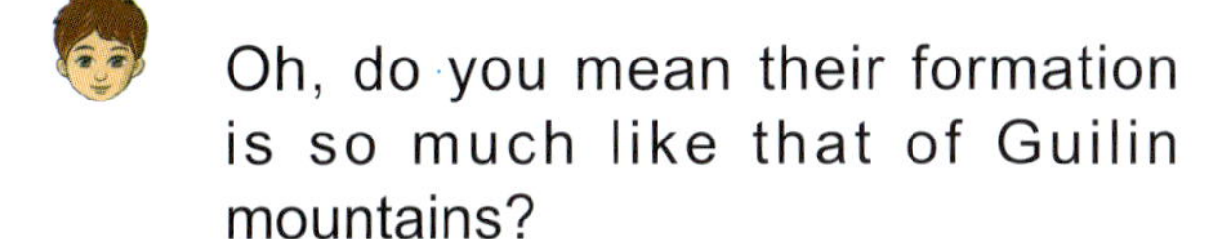

Yes. Guilin Mountains are a typical Karst landscape.

Grandpa, what is a Karst landscape?

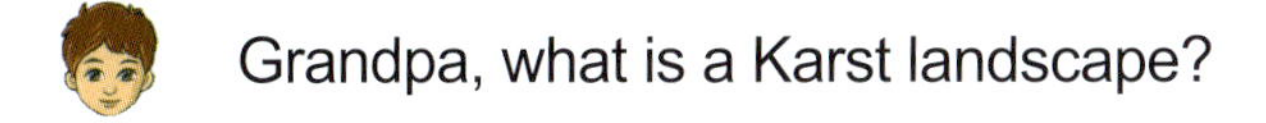

Karst is originally a place name in Slovenia, Europe, where there exists a large area exposed limestone. This area enjoys the most typical limestone dissolution characteristics with many well developed caves, sinkholes and underground rivers.

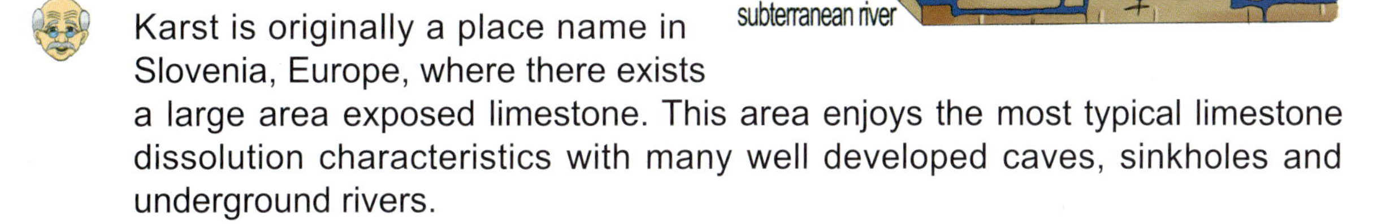

Oh, I see. It's just like Danxiashan in Guangdong and the corresponding term Danxia Landform for the unique and beautiful geological appearance.

Yes. In geology, Karst is a specific term for the phenomena of carbonate rocks. In fact, karst is the result of a kind of physical and chemical weathering, primarily chemical weathering.

Chemical weathering? Does it mean rocks are weathered through chemical reactions?

Yes. Chemical weathering is a long process where changes or decomposition of chemical compounds in the rocks decrease the hardness and density of rocks or change their volume, which promotes their dissolution after decomposition.

Do you mean dolomite is prone to chemical weathering, right?

Yes. Although it is not so easy for dolomite to form the typical Karst landscape as the limestones in Guilin, its chemical nature is likewise less stable. So dolomite is also prone to dissolution.

In terms of landscape, the dolomite in Shennongjia and the limestone in Guilin are really quite different. But in what way are the two different?

Xiaoming: You observe and think very carefully. The question you've raised is very good. Both dolomite and limestone are carbonate rocks, but they are different in chemical composition. The main composition of limestone is calcium carbonate while dolomite is composed of calcium carbonate and magnesium carbonate. Do you know the characteristics of calcium carbonate and magnesium carbonate?

No, I don't. Do they make a big difference in this respect?

Yes. They are rather different in chemical properties. Magnesium carbonate is not as active as calcium carbonate. Many people only know that they are both easy to corrode and prone to Karst phenomena, but they do not know the reason. Listen to me, Xiaoming. I'll explain it to you.

It must be a difficult subject.

It is not too difficult to understand really. First, Karst dissolution relates not only to the composition of rock, but also to the temperature, air pressure, water volume and water kinetic energy.

In the case of high temperature and air pressure, large volume of water and strong kinetic energy, carbonate is more easily dissolved. Because of different natural environments, the karst phenomena of Shennongjia can not be simply compared with those of Guilin.

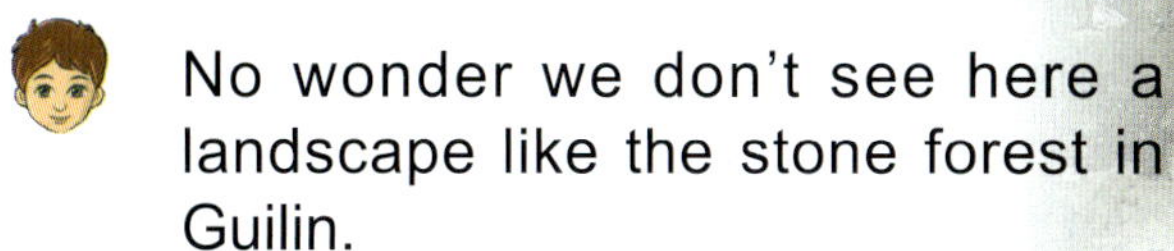

No wonder we don't see here a landscape like the stone forest in Guilin.

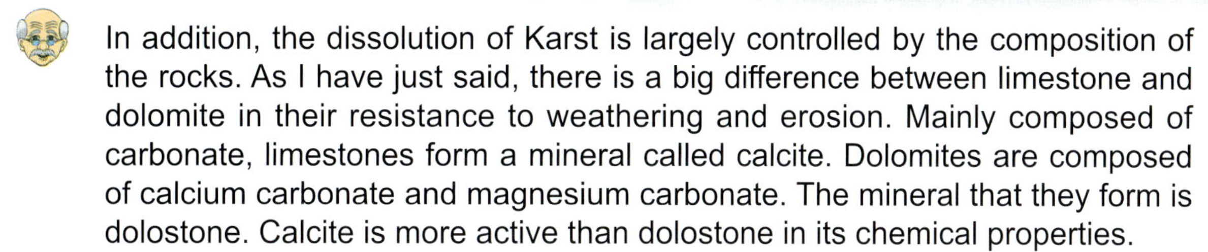

In addition, the dissolution of Karst is largely controlled by the composition of the rocks. As I have just said, there is a big difference between limestone and dolomite in their resistance to weathering and erosion. Mainly composed of carbonate, limestones form a mineral called calcite. Dolomites are composed of calcium carbonate and magnesium carbonate. The mineral that they form is dolostone. Calcite is more active than dolostone in its chemical properties.

That is to say calcite is more likely to be dissolved than dolomite, right?

Yes, you're quite right, but do you know how geologists identify them in the field?

They do it with a magnifying glass, I guess.

Yes. Tiny calcite crystals composed of calcium carbonate are often found in limestone. Dolomite crystals are usually not as good as calcite. Besides, geologists will use a knife to carve the stone.

What's the use of a knife?

The hardness of calcite is much softer than that of the knife, and it reacts readily to hydrochloric acid as well. Finally, geologists have the best trick to distinguish between calcite and dolomite.

Well, what is the best trick?

They will drop 5% hydrochloric acid on the rock and see whether the rock bubbles up or not. If there are bubbles, they can conclude that the main component of the rock is calcium carbonate and the rock must be calcite, that is, limestone.

Why is that?

Because only calcite, that is, the calcium carbonate, can react and form calcium chloride and carbon dioxide when it encounters hydrochloric acid. Sodium chloride is the main component of salt, which dissolves in water easily. Carbon dioxide is a gas, which produces the bubbles. The more intense the bubbles, more and purer the calcium carbonate. This is the way geologists determine the content of calcium carbonate in a carbonate rock in the field.

Geologists are intelligent indeed. Then what about dolomite?

Dolomite is mainly composed of dolostone. Magnesium carbonate is not as easy as calcium carbonate to react with hydrochloric acid. Drops of hydrochloric acid usually do not create bubbles. However, geologists have a solution.They often scrape some rock powder off the stone first, and then drop hydrochloric acid onto the powder. If there are bubbles, the

rock may be magnesium carbonate, that is, dolomite.

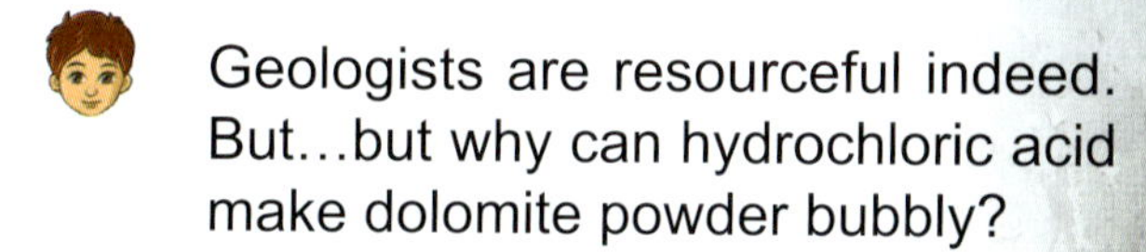

Geologists are resourceful indeed. But…but why can hydrochloric acid make dolomite powder bubbly?

To scrape some rock powder off the stone and then drop hydrochloric acid onto the powder is a way to increase the contact area for the chemical reaction. As each tiny powder particle fully contacts with the hydrochloric acid, the chemical reaction will be facilitated between the hydrochloric acid and magnesium carbonate.

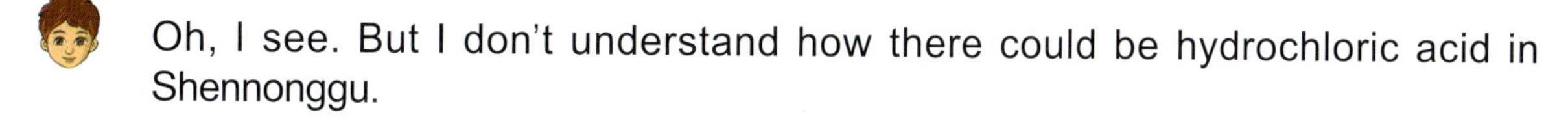

Oh, I see. But I don't understand how there could be hydrochloric acid in Shennonggu.

In fact, calcium carbonate and magnesium carbonate can react with any kind of acid, not necessarily the hydrochloric acid. If your floor is marble, for example, you must be very careful not to drop a vinegar bottle on the floor. Otherwise, there would be bubbles generated to ruin your floor.

Marble? Is marble also composed of calcium carbonate?

Sure. Marble is a metamorphic rock formed by limestone under high temperature and pressure. Like limestone, its main component is calcium carbonate, which will react with any kind of acid.

Oh, I understand. But what kind of acid is there in Shennonggu? Why could there be so much acid?

Do you know acid rain?

Yes, I have heard of it although I don't know exactly what it is.

Acid rain is produced by the combination of carbon dioxide and sulfur dioxide from factories with natural water. Their chemical constituents are carbonate acid and sulfuric acid. Carbon dioxide and sulfur dioxide are ubiquitous in nature. Animal and plant breathing, vehicle emissions, burning coal or other fuel, all this can produce a lot of carbon dioxide and sulfur dioxide.

Oh, I got it. As the carbon dioxide and sulfur dioxide rise in the air, they combine with the atmospheric water vapor to form acid rain. Ah! That's so terrible! The rain pouring from the sky might be acid rain. Is that so?

Yes. With the development of industrialization, the threat of acid rain is becoming more and more serious. These round pillars in Shennonggu were formed through millions of years of dissolution as a result of the long-term effect of chemical weathering.

Grandpa, will these pillars become thinner in the future?

Yes. They will become not only thinner, but also shorter until they eventually disappear altogether.

Really? What a pity, but why?

Don't you remember the geological thinking? I said that is the case millions of years later. However, if we pay attention to protecting the environment and reducing the emissions of carbon dioxide and sulfur dioxide, the process will be greatly slowed.

I think so. We should try our best to reduce the emissions of carbon dioxide and sulfur dioxide to protect the natural environment!

That is why we advocate "low-carbon emission reduction policy". If every one of us and every enterprise could do well in reducing carbon emission, our environment will completely improve and many geological relics can be saved for a longer time. So, for our own sake, and for the sake of future generations as well, we must do something to protect our earth!

Questions:

1.What is dolostone? In what way does it differ from limestone?

2.What is "Karst"? Can you name some tourist attractions in Karst landforms?

3.Explain why Shennonggu stones are relatively smooth according to your own understanding.

4.How is acid rain produced? What is the effect of acid rain on nature? And what should we do to achieve low carbon emission?

5.Do you understand the two chemical equations below? They constitute the theoretical basis for the karst phenomenon. $CaCO_3 + H_2O + CO_2 = Ca(HCO_3)_2$, $Ca(HCO_3)_2 = CaCO_3\downarrow + H_2O + CO_2\uparrow$

14

The senior professor resumed the journey with his group. The flying carpet moved slowly westward along the highway. Then Xiaoming suddenly found several huge footprints on a roadside stone. He could not help uttering a cry of surprise.

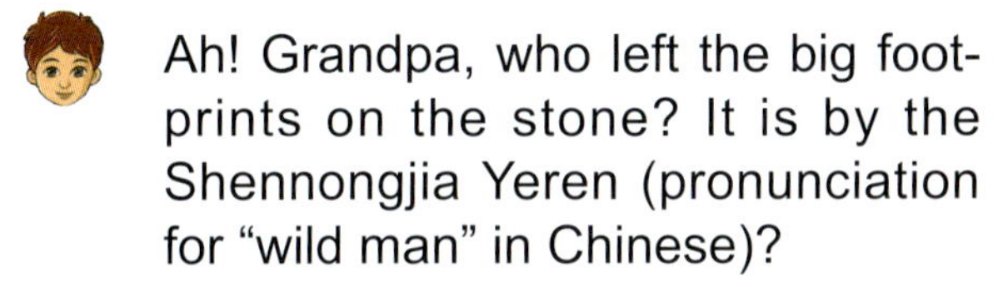
Ah! Grandpa, who left the big footprints on the stone? It is by the Shennongjia Yeren (pronunciation for “wild man” in Chinese)?

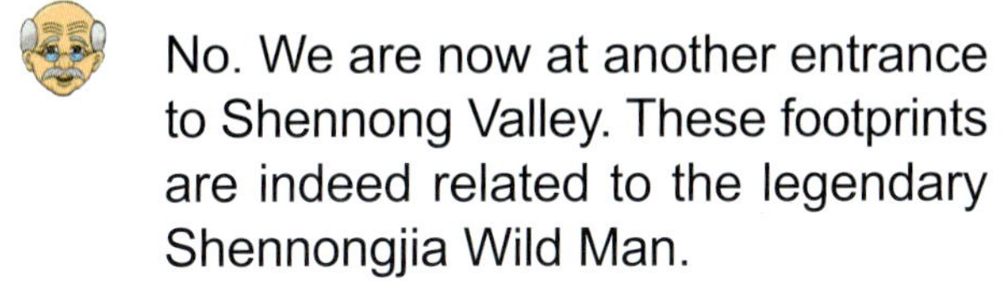
No. We are now at another entrance to Shennong Valley. These footprints are indeed related to the legendary Shennongjia Wild Man.

Well, what is it then?

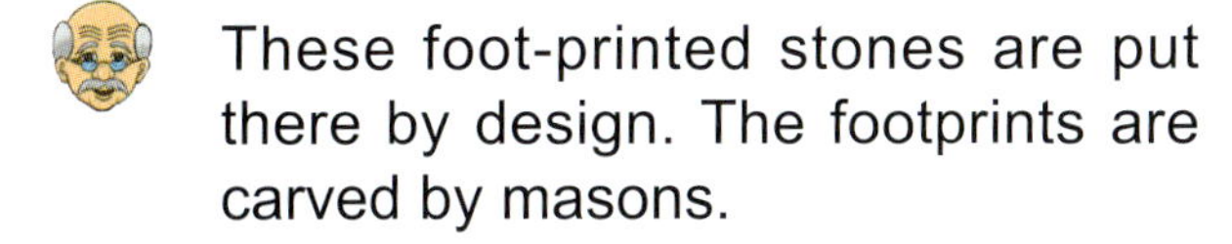
These foot-printed stones are put there by design. The footprints are carved by masons.

Why are those big footprints carved here?

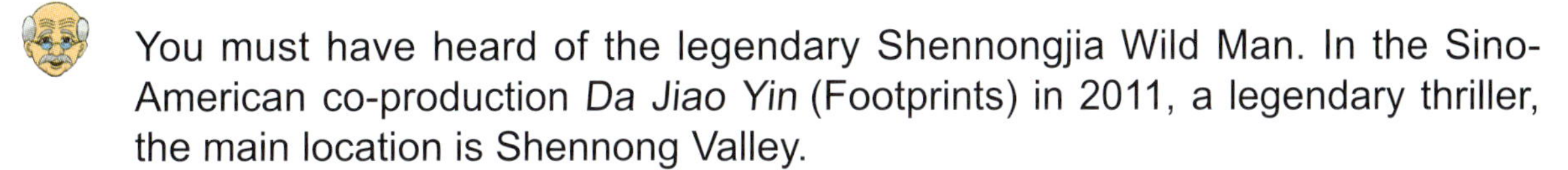
You must have heard of the legendary Shennongjia Wild Man. In the Sino-American co-production *Da Jiao Yin* (Footprints) in 2011, a legendary thriller, the main location is Shennong Valley.

It must be a wonderful movie with a very interesting story, I think.

Of course. Based on some famous legends of the Shennongjia Wild Man, the movie tells us a mysterious story about the unique local folk custom and the industrious, brave, and innocent lives of Shennongjia people. Because of the carved footprints at the entrance, Shennong Valley has been renamed as the “Bigfoot Canyon” by the Geopark.

No wonder there are so many footprints in here. Grandpa, do you think there really exist any wild man in nature?

There are quite a few legendary records about a so-called Wild man all over the world. For example, the ape-like creatures are said to have appeared in different places, for example, Yeti in the Himalayas, Qiuqiuna in Siberia, Cut Moss in Africa, Jove in Australia, and so on. In the Americas, tales about the "Bigfoot - Sasquatch" are popular locally. In fact, almost every continent in the world has reported a so-called trail of a "wild man".

Now then, is there really any wild man in the world?

With the passage of time, legends of Wild man have gradually disappeared in many places. For example, the witness reports about wild man in tropical primeval forest areas do not have any follow-ups in Yunnan, Guizhou and Guangxi. Only from Shennongjia there is constantly news about wild man.

I think there must be wild men in Shennongjia.

Quite a few explorers greatly interested in Wild man have come to Shennongjia for investigation. The Chinese Academy of Science and some civil society organizations like China Wild Animal and Plant Research Association, and the Exotic Animal Specialist Committee have organized a number of special investigations of the Shennongjia Wild Man.

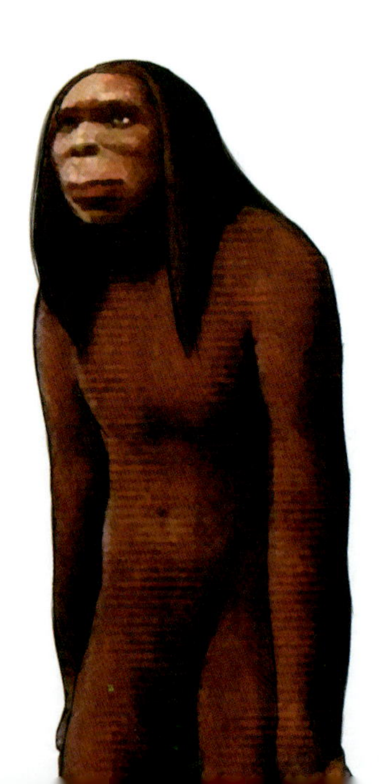

Have they found any wild man?

It is a pity that they haven't. Apart from some suspected sleeping nests, footprints, hair, feces and other indirect evidence of the wild man they have never caught or photographed a real live wild man to prove the true

existence of this legendary creature.

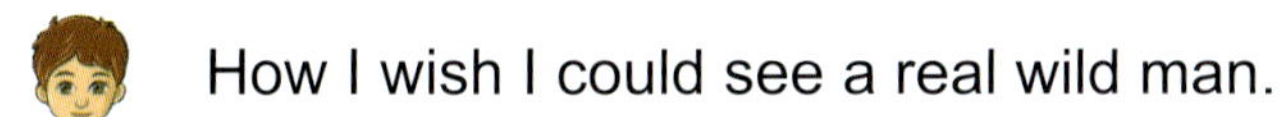

How I wish I could see a real wild man.

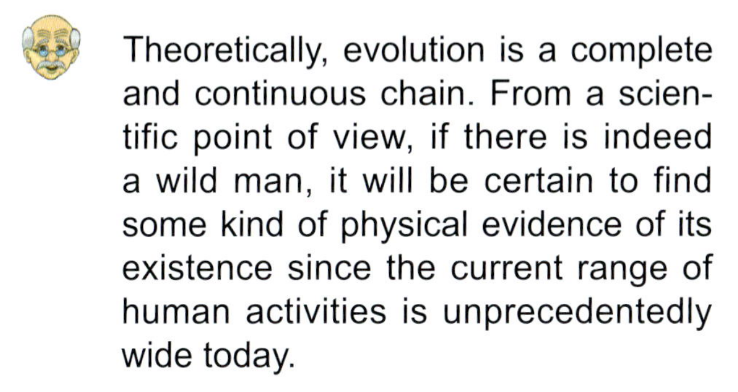

Theoretically, evolution is a complete and continuous chain. From a scientific point of view, if there is indeed a wild man, it will be certain to find some kind of physical evidence of its existence since the current range of human activities is unprecedentedly wide today.

Then, what can be said as the physical evidence?

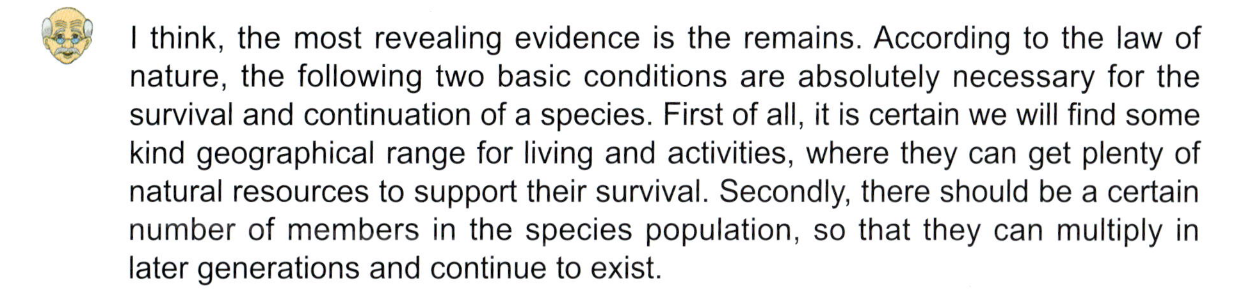

I think, the most revealing evidence is the remains. According to the law of nature, the following two basic conditions are absolutely necessary for the survival and continuation of a species. First of all, it is certain we will find some kind geographical range for living and activities, where they can get plenty of natural resources to support their survival. Secondly, there should be a certain number of members in the species population, so that they can multiply in later generations and continue to exist.

No doubt, without certain food and resources, the wild man would starve to death. Without a population, they would be quickly extinct.

If the Shennongjia Wild Man has a big population, it is impossible for them not to have left any exact physical evidence.

Has any evidence of such a kind ever been found in Shennongjia?

Unfortunately, not yet. The most important question, however, is no longer whether the wild man exists or not. What matters so much is that a unique Wild man culture has developed in Shennongjia, which is successfully promoting the local economy.

Yes. Today as long as Shennongjia is mentioned, people will naturally think of “Wild man”.

This is the fascination of the Shennongjia Wild Man culture. It can arouse people’s imagination and speculation, and stimulate their interest in scientific exploration.

Yeah, I’m always wondering what the wild men would do and eat every day if they were still living in Shennongjia. Could they sing? How would they do if they ran into wild boar or black bears? If I happen to come across them, especially a little one, would he play with me? Would they understand what I’m saying? So, I will observe the natural environment more carefully so as to find an ideal place for them to live in. How I wish I could!

The Wild Man culture stimulates people’s imagination and exploration. At the same time, it is also a source of many beautiful legends. It has made Shennongjia more attractive with boundless fascination. In the Tianyan Subarea of the Geopark, a special museum has been built with a fairly scientific and precise name: the Humanoid Animal Museum.

I like the name, too. Scientifically speaking, humanoid is appropriate name indeed. Come on, grandpa, there is a long footpath over there.

Yes, It's a footpath leading to Shennong Valley.

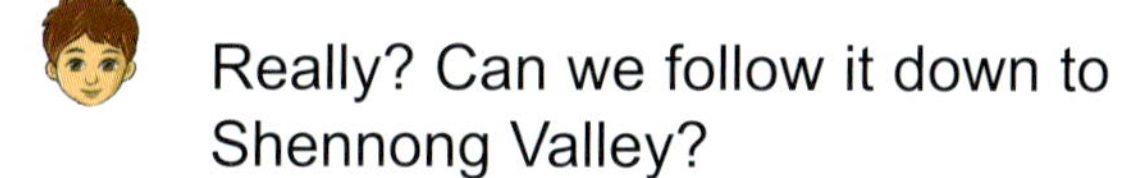
Really? Can we follow it down to Shennong Valley?

Of course. In order to make people fully enjoy the "Big Footprint Canyon", the magnificent natural landscape of Shennong Valley, the Geopark has built a 4km long "U" shaped sightseeing path. Starting at the entrance of the "Big Footprint Canyon", visitors can walk down along the winding path to the valley floor. Then, after a long walk through the forest, they can take the circling path upward until they come back to the ridge near the viewing platform where we just were.

Oh, it's quite a job to build the footpath.

Yes, indeed. The elevation at the entrance of the viewing platform is 2820m above sea level. The elevation difference from here to the lowest point of the path is 400m or so.

Wow, 400m! It's more than 100 storeys high. That's amazing!

To help tourists enjoy the wonderful scenery of the valley, the Geopark has overcome difficulties, spent a lot of human, material and financial resources to construct the footpath along cliffs. Full of twists and turns, the path winds through the beautiful Shennonggu mountains and rivers like a flying dragon in a picturesque landscape.

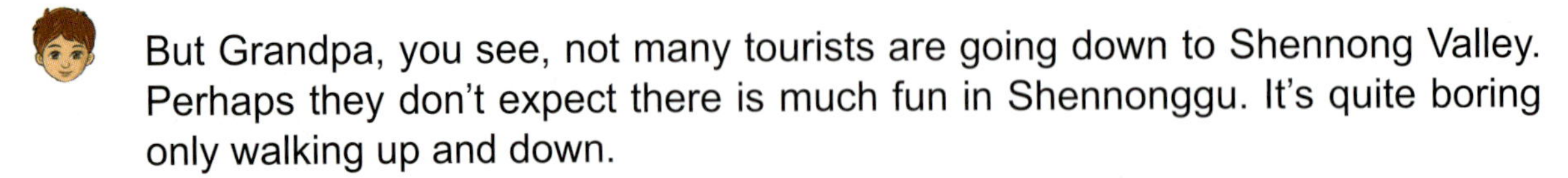

But Grandpa, you see, not many tourists are going down to Shennong Valley. Perhaps they don't expect there is much fun in Shennonggu. It's quite boring only walking up and down.

This may be true. Many people do not know how to enjoy Shennong Valley after a long walk of three hours or more. Most of the tourists are not prepared for such a long journey.

A tourist must be regretful if he fails to see Shennong Valley in his visit to Shennongjia.

He surely is. Shennong Valley is formed as a result of multi-geological activities such as extensional fractures, long-term weathering, ice split, erosion and collapse of vertical joints, and karst dissolution as well. It is characterized by steep rocks, numerous peaks and stalagmites. In such a fairyland you can appreciate a spectacular view of dense forests and a variety of beautiful scenes. As a graceful example of exogenous geological force landscaping,Shennong Valley enjoys the significance of scientific investigation and the high aesthetic value of the landscape as well.

Grandpa, look there. The path is stretching up between the narrow canyons like a ladder to the sky.

It's indeed a wonderful experience to walk through the canyons and view the beautiful scenes of the valley so closely. Looking over there, you can see clusters of stone pinnacles. The scene is particularly magnificent and beautiful!

Ah! The clouds are coming over.

Watching Shennong Valley clouds is one of the highlights in the tour. The drifting clouds add to the charm of the valley. Jagged peaks and stalagmites rise straight up in the changing clouds and mists. Various forms of cliffs and rocks appear and disappear from time to time with countless changes. Walking on the winding wooden plank road, we can appreciate the diverse and graceful natural landscapes.

Grandpa, shall we take a walk along the path?

Well, I'm afraid we are out of time today. We may do it next time. Let's go to the Banbiyan now. OK?

Banbiyan? Where is it?

That is one of the most popular tourist attractions in Shennongjia with a scene of another feature. You will know it soon.

Okay, let's hurry up to Banbiyan.

Questions:

1.From a scientific point of view, do you think there is a real wild man in Shennongjia?

2.What are the basic conditions for the survival and continuation of a species?

3.Why are the clouds viewed as a highlight of Shennong Valley?

4.Would you like to go down to Shennong Valley along the footpath for investigation? Why or why not ?

5.How do you introduce Shennong Valley to your friends?

15

Now the carpet continues to fly slowly over a highway. Built basically along the mountain ridges, the highway was stretching westward to Taiziya from Shennonggu. There are few trees on the mountainside , only alpine meadows, and clumps of bamboos, which have made the mountainside very messy. In spite of the monotonous scenery, visitors can enjoy something if they look far into the distance in a commanding position, where they will be overwhelmed by the broad imposing manner of the overlapping mountains, and feel quite relaxed and happy. You can find environmental protection tour buses traveling to and fro on the highway and lots of backpackers walking briskly on the mountain road. Very often the backpackers will stop to enjoy faraway scenes, or to take some pictures. Everyone seems charmed with the magnificent mountain scenery.

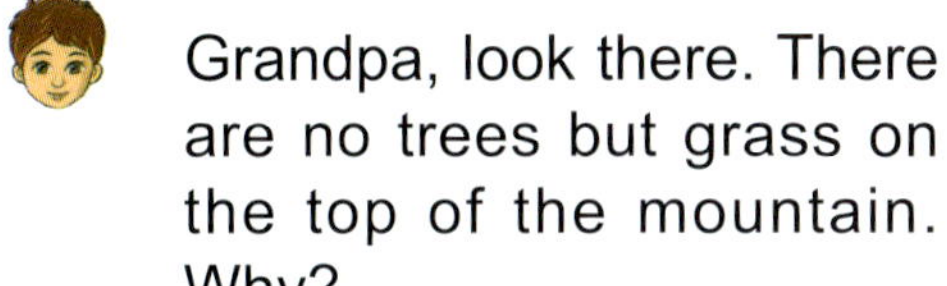

Grandpa, look there. There are no trees but grass on the top of the mountain. Why?

What we see are actually alpine meadows. Because of the large difference in geographical elevation and temperature change, there is a very interesting vertical vegetation zonation in the Shennongjia area...

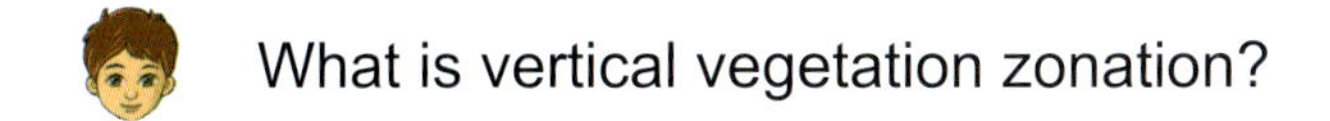

What is vertical vegetation zonation?

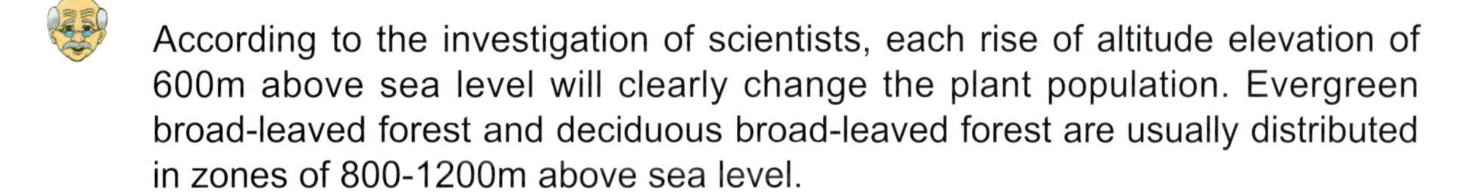

According to the investigation of scientists, each rise of altitude elevation of 600m above sea level will clearly change the plant population. Evergreen broad-leaved forest and deciduous broad-leaved forest are usually distributed in zones of 800-1200m above sea level.

What is the altitude of the alpine meadows we see now?

About 2400m above sea level. The vegetation here is represented by alpine meadows and bamboo forest. Westward in Taiziya the altitude is as high as 2600m above sea level. In addition to meadows and bamboos, there is a primary coniferous forest represented by a local fir. The alpine meadows, bamboos and firs are typical of the Shennongjia alpine zone vegetation.

Ah, are there still primary forests on the mountain top?

Yes, that's true. A 2km long foot path and 6 viewing platforms have been built in Taiziya primary fir forest, where people can enjoy beautiful woods and mountains, singing birds, fragrant flowers and fresh air of this quiet primeval forest close up.

Singing birds and fragrant flowers? Is that true, grandpa? The mountains are so high and it is so cold. How could there be fragrant flowers in the primitive forest?

It's true. In the alpine vegetation of firs, bamboos and meadows are also growing many alpine begonias and alpine rhododendrons!

Do you mean begonia and rhododendron?

No joke. Begonia is a very beautiful ornamental plant with red tender flowers. In the Taiziya fir forest you can see a lot of Begonia trees. In the flowering season, the green forest is embellished with the small red flowers of begonia. I once read a poem that praises the independent unassuming personality of Begonia.

Wow, how attractive the Begonia flowers are! Then what about rhododendron?

There are a lot of rhododendron trees in the mountain area of Shennongjia. Rhododendron is an evergreen plant of great adaptability, cold and heat tolerance. It is therefore praised as the "royalty of the green world". According to statistics, among the 58 kinds of rhododendrons in Hubei Province, 35 are growing in Shennongjia, are distributed in the fir forest or on meadow mountainside at 1800-3100m above sea level. In the flowering season every spring, the pink and bright white rhododendron flowers will blossom and the mountainsides will be breathtakingly beautiful.

Wow, it's so attractive! Then, what's the best season to see rhododendron flowers in Shennongjia? I really want to see for myself the beautiful scenery of rhododendrons in full bloom.

It's in May. On the green mountainside in spring, you can enjoy clusters of red and white rhododendron flowers like bright flames and white clouds, which makes the mountainside a world teeming with life and colors.

How beautiful the green mountains are with so many white and pink rhododendron flowers.

Rhododendron is known as the "Belle of Flowers". A famous poem by Bai Juyi praises the great beauty of rhododendron: "Looking back, I find peach and plums are colorless, and hibiscus is not a flower at all."

Wow! I can't imagine that rhododendron in the wild is so beautiful. Let's come to Shennongjia to see rhododendron in May. The flowering rhododendron of May!

OK. In spring, the Shennongjia area will become a colorful world.

There're so many valuable plants here! Grandpa, what other fascinating plants are growing here?

Oh, quite a lot. Let me introduce you to the rare plant of red birch.

Red birch? I have some idea about white birch. This is the first time I hear of red birch.

Professor: Due to the pink bark, red birch is particularly striking in green trees. Its bark falls off once a year. The fallen bark is so smooth that young people would write letters on it to express love. So the red birch has got a nice name "Love Tree" in Shennongjia.

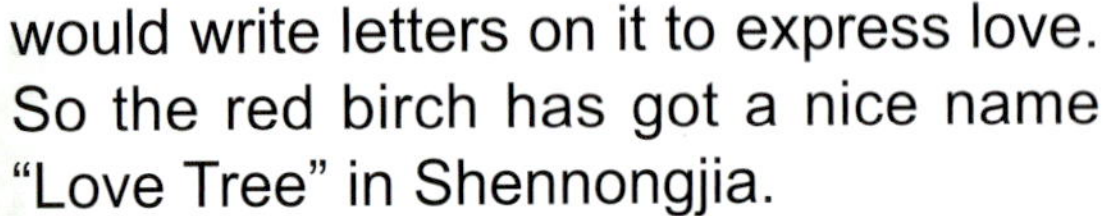

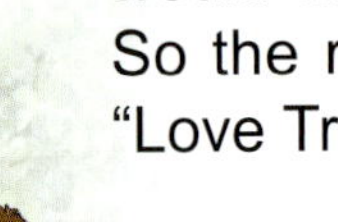

Ah, what a romantic name, "Love Tree"!

Red birch also secretes a kind of juice, called red birch juice. It's very sweet, containing more than 10 kinds of essential amino acids. In the past, people who lived in the mountain would drill holes in red birch trees and get the juice for a very valuable drink. But since the natural forest protection measures are implemented, nobody is supposed to do so in Shennongjia.

Ah, there are so many treasured objects in Shennongjia mountain forests.

Yes. With the forest coverage up to 96%, the Shennongjia area preserves a complete subtropical forest ecosystem. Up to now, more than 3400 kinds of higher plants have been identified, including 116 species of native plants and 33 new species. Besides, there are more than 1800 kinds of medicinal plants. Many of them are valuable Chinese medicinal herbs like “a pearl on head”, “a bowl of river water”, “a pen of King Wen”, “Aescin a flower”, etc. According to legend, Emperor Yan, Shennong, is said to have collected Chinese medicinal herbs here in order to cure people and compile his famous book *Shennong's Herbal Classic* as well.

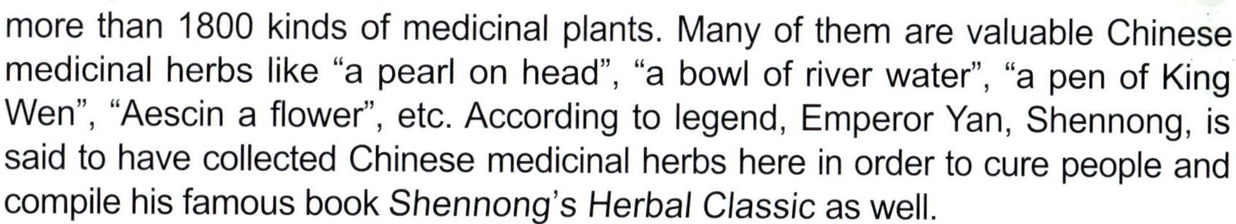

No wonder I once have read the laudatory names for Shennongjia: the Herb Garden and the Green Treasure House.

I think these names are very appropriate. Shennongjia's favorable geographical location, dense forests and rich vegetation are indeed extremely rare in the world.

I hope Shennongjia always beautiful and green.

The Shennongjia forest region was first established as an administrative area in China for the purpose of cutting wood. Since 2000, Shennongjia has completely given up deforestation. Forest protection and afforestation have become the general principle, especially since the establishment of the Geopark.

Oh, what a great change!

Yes, a very great change. Shennongjia in its history has witnessed great progress in our understanding of nature. Today, people are paying more and more attention to the protection of the natural environment of the mountain forest. I believe that Shennongjia will always stay green and become more and more beautiful in the future.

Questions:

1.What is the vegetation vertical zoning phenomenon?

2.What plants represent the vegetation of the alpine region in Shennongjia? What are the main features?

3.Why are there singing birds and fragrant flowers in the fir forest of Taiziya?

4.Can you give a brief account of the Shennongjia red birches?

5.What is the significance of the change from deforestation to afforestation and the forest protection policy of the Shennongjia forest region?

16

In speaking of the protection of forests and environment, the senior professor and Xiaoming found themselves lost in deep thought. The carpet was flying smoothly over the verdant mountains covered with vast meadows and bamboos. Everything was quiet except the gentle breeze and intermittent bird singing in the distance. Two eagles were soaring high in the sky. Suddenly, with a loud scream, one of the eagles swooped down like an arrow. After skimming across the bamboo forest, the eagle vanished behind a hill, followed by the loud repeated barks of Wangwang.

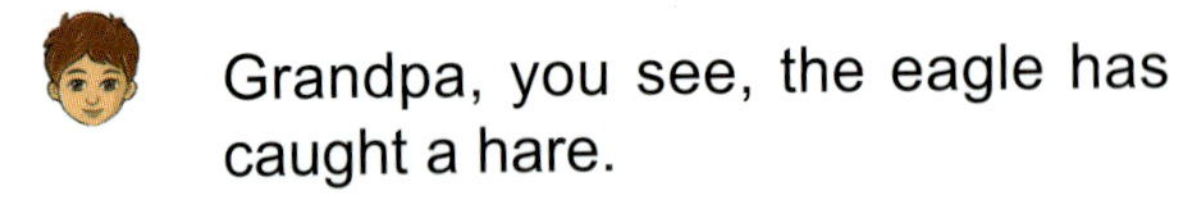

Grandpa, you see, the eagle has caught a hare.

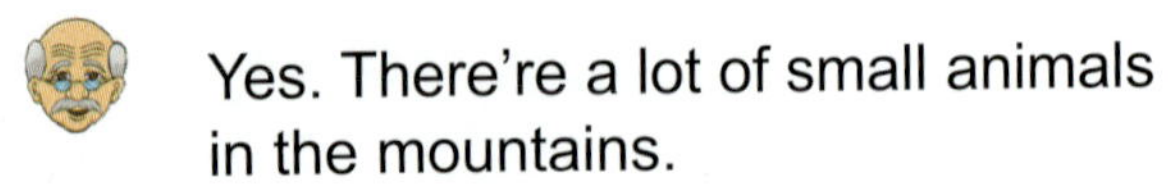

Yes. There're a lot of small animals in the mountains.

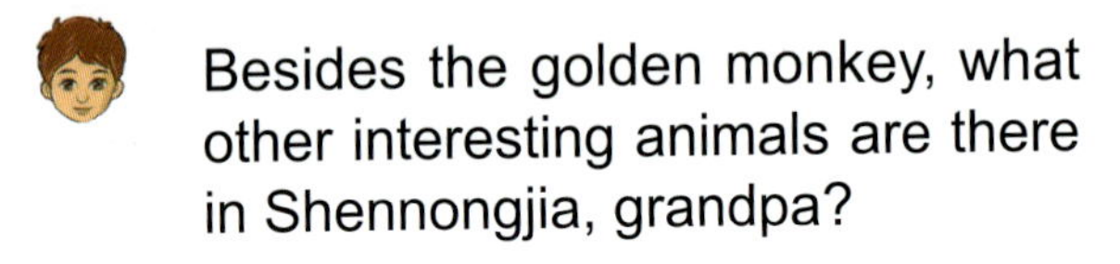

Besides the golden monkey, what other interesting animals are there in Shennongjia, grandpa?

Shennongjia is not only a treasure house of plants and forests, but also a home for various wild animals.

There must be many kinds of animals in the forest, I'm sure.

For the very rare biodiversity in the northern hemisphere, Shennongjia was accepted as a member of UNESCO's the "Man and the Biosphere Protection Program" (MAB) in 1990.

What is the "Man and the Biosphere Protection Program"?

In response to the challenges of growing population, and demands on resources and the environment, this research program of the United Nations is designed to protect the global environment that humans live in.

Then, what's the role of the MAB program?

The main purpose of MAB is to coordinate the relationship between man and the biosphere. "Biosphere reserve" is the core of MAB, which provides multiple functions of protection, sustainable development, scientific research, as well as teaching, training, monitoring, and so on. Since its identification as

a "biosphere reserve" of MAB by UNESCO, Shennongjia has been carrying out such functions for many years.

That's great! Shennongjia is really a treasure.

Yes. The Shennongjia area is described as a "natural zoo" and a "species gene pool". Here you may find 79 kinds of national key protected wild animals including the first-grade State protection animals like golden monkey, leopard, white stork and golden eagle, and 74 kinds of the second-grade State protection animals like golden cat, musk deer, yellow-throated marten, vulture, large viverra, golden pheasant, giant salamander, and so on.

Giant salamander? It is the so-called Wawayu in Chinese, isn't it?

Yes. Giant salamander is the largest and the most precious amphibian in the world. Because it sounds like a baby crying, Chinese people also call it "Wawayu".

I've seen giant salamander only in the zoo. But they are not wild.

You can see many of them in Shennongjia mountain streams. There are over 200 giant salamanders in the Giant Salamander Breeding Research Center especially set up by the Geopark.

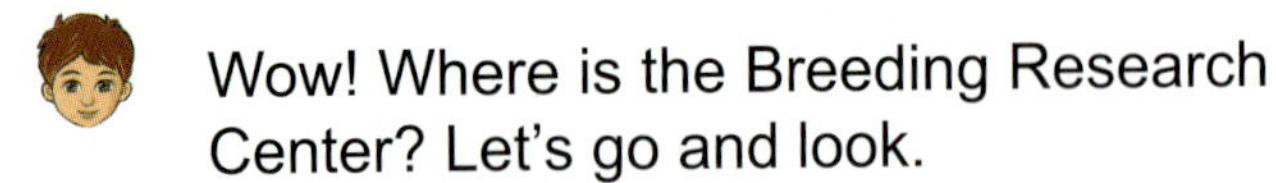

Wow! Where is the Breeding Research Center? Let's go and look.

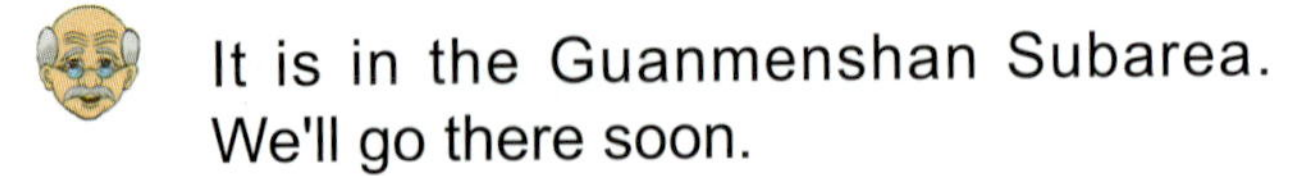

It is in the Guanmenshan Subarea. We'll go there soon.

Great! How I want to see the wild giant salamanders!

The complex climate and rugged terrain have made Shennongjia a paradise for wildlife, including a large number of rare species. Shennongjia is now a shelter for endangered animals.

I think so. Listen to the songs of insects and birds. They are bringing vigor and vitality to the whole area.

But Xiaoming, do you know that Shennongjia possesses more than 600 kinds of wild mammals, birds, fish and amphibians. Among them more than 259 kinds of wild animals are protected by the State for their important economic and scientific research value. The number of insects alone is more than 4000 kinds.

Oh, my God. What a large number! It's simply amazing.

More interestingly, there are more than 30 kinds of albino animals here.

Albino animals?

Yes. There're albino animals like white musk deers, white serows, white snakes, white hedgehogs, and so on.

Animals like polar bears, white swans, arctic foxes are all albino animals, aren't they?

No. Polar bears, white swans, and egrets are not albino animals. Their white color is just the normal expression of their dominant genes while an albino animal is the homozygous product of a recessive gene. Ah, as for genes, I have to emphasize that it is a very complex scientific problem, and you have to learn a lot before you can understand it.

But the problem is really interesting. I will look up information and make a study of it.

I like your serious attitude towards learning. The cause of the color variation of animals is very complicated. In addition to genes there must be some other reasons for the existence of so many albino animals in Shennongjia, to name but a few, the relative isolation and diversification of geographical ecological environment.

I guess so, too. It's amazing to turn Shennongjia into a paradise for animals and plants!

It is more than a paradise. In the case of serious deterioration of the earth's environment, Shennongjia may be the last shelter for some biological species like the "ark" in the Bible. Therefore, the most important is to protect the natural environment of the Shennongjia area.

Is the problem so serious, grandpa?

Yes. The situation is indeed very serious. For example, “Chinese bees”, the hard-working and lovely little creatures were once seen almost everywhere in our country. But today they are almost extinct in many places. Now we can see them only in a few inaccessible places like Shennongjia.

Why?

Because their living space is being occupied by bees introduced from Italy. The newcomers are crowding out the small native bees. Of course, their extinction is also due to the impact of environmental pollution and climate change and other factors.

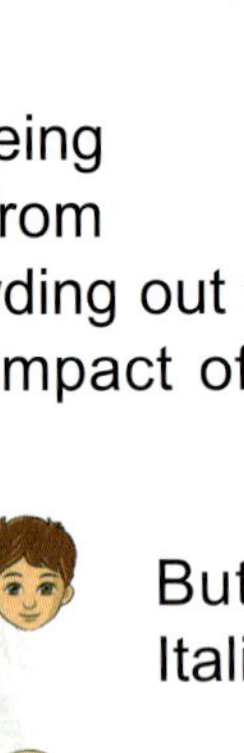

But then, why do we have to introduce Italian bees? Can't we get rid of them?

The Italian bees are more productive but they can hardly adapt to the cold climate. Chinese bees, on the other hand, are not afraid of cold. They can collect honey outside in the temperature of 4°C. High mountain plants in Shennongjia are often pollinated by Chinese bees for reproduction. Without them, Shennongjia would become bare mountains in no time.

Wow! That's so serious. I didn't realize the great importance of the small Chinese bees for the natural environment.

We should know that nature is a complementary and interlocking organism. Absence of any link in the chain would cause unimaginable consequences and even a series of irreversible malignant changes and hasten the collapse of the local environment. Therefore, to protect the natural environment and save the endangered species is one of the most serious challenges facing earthlings.

Yes. It is an urgent task for us to protect animals, plants and the global environment without delay.

For our future generations, we must protect all the species on earth and the environment that they need to survive.

Questions:

1.What is the "Man and the Biosphere Protection Program" (MAB)? What role does it play?

2.Why is Shennongjia described as a "natural zoo", a "species gene pool" and an "endangered animal shelter"?

3.Have you ever seen an albino animal? Why does the Shennongjia area possess so many albino animals?

4.Why did the professor say that without the little Chinese bees Shennongjia would become bare mountains in no time?

5.How to protect our plants, animals and the natural environment? What can we do for that?

17

The carpet resumed its flight along the highway on the mountain ridges. The gentle mountainside is variegated and disordered by scattered alpine meadows and tall bamboos. The bamboo forest is rippled by the mountain breeze, rising and falling like the sea. As the odd shapes of some dark grey rocks came into view on distant mountain sides, the professor had the carpet land in a big parking lot near the rocks. Then he took Xiaoming and the little dog to a stone sign with three big Chinese characters "板壁岩"(Banbiyan) (literally, a wall of rocks. So, it is also called "Wall Rocks").

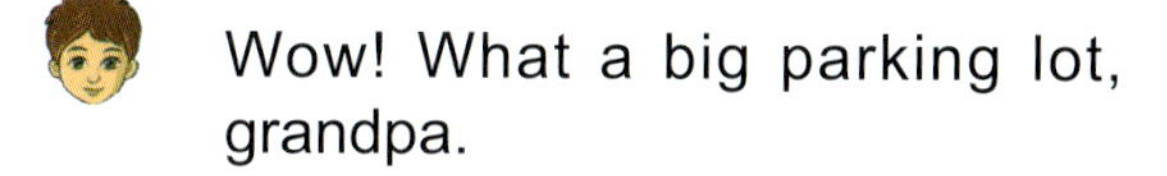
Wow! What a big parking lot, grandpa.

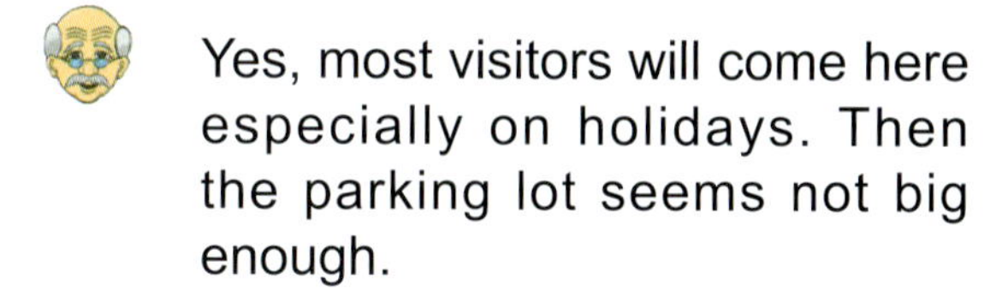
Yes, most visitors will come here especially on holidays. Then the parking lot seems not big enough.

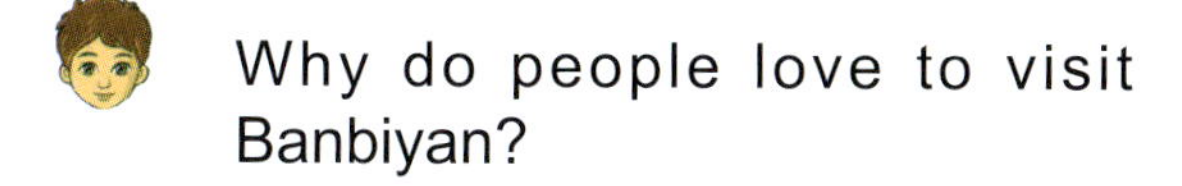
Why do people love to visit Banbiyan?

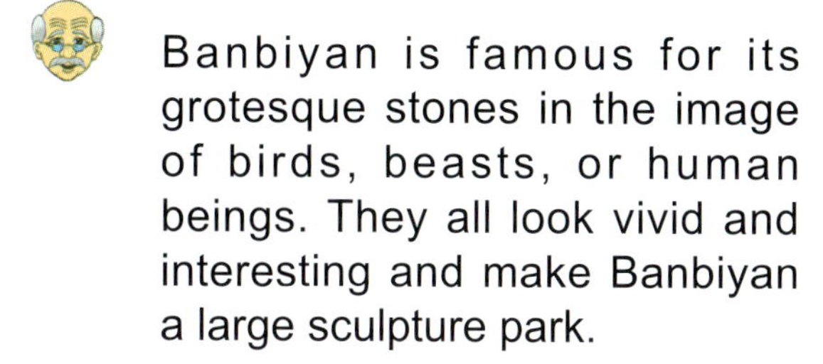
Banbiyan is famous for its grotesque stones in the image of birds, beasts, or human beings. They all look vivid and interesting and make Banbiyan a large sculpture park.

Grandpa, why is this place called "Wall Rocks"?

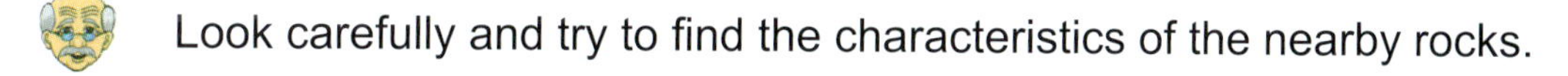
Look carefully and try to find the characteristics of the nearby rocks.

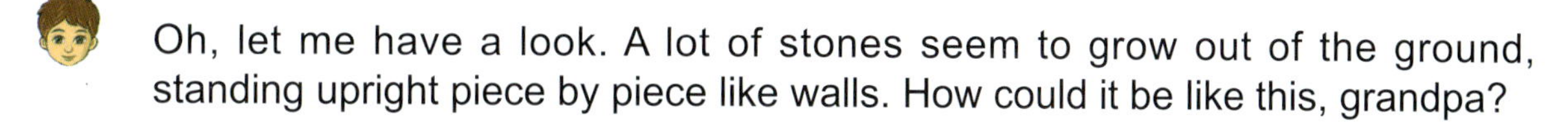
Oh, let me have a look. A lot of stones seem to grow out of the ground, standing upright piece by piece like walls. How could it be like this, grandpa?

Rocks in this area are squeezed and deformed like thin plates, extending along a certain direction, exposing themselves to the earth surface almost upright. That is to say, the attitude of the rock formation is almost vertical. As a result of weathering and erosion, they are varied in heights just like the uneven stone buds sprouting from the alpine meadows. Some rocks are

standing vertically like a door plank or wall. So people give them a very vivid name of Wall Rocks, or Banbiyan in Chinese.

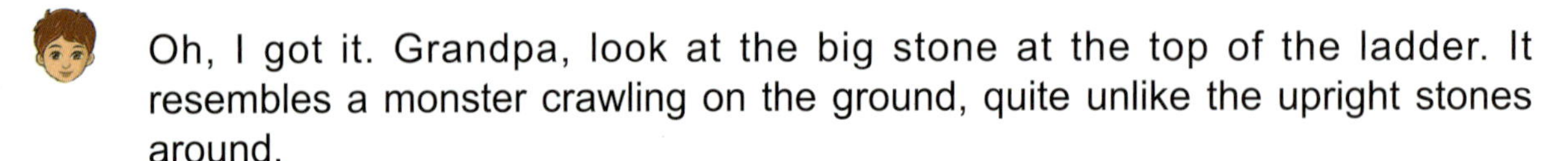

Oh, I got it. Grandpa, look at the big stone at the top of the ladder. It resembles a monster crawling on the ground, quite unlike the upright stones around.

I'm glad you're very observant. The question you raise is very good. I'm sure you will be able to answer by yourself if you look and think carefully.

Well, let me see. Oh, I don't think this stone belonged in here originally. Maybe it rolled to this place from somewhere else.

That's right. Stone of this kind is geologically called a Rolling Stone while those exposed in situ are called Bedrocks. Now can you distinguish between rolling stones and bedstones?

Well, I think both bedrock and rolling stones may belong to the same kind of stone. The bedrock shows clearly the original rock attitude.

Quite right. The rolling stone, however, only provides the nature and characteristics of the stone itself, such as the category it belongs to and main minerals it contains, etc. But we cannot trace the attitude of the rock stratum through a rolling stone. That is to say, we cannot tell in which direction it tilts and extends, and so on.

Oh, Grandpa, look there! That big rock is crumbling as if to roll down soon. Who put it up there? It's dangerous indeed.

Nobody could do that.

Do you mean it was caused by the earth's exogenic force?

Yes. Just look carefully and try to figure out how it should have been so.

Well, it doesn't beat me. You see, the shaky stone above and the big stone seat below are almost at the same attitude. From this I can deduce that the two stones should have been linked together at first. And then, as a result of weathering and erosion along the rock layer and cracks perpendicular to the layer, the stone above became hollowed out around. Finally it separated from the stone seat although it is still in the original position.

Fine. You've done a very good analysis. Every time you see a strange geological phenomenon in the wild, you should think about it and analyze the causes of this phenomenon carefully. This will make your field trip much more interesting.

Thanks, grandpa. Do you think this rock will slide down in the end?

What's your idea according to the "geological thinking"?

Oh, I understand! The stone will slide down some day. But if we try to implement the "low carbon emission reduction" to protect the environment, this day will be greatly postponed.

Very well said! Come on, Xiaoming, what is the stone like in the distance?

It looks like a chicken head.

You're right. So it's got a good name "Golden Rooster Heralding the Daybreak" for its image of the rooster, which roars out a heroic song to announce the morning. It's a very vivid metaphor.

Look, grandpa. There seems to be a stone bird over there.

Yes, the stone is like a newly hatched bird that is crying piteously for food. It is therefore described as "Chufengdaibu" (a young phoenix waiting for feeding).

Wow! These stone sculptures are all interesting indeed.

Banbiyan is actually a natural sculpture park. Here you can appreciate a lot of beautiful works created by nature. Many of them have got vivid apt names such as "Wukong Escorting", "Peacock Stone", " Beauty in the Mirror", "Root of Life", "Galloping horse".

Nature is really a skillful craftsman, whose creation is always fantastic.

Yes. Nature is omnipotent. The high mountains, cliffs, gullies, and hills we see today are all created by the internal and external forces of nature. More wonderfully, they are always in a changing, dynamic balance. For example, the shaky,

dangerous stone you just mentioned will really slide down some day. By then, the landscape of Banbiyan would become a more beautiful sculpture instead.

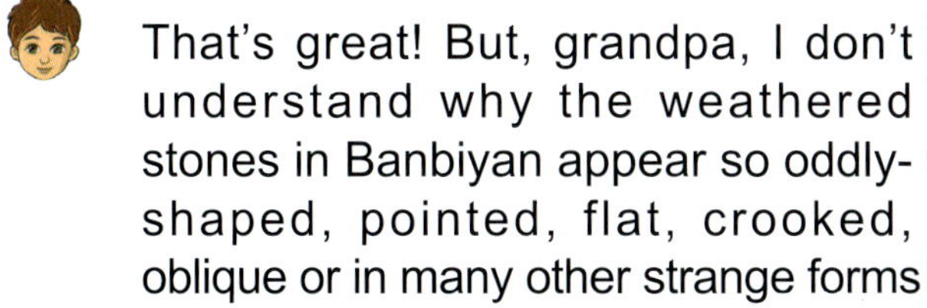

That's great! But, grandpa, I don't understand why the weathered stones in Banbiyan appear so oddly-shaped, pointed, flat, crooked, oblique or in many other strange forms as they are supposed to be erect like stone tablets, door planks or bamboo shoots. Why are they so?

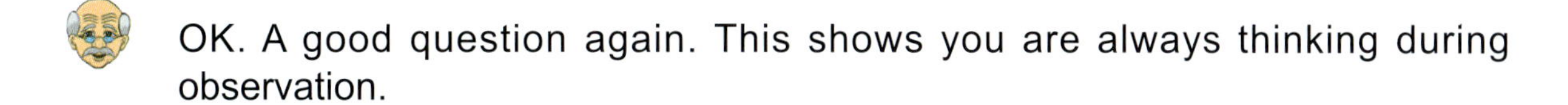

OK. A good question again. This shows you are always thinking during observation.

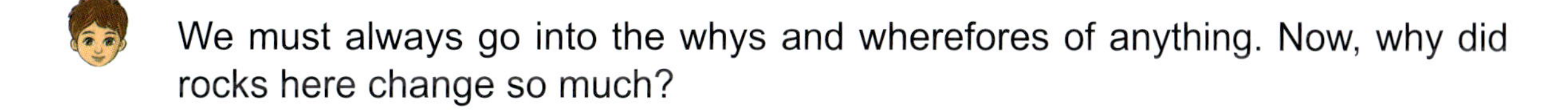

We must always go into the whys and wherefores of anything. Now, why did rocks here change so much?

There are two main reasons, I think. First, in the long geological history rocks have naturally produced a lot of cracks, or "joints" as defined in geology.

Yes. You told me that the natural cracks in rock are called joints.

The growth of joints is very uneven. They may be quite dense in some places but relatively sparse in some other places. When weathering and erosion occurs, the dense part of joint will be eroded soon. Because of the irregular extension of joints, the weathered rocks could become oddly shaped afterwards.

Oh, no wonder we find the rock is covered with grooves of different sizes, length and depth. Rocks are weathered and eroded along these joints.

Secondly, the composition of the rock itself is also uneven with higher calcite content in some places or layers and higher dolomite content in some other places. Calcite and dolomite are quite different in resistance to weathering.

Oh, I understand that calcite is more likely to be weathered and eroded to form pits and grooves.

Yes. That's why weathered stones are always jagged. In geology, this phenomenon is called differential weathering.

Differential weathering? Do you mean the different resistance to weathering has produced the uneven surface of weathered rocks?

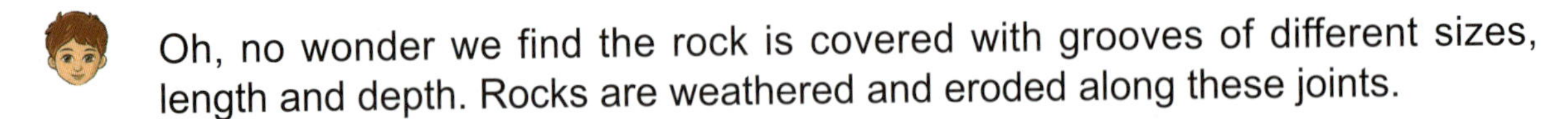

Yes, You have drawn a very good conclusion. Rocks here often contain thin silicon layers, silicon bands, silicon masses, and so on. These objects are more difficult to be weathered. They often rise in bulges on the rock surface.

I think so. No wonder I see a lot of uneven beddings and irregular bulges on rocks.

Differential weathering of rocks often results in very beautiful

ornamental stones. In addition, calcite and quartz veins are often found in rocks.

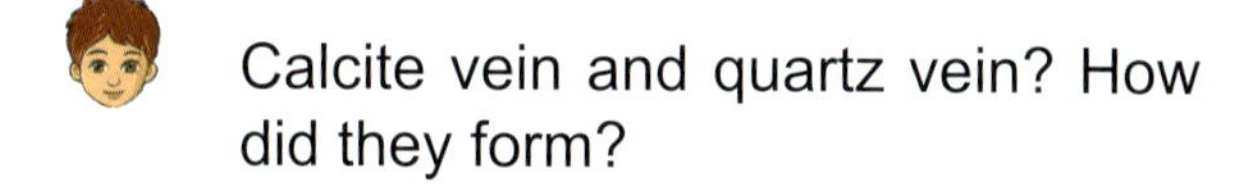

Calcite vein and quartz vein? How did they form?

When liquids rich in calcium carbonate flow through rock cracks, they will, under conditions of certain temperature and pressure, gather and settle down, or fill in the crack until they become calcite veins at last.

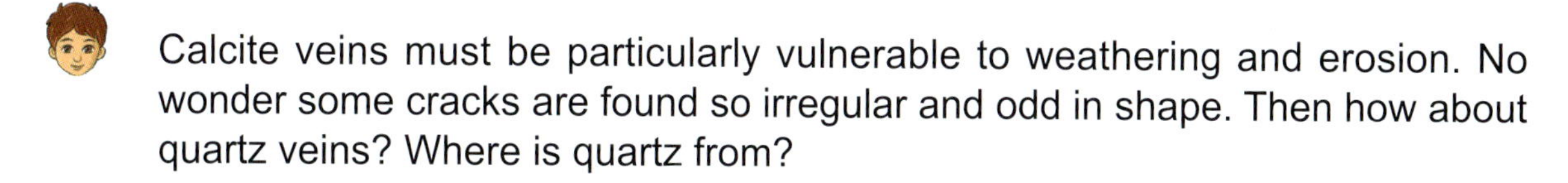

Calcite veins must be particularly vulnerable to weathering and erosion. No wonder some cracks are found so irregular and odd in shape. Then how about quartz veins? Where is quartz from?

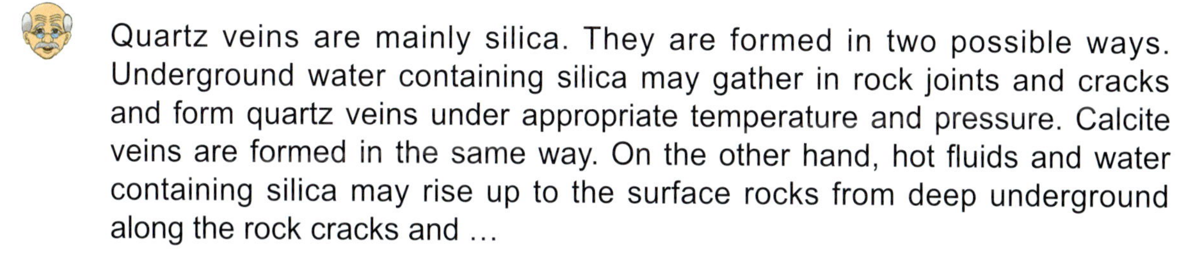

Quartz veins are mainly silica. They are formed in two possible ways. Underground water containing silica may gather in rock joints and cracks and form quartz veins under appropriate temperature and pressure. Calcite veins are formed in the same way. On the other hand, hot fluids and water containing silica may rise up to the surface rocks from deep underground along the rock cracks and ...

What is meant by “hot liquid and water containing silica rise from deep underground”?

Oh, they are often associated with the magma activity deep in the earth.

Magma? Do you mean there is magma here? Isn’t that dangerous?

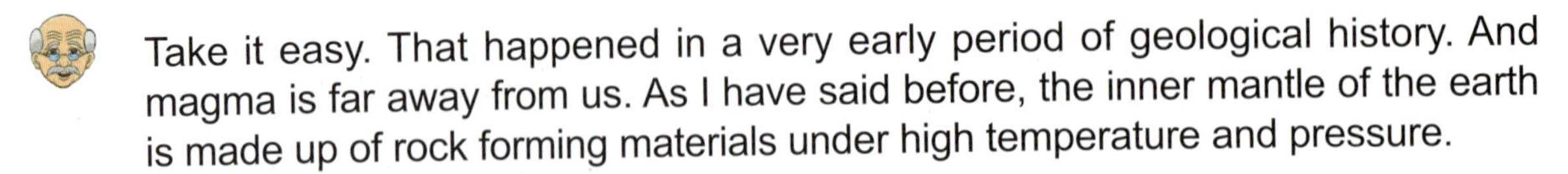

Take it easy. That happened in a very early period of geological history. And magma is far away from us. As I have said before, the inner mantle of the earth is made up of rock forming materials under high temperature and pressure.

Then what are the rock forming materials?

With common minerals like feldspar, quartz, hornblende and biotite as the main contents, the so-called rock forming materials are actually ordinary substances that make rocks.

Oh, I know. These minerals are often found in granite rocks.

Yes. Many of these minerals are rich in the soft flow layer on the top of the mantle. A large number of radioactive elements are also found there in the soft flow. As they disintegrate and release heat, they will make rocks softened and partially melted at high temperature to form magmas, producing lots of hot fluids and gases containing metallic and nonmetallic elements.

But how could the hot fluids and gases come to the wall rocks?

This is because changes in the geological structure have spread faults and fractures all over the earth's lithosphere. The hot fluids and gases constantly migrate upward along rock fractures. Rocks formed in this way are known as "hydrothermal deposits" in geology. The change of

temperature and chemical environment, the metallic and nonmetallic elements contained in the hot fluids and gases are deposited as useful minerals like copper, iron, phosphorus, boron, fluorine, selenium, etc.

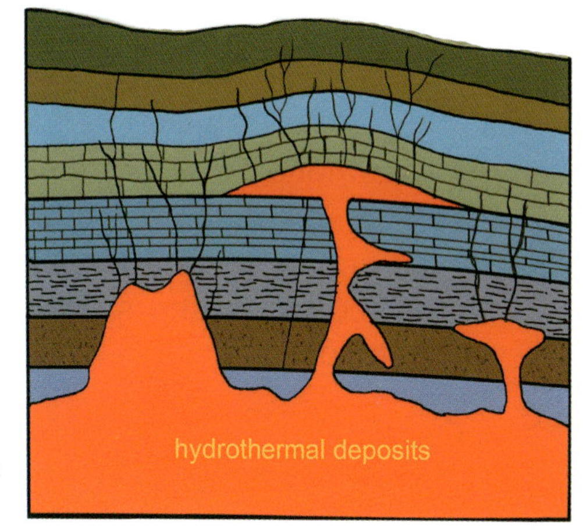

Well, I understand how silica rises to the rock wall along with the hot fluids and gases.

On the other hand, due to the lowering of temperature and the change of chemical environment, the silica attached to the surface of the rock fracture will deposit, crystallize and even fill the entire fracture to form quartz veins. Did you know that this process may form a very beautiful quartz geode? Under specific conditions, quartz crystals in the geode may be as huge as several meters long.

Wow, it's as beautiful as a fairy tale world, isn't it? Hot fluids or gases in the magma rise along the rock fissures and settle down at the right place.

Yes. Thus the earth is full of awesome scenes.

Oh, what an amazing earth we have! Grandpa, will you tell me how to distinguish a quartz vein and silica layer? They are both mainly made of silicon dioxide.

Xiaoming, you're really an enquiring boy. But you should choose the correct perspective for a question carefully before you can answer it.

Well, let me see. One is vein and the other is layer. Oh, the causes and processes of their formation are not the same!

You are so smart. Quartz veins and silica layer are different in shape because of the different ways and mechanisms of their formation. The silica layer is a thin layer formed by the precipitation of silica in the containing sediments. Therefore, it is consistent with the occurrence of rock bedding. However, the quartz veins are formed by the filling of fissures in the rock, which gives the quartz vein an irregular shape.

Thank you. I have really learned a lot here in Banbiyan. I have got a better understanding of the natural changes under various geological processes. And I can confidently tell how a pictographic rock sculpture is formed in Banbiyan.

That's great progress. Now can you explain the formation of the vertical rock named the Root of Life?

It's easy. You see, the rock stands there like an upright pillar. It should have been a part of the vertical cliffs originally. Then lots of vertical joints occurred around to provide passages for weathering denudation, including both chemical and physical weathering. As the broken rocks were carried away by wind and runoff water, cracks

consequently became wider and wider. At last, the weathered and eroded rock was reduced into a single upright pillar. I'm convinced that this pillar Root of Life will completely collapse some day.

You've summarized well. We should gain something new from our travel. Oh, it's getting late. We have to leave for Shennongtan.

Okay, let's go, Wangwang. Bye-bye, Banbiyan!

Questions:

1.What affects the weathering and erosion of rock?

2.How are quartz veins and silicon layers formed respectively? What's the difference between them?

3.What is differential weathering? Can you identify it in the field?

4.Can you cite a few instances of Banbiyan to illustrate how pictographic stones are formed in terms of weathering and denudation?

5.Please inquire and sum up the formation of hydrothermal deposits associated with magmatic activities.

18

The flying carpet continued its journey eastward over the undulating mountains, floating white clouds and winding streams. The dark green tea field on the slope is dotted with colorful tea pickers. Occasionally some black- tile and white- wall houses swung into view together with the small blue ponds in front of them. This landscape is beautiful enough to induce fantastic reveries. The carpet came down quickly through the Xiangxi Valley(Xiangxigu) and landed beside a huge statue—the iconic statue of Emperor Yan. Alighting from the carpet, the professor led his group onto the steps in front of the statue, where two flights of flag-decorated steps were found leading to the famous square for annual ritual activities.

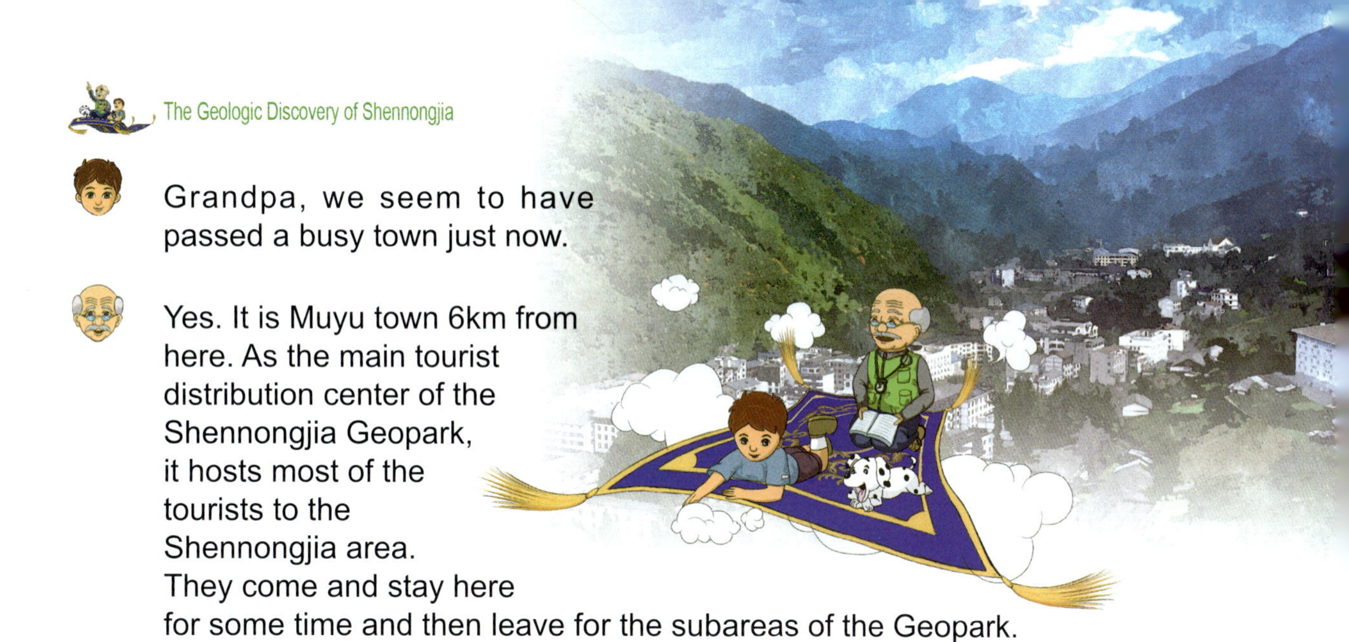

Grandpa, we seem to have passed a busy town just now.

Yes. It is Muyu town 6km from here. As the main tourist distribution center of the Shennongjia Geopark, it hosts most of the tourists to the Shennongjia area. They come and stay here for some time and then leave for the subareas of the Geopark.

Muyu town! What a good name that sounds both beautiful and meaningful. I will check it. There must be a story about it, I think.

Yes, the town is said to be related to a beautiful love story.

Really? The story must be interesting. Where are we now, grandpa?

This subarea is called "Shennong Altar"(Shennongtan), the south entrance of the Geopark. It is a scenic area of the Shennong culture.

Oh, no wonder there stands the giant statue of Emperor Yan. The commanding position makes it extremely magnificent.

The altar is built precisely in Chinese royal garden style. In front of the Shennong statue is Heaven Altar (Tiantan) reserved specially for emperors in rituals; below it there is Earth Altar (Ditan) , which is the place, by a strict hierarchy, for ordinary people in rituals.

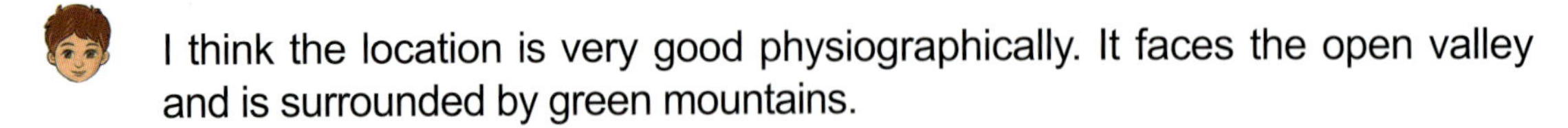

I think the location is very good physiographically. It faces the open valley and is surrounded by green mountains.

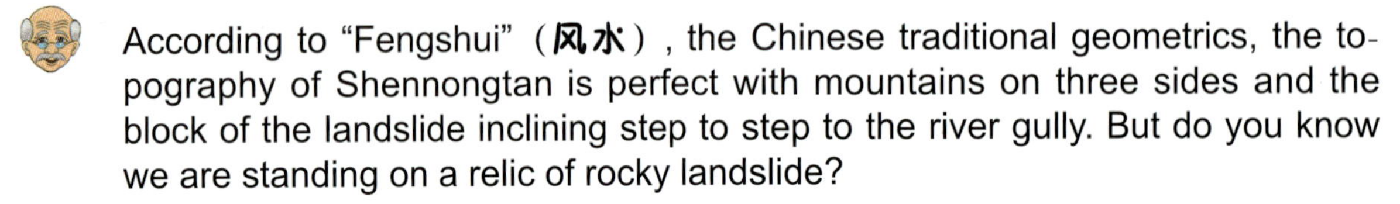

According to “Fengshui” (风水), the Chinese traditional geometrics, the topography of Shennongtan is perfect with mountains on three sides and the block of the landslide inclining step to step to the river gully. But do you know we are standing on a relic of rocky landslide?

What? Do you say a landslide? That must be very dangerous, I’m frightened.

Don’t be afraid. This is an old rocky landslide. It’s stable now with no danger for the time being.

Then, what caused the landslide?

A Landslide is a common geological hazard. It simply means the downward movement of the rock soil from the slope along a certain weak belt or surface under the action of gravity due to underground water or earthquake. It’s quite destructive.

We are now standing on a big landslide, aren’t we? It must have been terrible when the landslide happened!

Sure. The place we’re standing on and both sides of the slope are a huge landslide. During the sliding, the earth must have trembled and the mountains swayed as if to shake heaven and earth.

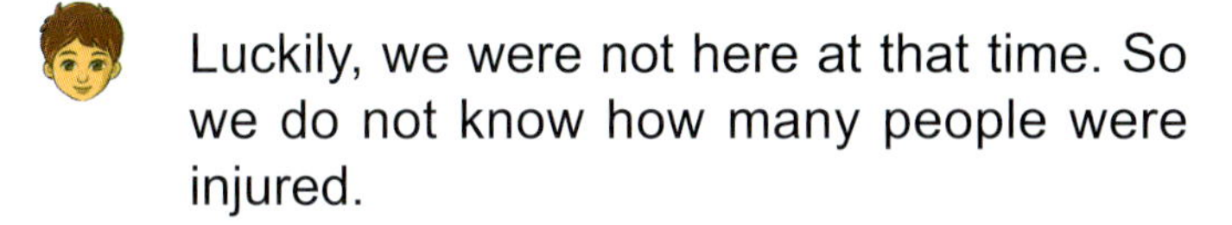

Luckily, we were not here at that time. So we do not know how many people were injured.

before landslide

It is assumed that no people were living here when the landslide occurred long ago. Anyway, landslides, debris flows, landslips are great geological disasters that cause great damage to our lives and property.

after landslide

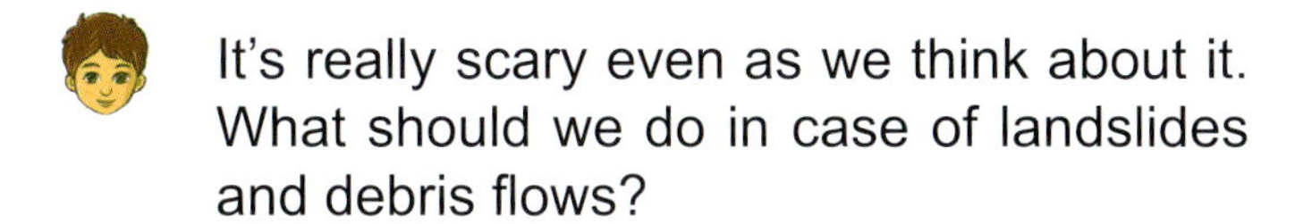

It's really scary even as we think about it. What should we do in case of landslides and debris flows?

Don't stay long in high incidence areas like steep slopes, cliffs, gully mouths during the rain. In case of a landslide or debris flow happens, people should evacuate to either side of the slope as soon as possible.

Grandpa, hill slopes in this place must have been very steep before the landslide, I guess.

Yes. They were relatively steep and very unstable at that time.

But now everything here seems so quiet and peaceful as if nothing had happened. And, you see, the slope is quite gentle. Looking down from here, we can appreciate a beautiful landscape in the valley below. It's quite enjoyable.

Yes, something negative may turn out positive. Now people have made full use of the natural terrain of the landslide and built the altar and the Shennong

statue on the hillside for people to pay their highest tribute to this earliest forefather of Chinese agricultural civilization.

Yeah, they are both dignified in appearance.

Xiaoming, just look at the giant statue on top. It's 21m tall, 35m across, together the total is 56, a symbol of great unity of the 56 ethnic groups in China.

Oh, a very good design. China has a long history of farming civilization. As people gather here for ritual activities every year, they will look up at the Shennong statue with love and worship, and pray for harvest, peace and prosperity. Meanwhile, they will take the opportunity to express their determination to carry on the farming civilization.

You seem to be well prepared for this visit.

Yes, but how can I come here without knowing anything about Emperor Yan?

Well then, can you summarize the main achievements of Emperor Yan in a few words?

OK. As one of the founders of Chinese farming civilization, Emperor Yan achieved a lot of miracles. For example, he divided day and night and set the lunar calendar. He instructed people how to use tools and sow grains according to season. He collected Chinese medicinal herbs as a doctor. He showed love, grace and wisdom, fired some

pottery and pottery paintings. He taught people to make arrows for hunting and musical instruments for entertainment, and to practice dancing for health.

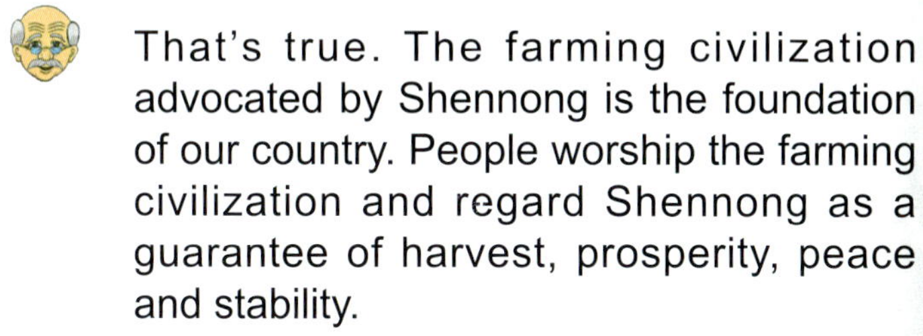

That's true. The farming civilization advocated by Shennong is the foundation of our country. People worship the farming civilization and regard Shennong as a guarantee of harvest, prosperity, peace and stability.

When are the rituals in honor of Shennong?

The solemn rituals are held here every year by the Geopark to celebrate the birthday of Shennong on Lunar April 26th.

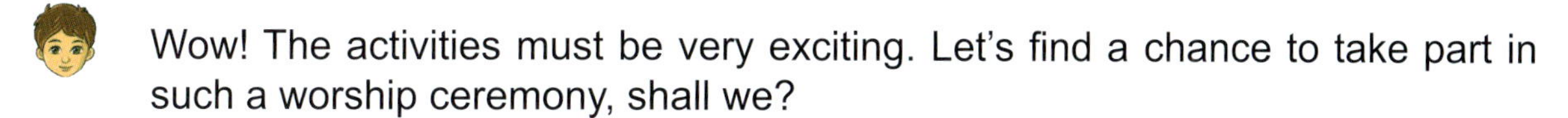

Wow! The activities must be very exciting. Let's find a chance to take part in such a worship ceremony, shall we?

Of course. Visitors to Shennongjia should come to see how people pay tribute to this earliest forefather. In addition to the worship rituals, Shennong Altar processes many other things related to Emperor Yan.

Is that so? Please show us around.

Do you see the big tree beside the square? It is 48 m tall, and 9 m in circumference. The tree is so huge that it takes 6 adults to embrace, hence the name “QiannianShanwang”, or the King of Firs, which is said to have grown for more than one thousand years.

Oh, it's huge indeed and also sturdy with red-coloured strips. But is it really a thousand years old?

No doubt. In the long history of more than 1200 years, the tree has survived the dynasties of Tang, Song, Yuan, Ming and Qing. Old as it is, the tree is still growing well in leafy profusion. During festivals, people come to worship the Divine Tree and ask for blessing. Sick people from afar often come to pray for health.

Wow, it's really a tree of good fortune.

One reason for the long-lasting growth of the tree may be that the broken earth rocks of this landslide body enable the roots to extend underground for more food and water.

This once again illustrates that negative things can turn out positive. Things in the world are always complementary and interrelated.

That's the truth. We should try our best to translate natural disasters into beneficial conditions for mankind. Xiaoming, do you see the small dome-roofed houses over there?

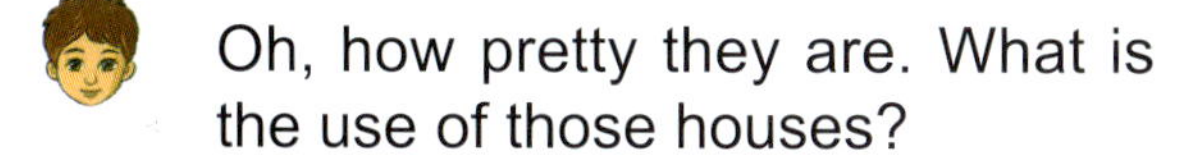

Oh, how pretty they are. What is the use of those houses?

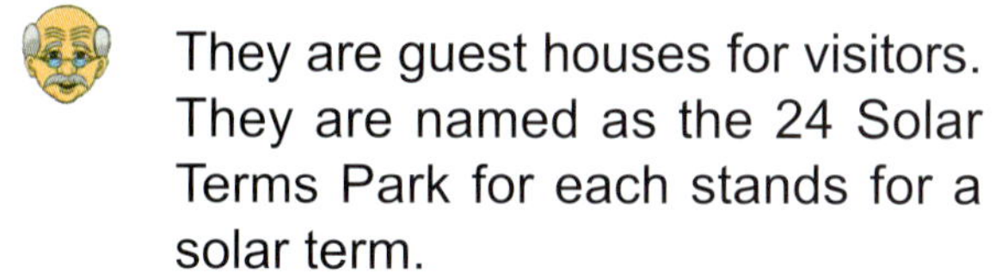

They are guest houses for visitors. They are named as the 24 Solar Terms Park for each stands for a solar term.

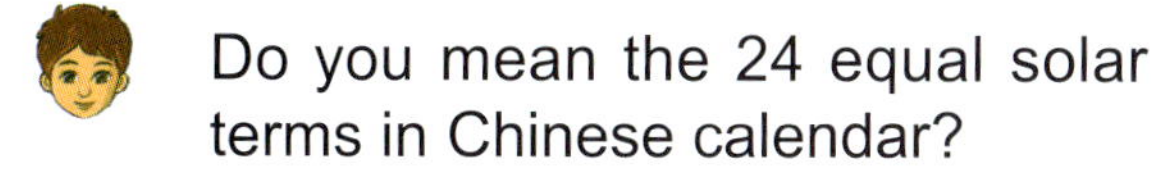

Do you mean the 24 equal solar terms in Chinese calendar?

Yes. The 24 solar terms system is believed to have been first created by Shennong and other ancestors of Chinese people. Based on the ancient Chinese people's constant observation and exploration of astronomy, weather and phenology, this system has constituted a very valuable scientific heritage to reveal the natural order of things.

And it's a good idea to build the 24 Solar Terms Park as visitor' guest houses.

Yes, it's a brilliant idea. At the back of the king of Fir lies a special garden, the Shennong Herbs Garden.

Oh, let's go and have a look.

You may know that Shennongjia is named after the legendary emperor, Shennong, who once collected herbs here in the mountains and forests. *Shennong's Herbal Classic* is the earliest extant pharmaceutical monograph in China with a collection of 365 kinds of medicinal herbs.

Wow, that's amazing!

Today more than one hundred common medical herbs are planted in the garden for the purpose of cultivating precious herbs and studying their usage in Chinese

medicine. In this sense, Shennongtan is an important place for the revival and promotion of traditional Chinese medicine.

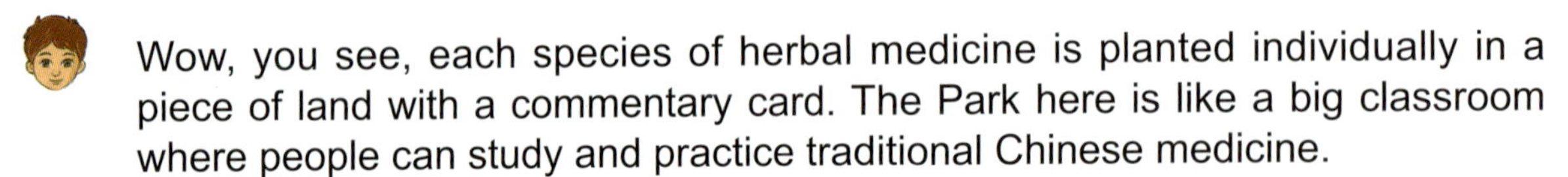

Wow, you see, each species of herbal medicine is planted individually in a piece of land with a commentary card. The Park here is like a big classroom where people can study and practice traditional Chinese medicine.

Walking up, we'll see the Shennong Tea Park. When it comes to tea, I recall a legend of Shennong. Once when he tasted herbs, Shennong mistook a rare poisonous weed and fainted right here. Thanks to the dew drops dripping into his mouth from a tea tree, he slowly came to his senses. Thus he found the medicinal value of tea and began to teach people tea planting.

According to some literature, the roasted mountain tea Chaoqing is produced in Shennongjia. Cheap and yet fine, this famous brand of tea proves to be a good health drink.

That's true. As the saying goes, "mountains and mist produce good tea". The Shennongjia area is at a higher elevation. It is mist-shrouded all the year round and free from chemical pollution by pesticides and fertilizers. This provides the area with pure air and fertile soil rich in organic matter. During the day, sufficient sunlight enhances photosynthesis to create a wealth of organic matter; while at night, low temperature slows cell respiration and reduces the consumption of these organic compounds. Growing up in such good conditions, the tea of Shennongjia always grows with fat shoots and soft leaves, which ranks it as the best of teas.

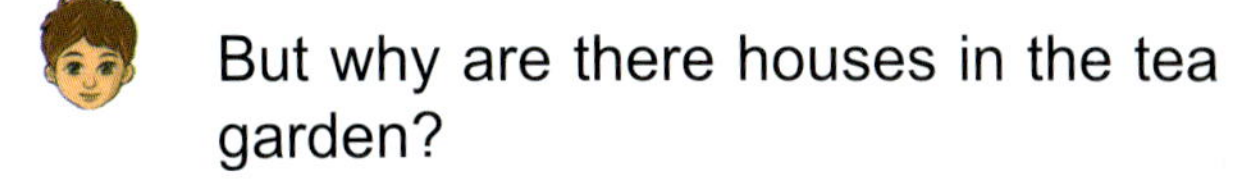

But why are there houses in the tea garden?

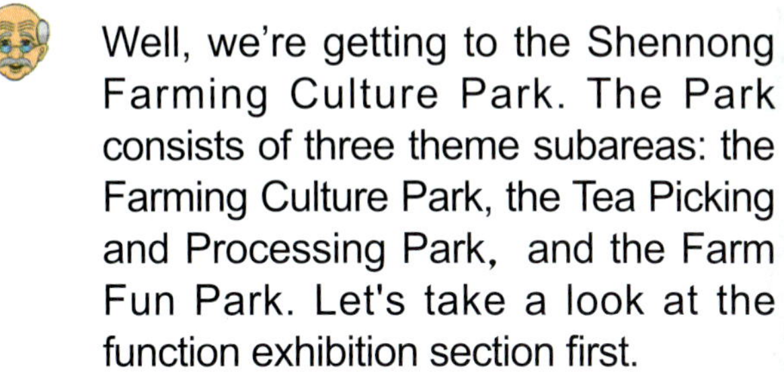

Well, we're getting to the Shennong Farming Culture Park. The Park consists of three theme subareas: the Farming Culture Park, the Tea Picking and Processing Park, and the Farm Fun Park. Let's take a look at the function exhibition section first.

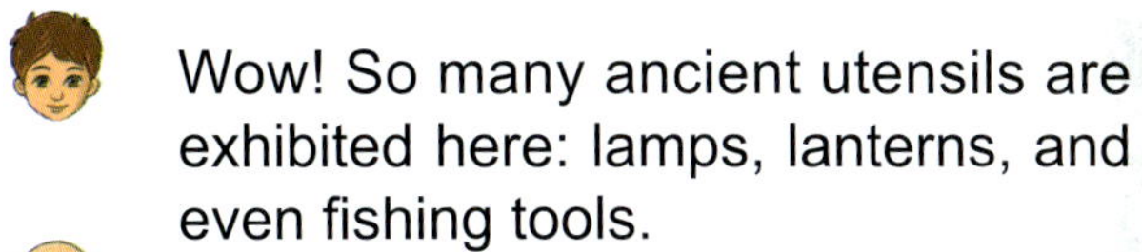

Wow! So many ancient utensils are exhibited here: lamps, lanterns, and even fishing tools.

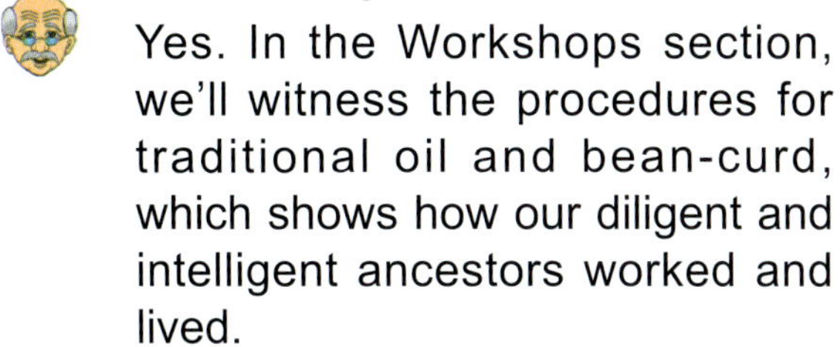

Yes. In the Workshops section, we'll witness the procedures for traditional oil and bean-curd, which shows how our diligent and intelligent ancestors worked and lived.

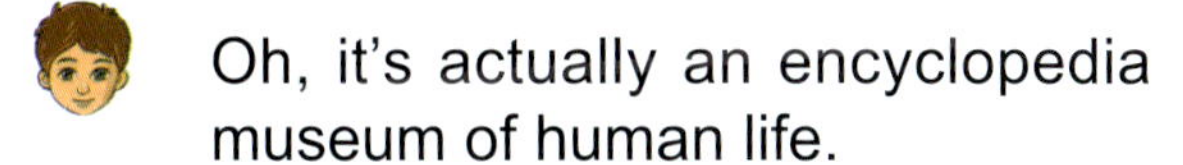

Oh, it's actually an encyclopedia museum of human life.

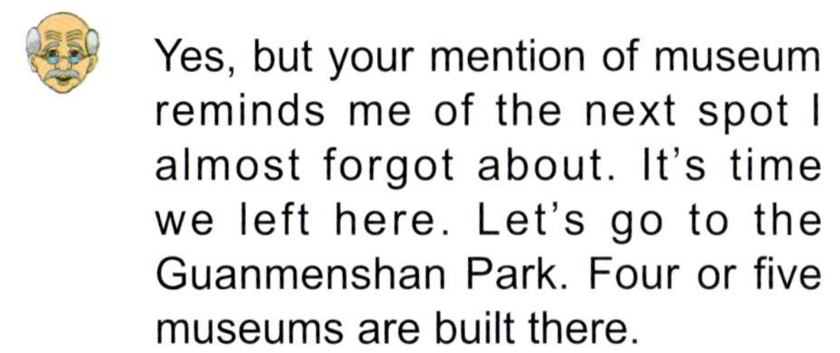

Yes, but your mention of museum reminds me of the next spot I almost forgot about. It's time we left here. Let's go to the Guanmenshan Park. Four or five museums are built there.

Really? Shall we hurry up? I like visiting museums best.

Questions:

1.What is a landslide? How do you avoid disasters caused by a landslide?

2.Why is a grand solemn ritual ceremony held every year in Shennongjia to commemorate Shennong? Can you give a detailed account to the main achievements of Shennong?

3.What are the 24 solar terms and their implications? How are they related to agricultural activities?

4.Why is the Shennong tea reputed to be the best of teas?

5.What facilities are associated with Emperor Yan in the Shennong Altar? What have you learned from your visit to the Altar?

19

Over a lush green ridge, the carpet flew upstream along the gully. Looking down, the professor and his group enjoyed the clear water lapping against rocks, sending up a pearly spray. In the broader section of the river, the rippling water runs smoothly downstream. Sightseeing battery-driven cars shuttle back and forth on the winding road along the river. Obviously, this is a great scenic spot.

Along the river valley stretches a dark wooden path built on the mountainside. Steep cliffs stand vertically on both sides of the river. The river valley becomes wider suddenly in an “S” shape corner, which gives a clear view of some pretty buildings in the green thicket. Now the carpet is landing in front of the “Landform Hall”, a beautiful building with a large staircase.

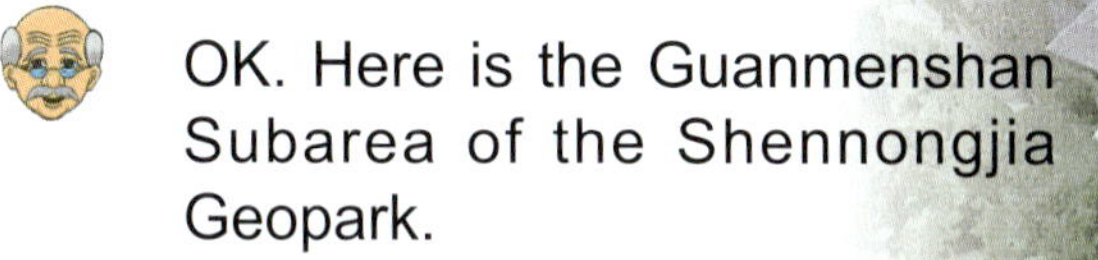

OK. Here is the Guanmenshan Subarea of the Shennongjia Geopark.

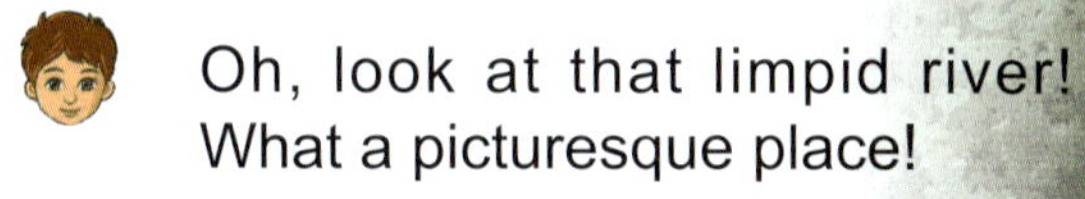

Oh, look at that limpid river! What a picturesque place!

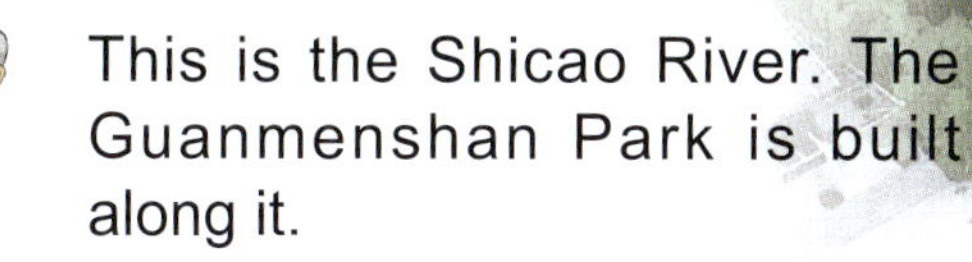

This is the Shicao River. The Guanmenshan Park is built along it.

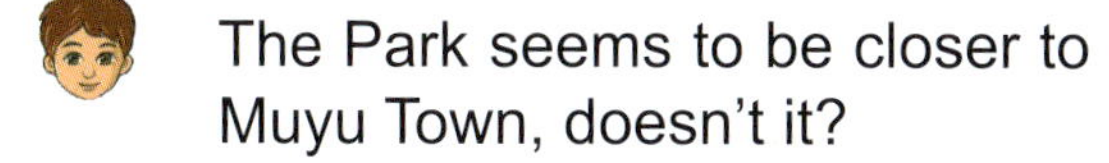

The Park seems to be closer to Muyu Town, doesn't it?

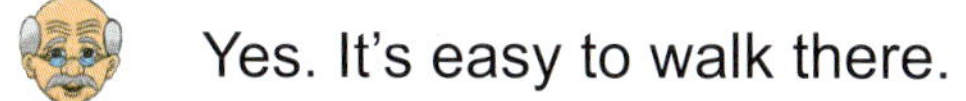

Yes. It's easy to walk there.

Great. It's very convenient.

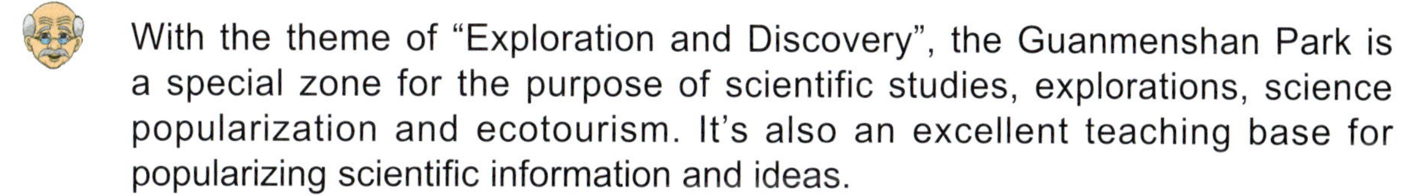

With the theme of "Exploration and Discovery", the Guanmenshan Park is a special zone for the purpose of scientific studies, explorations, science popularization and ecotourism. It's also an excellent teaching base for popularizing scientific information and ideas.

Well, what facilities for scientific research and science popularization are there in the Park?

A number of museums have been built here, such as the Geo-museum, the Wonder Stone Gallery, the Animal Museum, the Plant Museum, the Scientific Exploration Museum, the Folkway Museum, plus a 4-D Cinema and many animal and plant ecological cultivation and breeding parks.

Wow! It's a museum complex, isn't it?

No doubt. In fact, they are the specialized sections of the Shennongjia Nature Museum, which started to be built in 2008, and was completed and fully

operational in 2013. With an investment of 65,000 RMB, this comprehensive museum covers an area of nearly $30,000m^2$ and construction area of $15,000m^2$. Its completion and production has turned the whole valley into a large classroom for science popularization.

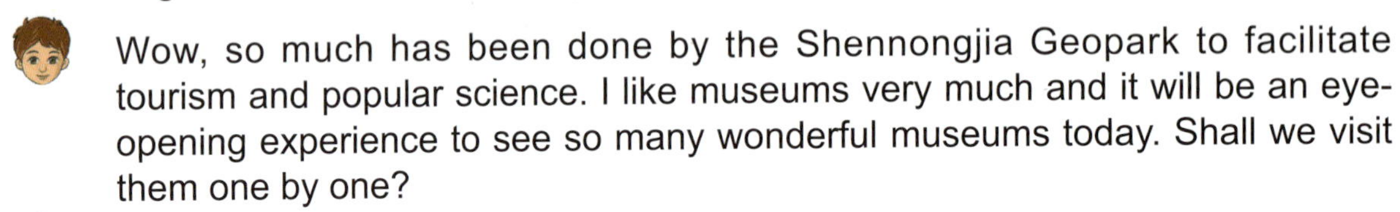

Wow, so much has been done by the Shennongjia Geopark to facilitate tourism and popular science. I like museums very much and it will be an eye-opening experience to see so many wonderful museums today. Shall we visit them one by one?

Of course. Let's begin with the Landform Hall.

The Landform Hall? We saw the landscape very clearly from the carpet, didn't we?

Yes, but it's not the same. In the Hall, you can see a topography sand table model, on a scale of 1:10,000. It serves as a big map of the Shennongjia area. Besides, the Hall provides facilities for consultation, complaints, rescue, and even special services for the disabled.

Wow! The Landform Hall is so well equipped.

Come on now, Xiaoming. Look at the sand table model. It clearly displays the river systems of Shennongjia and gives you an overall idea of the whole area. You can see that the watershed of the Yangtze River and the Han River is constituted by Shennong Peak and the related mountain ranges. The river basins are all clear in the sand table model.

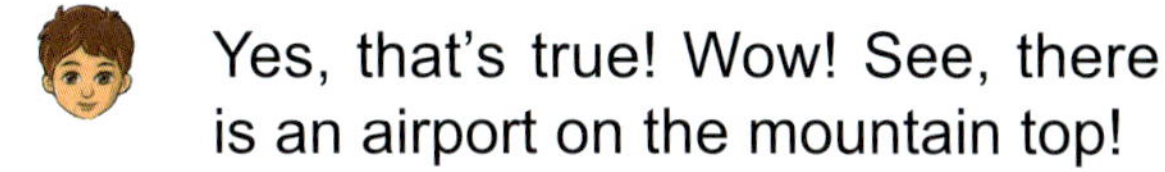

Yes, that's true! Wow! See, there is an airport on the mountain top!

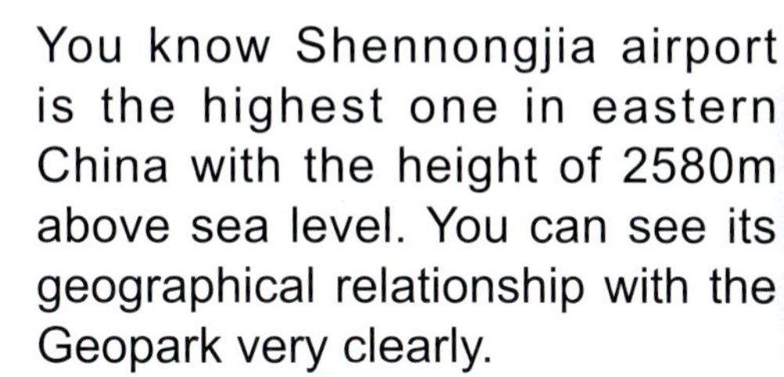

You know Shennongjia airport is the highest one in eastern China with the height of 2580m above sea level. You can see its geographical relationship with the Geopark very clearly.

This model is wonderful to tell us that we have visited so far is only a small section of the whole.

That's true. The Shennongjia area is actually a huge geology park with an area of more than 1000km^2. We have visited only a small part of it. Well, it's time we went on to the next museum.

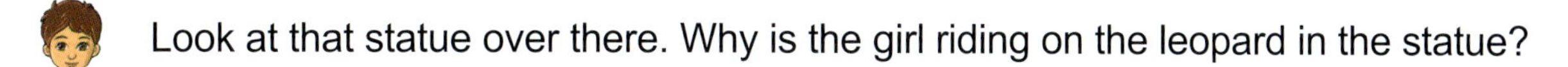

Look at that statue over there. Why is the girl riding on the leopard in the statue?

Oh, this statue tells a story about a 20-years-old girl named Chen Chuangxiang. One day in the summer of 1957, the brave girl risked her own life to save a little boy of 3 years old from the mouth of the leopard. She fought the ferocious animal and killed it at last.

Wow! She's really something to kill a leopard. She must have been strong enough, I guess.

When she was fighting the leopard, the girl suddenly thought of what senior people had told her of the leopard. It is said that a leopard has a copper head, iron tail and a weak waist like a hemp rod. So she rode on the leopard's back and tried her very best to crush the animal's waist with all her efforts. She succeeded at last.

Oh, the girl is both brave and resourceful in the face of danger.

Yes. The story also tells us that the industrious and brave locals know much about the animals in the forests. Today, we can use their knowledge to improve the protection of these animals.

Oh, we are already here in the Geo-museum.

Yes. It's one of the chief sections of the Shennongjia Nature Museum.

The design of exhibition is quite unique. You see, the domed ceiling looks like the deep blue sky of stars, vast and fantastic. Panels and photos on the walls support the corresponding rock and mineral samples in the glass cabinets seem to narrate the geological history of Shennongjia in the past hundreds of millions of years.In this way, visitors feel themselves transcending time and

space. This is a wonderful experience indeed.

Through the display of specimens, photographs and commentary, the exhibition is a visual introduction to the geology of Shennongjia, showing how this area has risen to become "the Roof of Central China" through a lot of sea-land changes in the long geological history of more than a billion years. The Wonder Stone Gallery just lies downstairs. It's also a favorite spot for many tourists.

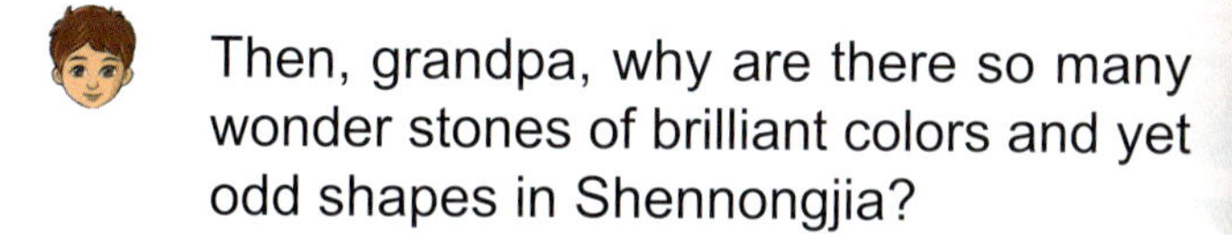

Then, grandpa, why are there so many wonder stones of brilliant colors and yet odd shapes in Shennongjia?

As I said, nature is a magical creator. Wonder stones refer to those naturally formed stones either of strange shapes or peculiar colors. With unusual materials, shapes, colors or figured patterns, wonder stones satisfy people's curiosity or aesthetic appreciation. Therefore, they are especially suitable for collecting, watching and playing with.

Yes. The creativity of nature is really amazing. Wow, grandpa, look at the figured pattern of this stone. It's almost like a landscape painting. How was such a stone formed?

Do you mean this stone? It is an example of ornamental marble stone. I did tell you that marble is produced where limestone is metamorphosed at high temperature and pressure.

Well, what change would take place when limestone is metamorphosed into marble?

In this process, the calcite in limestone is often recrystallized. The original small crystals would be recrystallized into larger ones to form a crystalline stone. The white crystalline marble, known as the White Marble, are sparkling and crystal clear. It is particularly employed as the finest decoration material for magnificent constructions like Huabiao, ornamental columns on Tian'anmen Square, and railings of the

Jinshuiqiao Bridges in front of Tian'anmen.

The white marble is beautiful indeed. Metamorphism means recrystallization of rocks, right?

Yes, but recrystallization is just one kind of metamorphism for some new minerals are produced during metamorphism. These products are named as "metamorphic minerals". Material contained in a metamorphic rock can form different metamorphic minerals under different temperatures and pressure. That's why geologists often determine the intensity of metamorphism according to the combination of metamorphic minerals in rocks.

Geologists are great. But how can we know whether there are metamorphic minerals in a marble since it is so delicate and smooth, and with no signs of recrystallization.

Well, this is what I'm going to talk about in the next topic about the third type of metamorphism. Metamorphism may also bring about changes in rock structure under high temperature and pressure. These changes will cause mineral fragmentation or directional re-arrangement in the rock and make it foliated or gneissic. This can be seen more clearly under a microscope. However, some rocks become as soft as dough in the same conditions. The original rock must be a limestone with dark carbonaceous thin layers. In the process of metamorphism under high temperature and high pressure, the rock twists and crumples like the dough being kneaded to create decorative patterns of landscape paintings on the rock.

Oh, this seems like the way we make bread rolls.

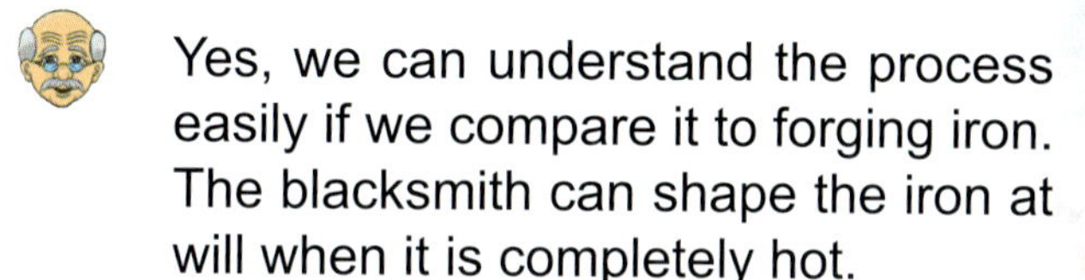

Yes, we can understand the process easily if we compare it to forging iron. The blacksmith can shape the iron at will when it is completely hot.

This comparison is accurate to illustrate and understand.

In the flexible state, the dark carbonaceous layers of the original rock are twisted, crumpled and deformed to produce beautiful patterns of landscape painting. Marbles of this type belong to an important category of ornamental stones.

Nature is really a skillful craftsman. Oh, grandpa, look at the bamboo leaves drawn on this stone. They look so beautiful!

Well, this is a typical bamboo leaf limestone.

Bamboo leaf limestone? How did bamboo leaves come onto a limestone ?

They are not real bamboo leaves at all. They are so named because they have the similar shape with bamboo leaves.

Then, how are the leaves produced? They look like real bamboo leaves more than the drawings by artists. What's more, they are 3-D.

Bamboo leaf limestone is a common type of syngenetic conglomerate.

What does syngenetic conglomerate mean?

Conglomerate normally refers to pebbles and gravel blocks, which had been carried by streams and accumulated in lakes or oceans. Then they were gradually solidified and cemented by sediments or other chemical substances in the rock's long geological history. Of course, it would take a long time to develop from the deposition to the consolidation and cementation of rocks. But this is not the case of syngenetic conglomerate.

Why? Doesn't it take only a short time to solidify and cement pebbles into rocks?

The process cannot be so short. What we can say is that the gravel and the solidification of syngenetic conglomerate occurred almost at the same time. This process is as follows: at first, before full solidification, the thin layers of limestone deposited in the shallow water will be first broken, washed and worn away into the size and shape of bamboo by the sea waves. Then they will be solidified and cemented by the deposited calcareous sediments until they turn into the beautiful bamboo leaf limestones at last.

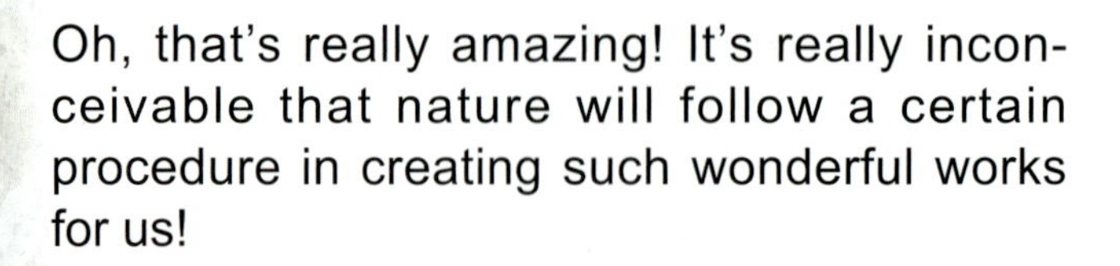

Oh, that's really amazing! It's really inconceivable that nature will follow a certain procedure in creating such wonderful works for us!

That's right. Nature is a magical creator indeed. From its creations, we can understand the formation of these stones and even stories behind them. Come on, Xiaoming, can you analyze the environmental changes in the formation of the bamboo leaf limestone?

Okay, let me try. I think, at first, the sea should be relatively calm to deposit a number of thin-layered limestones. Then before they were solidified into rocks, they suffered strong winds, waves, and even a storm or a typhoon probably.

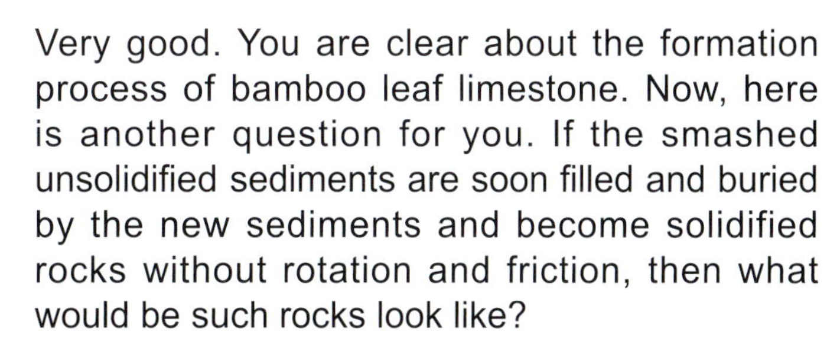

Maybe a tsunami, right?

Yeah, maybe. The strong wind and waves would break up those thin layers that had not been smashed already. Due to rotation and friction waves, the rock fragments had their edges worn away and became thin pieces like bamboo leaves. When the sea was calm again, the new deposits got the bamboo-leaf-looking limestone fragments cemented, which would be further solidified and fossilized until they became bamboo leaf limestones many millions of years later.

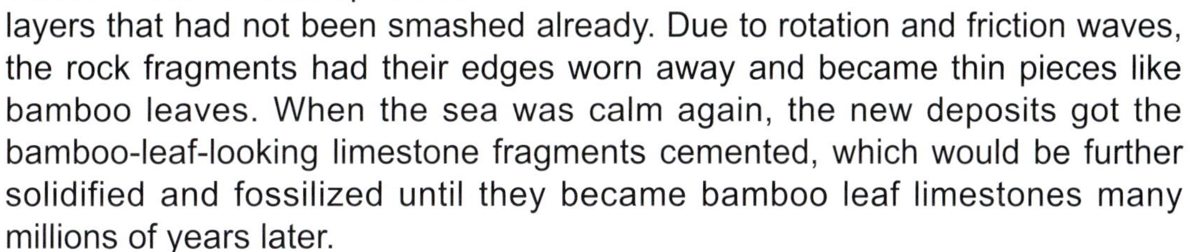

Very good. You are clear about the formation process of bamboo leaf limestone. Now, here is another question for you. If the smashed unsolidified sediments are soon filled and buried by the new sediments and become solidified rocks without rotation and friction, then what would be such rocks look like?

Do you mean the rocks are solidified and cemented as soon as they are smashed? I think, this kind of rocks should have sharp rather than round edges like bamboo leaves. Am I right, grandpa?

Yes, you're right. They are sharp edged and named as sedimentary conglomerates consequently.

Wow, it sounds interesting that a simple glance at a rock can reveal its formation. Geology is really a fascinating science.

Geologists analyze and judge the events and environmental changes in the geological history in this way. Now you've learned something about the important and useful principle in geological analysis, namely, uniformitarianism.

But what does the serious term mean, grandpa?

Uniformitarianism is a geological principle advanced by J. Hutton in 1788. According to him, geologists can analyze the past events and phenomena, and speculate the possible reasons according to the similar natural phenomena we see and learn today. This principle is the basic belief that "The present is the key to the past".

Oh, The principle is so interesting and useful. I'll study it well so that I can tell interesting stories behind the geological phenomenon, whatever and wherever they are.

Yes. The more you learn about the phenomena, the better you are at the application of uniformitaranism and the more correct the conclusions you can make.

I think so. From now on, I should carefully observe and study the natural

phenomena around us to accumulate as much knowledge as possible. Thus, I can tell stories about the formation of and reason for any wondrous geological phenomena I encounter.

There are so many amazing things in nature that are waiting for us to study, explore, and analyze.

Yes. Every stone has an interesting story behind it. I want to be a good story teller of stones about their origin and experience. Wow! What fun it will be!

OK. I'm sure you will be an excellent story teller as long as you pay close attention to accumulate knowledge.

Questions:

1.What do you think of the Guanmenshan Subarea of the Shennongjia Geopark? How do you introduce it to your friends?

2.How is the beautiful patterns of landscape painting formed in marble?

3.What changes are brought about in rocks during metamorphism?

4.Can you describe how bamboo leaves limestone is created? In what way does it differ from sedimentary conglomerate?

5.What is uniformitarianism? Can you use it to explain why some rivers are found with gravel beaches and others with sandy beaches?

20

The professor chatted with Xiaoming cheerfully over the past and upcoming attractions. Wangwang was also busy. It ran forth and back around the exhibition cabinet, hoping to see something inside. As it failed again and again, the dog began to feel a bit bored until it entered the Animal Museum, where it was excited by various wild animals. It was so happy that it jumped and barked. Sometimes it got down on all fours as if to mutter something to the animals. When it realised that the animals inside would not play with it, the dog became lazy and dozed off at the entrance.

Walking out of the museum, the professor took Xiaoming and his dog across a small bridge and went toward the plank road along the river. There are steep cliffs on one side of the plank road, and a flowing stream on the other. The breeze from the woods was cool, brisk and invigorating. The sun shone warmly through the forest, shedding shimmering light spots on the plank road. Wangwang chased the light spots and falling leaves excitedly, running and jumping with great joy. This place seems much more fun for the dog. Upon seeing a ladder leading to the river, it would jump down and smell the stones in the river, looking for something that is of interest to it.

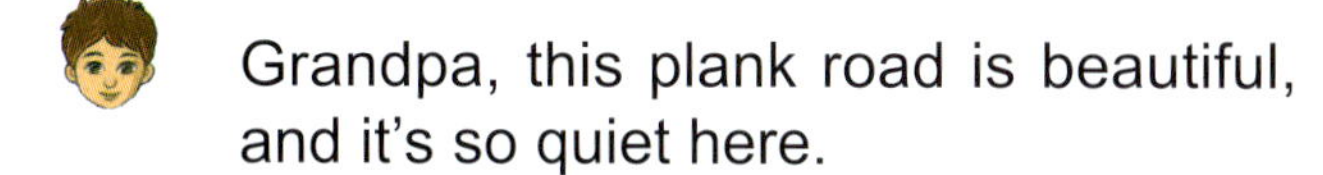

Grandpa, this plank road is beautiful, and it's so quiet here.

Yes. The plank road built along the river bank is 8km long from the Geopark entrance. It's a green passageway for visitors to take a close look at a diversity of rocks, animals and plants in nature.

Yeah. Walking on such a passageway we can appreciate not only green plants, gurgling streams, singing birds, fragrant flowers, but also clean fresh air. We are quite refreshed, calm and joyful.

Actually, as we walk on this passageway, we can enjoy both the beautiful environment and value of true nature.

Grandpa, I don't understand what you said just now. Isn't that a road? How does it make us see more of true nature?

Listen to me, Xiaoming. The plank road has been built by the Geopark to show the rich biological diversity, geological relics, and the beautiful natural environment as well. In this sense, the plank road is not only a road, but also a long corridor, where visitors feel much closer to nature and experience nature for themselves.

Well, what can we see as we walk along it?

Along this road, the Geopark possess several gardens of flowering plants and fruit trees, such as the Tea Garden, the Medicine Garden, the Orchid Garden,

the Wintersweet Garden, the Red Maple Garden, the Azalea Garden and the All-kinds-of-fruit Garden in addition to the Vicissitudes Park, the Celebrity Trees Park, the International Park, the Honey Bee Park and the National Nature Reserves Union Forest, China Geoparks Development Forest, and so on. Furthermore, a panda house is being built on the riverbank of Longtou village. It will be a new home of the national animal in Shennongjia.

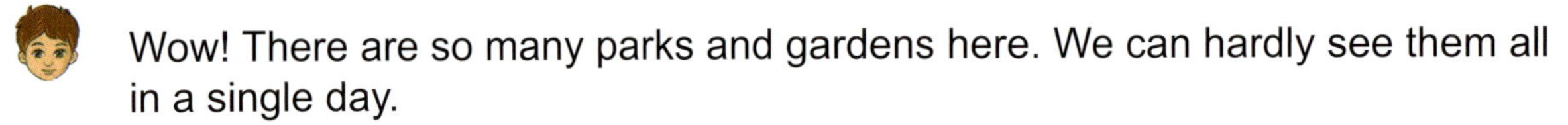

Wow! There are so many parks and gardens here. We can hardly see them all in a single day.

Apart from these parks and gardens, a number of interesting geological relics can be found along the plank road.

Oh, is that so? What are they, grandpa?

For instance, underground rivers, stromatolite fossil remains, named places of important stratigraphic units and odd outlets, etc.

What is odd outlet? It sounds interesting.

It is actually the mouth of a strange bottomless karst cave with the wind whistling in and out all the year round.

Really? How can this be?

It's simple to understand. As the cave is deep underground, the air temperature inside is relatively stable throughout the year, while the outside temperature changes with the seasons. Therefore, there is always a certain difference between the inside and outside temperatures, which consequently causes air circulation.

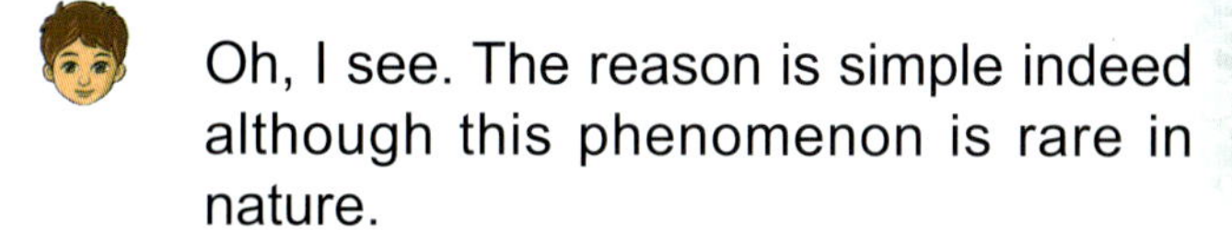

Oh, I see. The reason is simple indeed although this phenomenon is rare in nature.

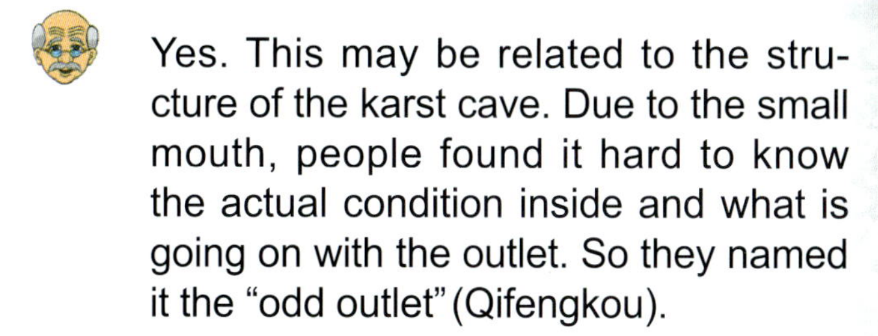

Yes. This may be related to the structure of the karst cave. Due to the small mouth, people found it hard to know the actual condition inside and what is going on with the outlet. So they named it the "odd outlet"(Qifengkou).

Oh, such is nature, which is always creating challenges for thought.

Yes. With all these interesting things, this plank road is worth exploring. Only a few visitors to the Guanmenshan Park know much about this plank road. They think the Guanmenshan Park is noted only for museums and cinemas.

In fact, the museums, cinemas and the plank road are mutually complementary. Visitors can look at stuffed examples in museums and enjoy the objects along the plank road. Objects are purely natural and more interesting to visitors.

That's right. A special park, the Reconstructed Logging Field, has been built in the mountain gully on the other side of the river near the International Park.

The Reconstructed Logging Field? Was it a logging site in the past?

Yes. The reconstructed field is a vivid exhibition of the development of the Shennongjia forest region, which was established in 1970 mainly for logging.

In order to provide the timber resources for the national construction, a lot of woodmen came to this beautiful forest. Many huge age-old trees were cut down and transported outside.

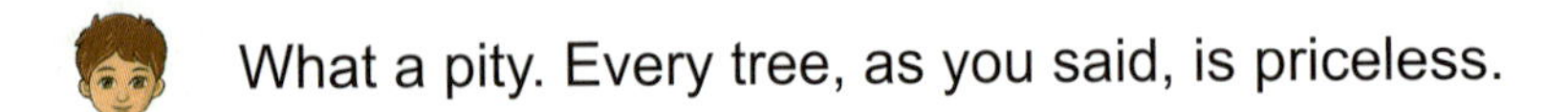

What a pity. Every tree, as you said, is priceless.

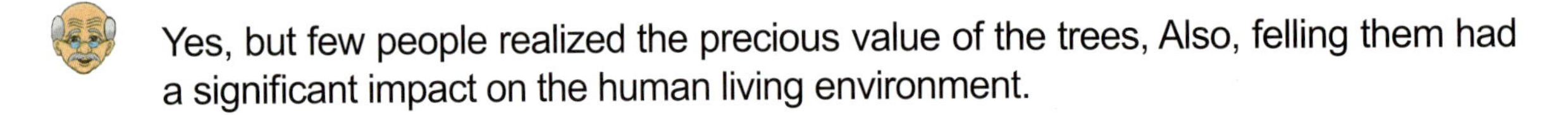

Yes, but few people realized the precious value of the trees, Also, felling them had a significant impact on the human living environment.

What should we do then? It's really worrying to see so many trees were chopped down.

To meet the increasing demand to protect the environmental, Shennongjia has completely given up deforestation this century, and gone over to continuous large-scale afforestation.

That's great.

The great change of people's ideas from deforestation to afforestation is also a strategic shift in the protection of environment and resources. So the Reconstructed Logging Field has both important educational significance and tourism value. A reflection of the temporary detour will inspire our love of the forest and nature, and ensure our awareness of environmental protection.

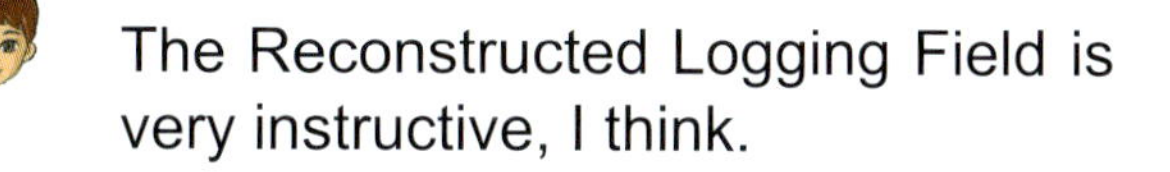

The Reconstructed Logging Field is very instructive, I think.

If you walk along this experience gallery to have a close look at nature, you can learn a lot. On fine days, you can enjoy gurgling streams and fragrant flowers and plants. It's a wonderful experience indeed.

Shh, Grandpa! Look at the bird over there. It's really so beautiful.

This thick forest has lots of insects plus streams, and provides beautiful birds with an ideal habitat. Therefore, here is regarded as a birdwatchers' paradise.

Oh, fine spring days are around us. It's so comfortable to walk on this road with various birds and flowers. And we can learn a lot during the visit.

Yes, it's great to be here. The Azalea Garden is a ready example. Built in 2008, the garden covers an area of over 6000 m^2. There are more than 50 species of Ericaceae and associated trees in and around Shennongjia.

Wow! There are so many kinds of azalea. They must be extremely beautiful in the flowering season.

Yes indeed. As one of the world's three precious Alpine ornamental flowers, azalea has been praised since ancient times as the only beautiful flower like the great beauty Xishi in the human world. The azalea of Shennongjia may be red, yellow, purple, white or other colours. Every April and May, the whole area is gorgeous with the flowering azaleas, offering people a feast for the eyes.

Oh, this is not unattractive. We are coming to see the blooming azaleas next time.

Well, Xiaoming, you can see flowers right now if you want to. The Orchid Garden is just ahead.

Are we going to see orchids? That's great. I like orchids very much.

Many people do. In China, the orchid symbolizes elegance and nobleness and in traditional Chinese paintings it is regarded as one of the so-called "Four Gentlemen" like plum blossom, bamboo and chrysanthemum. Ranking in China's top ten flowers, the orchid also stands for constant pure love.

The orchid is noble in spirit, I believe. Growing in the excellent natural conditions, orchids here must be particularly beautiful, shapely with specific fragrant.

The Orchid Garden was built in August, 2007 on a mountainside at 1280m above sea level. It covers an area of more than 16,000m^2 with over 10,000 orchids plants including 36 genera and over 90 species.

Wow! The Garden is so beautiful. Grandpa, you see, the mountainside is covered with orchids. I like the white and purple orchids best.

The yellow and pink orchids are also very nice. Come on, Xiaoming. Just look at the flagstones carved with flowers.

Well, what flowers are carved?

I won’t tell you now. You try to identify them for yourself.

Okay. I’ll look carefully. …Hey, grandpa, is this what you mentioned? There’re some circles on it. How fine it looks! What is it?

It is a fossil, called a stromatolite.

Stromatolite? It’s a good name for rocks that have layered stacks on them. Come on, Grandpa. I'll find more of them on flagstones and beside the road.

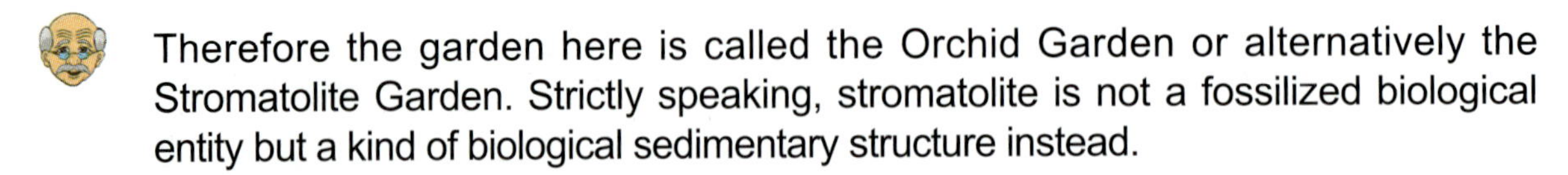

Therefore the garden here is called the Orchid Garden or alternatively the Stromatolite Garden. Strictly speaking, stromatolite is not a fossilized biological entity but a kind of biological sedimentary structure instead.

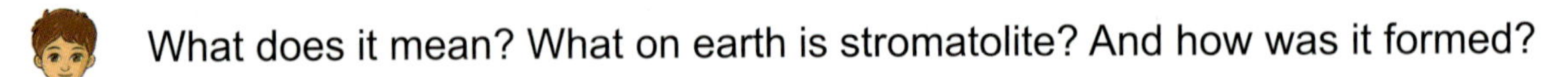

What does it mean? What on earth is stromatolite? And how was it formed?

A stromatolite is a special sedimentary structure built (formed) by primitive living marine unicellular algae and were one of the most ancient organisms on earth living one billion and three or four hundred million years ago.

Oh, they were so ancient! It's really a miracle to see them today.

Certainly. More importantly, the algae are a miracle creator to the earth.

Really? What miracles have they created?

They have made very important contributions to the birth and multiplication of a large number of lives on this blue planet.

Are they so important?

No doubt. The early earth was filled with carbon dioxide, methane, nitrogen, hydrogen sulfide and ammonia. There was no life on it at all. Then, through photosynthesis,

the primitive algae made a lot of oxygen. After tens of millions of years of accumulation, oxygen gradually changed the composition of the earth's atmosphere and created an indispensable oxygen enriched environment for later biological reproduction. That's why our Earth enjoys a great biological diversity today.

Wow! That's really a great contribution the algae have made to the earth. Grandpa, how can they grow in layers and circles?

The algae that form stromatolites grow on the surfaces of rocks. During the course of growth from generation to generation, they constantly add tiny sediment particles from the sea and construct a fossil structure that is concentric layer upon layer.

Oh, it's so interesting to know the real formation of stromatolites.

What is more interesting is that scientists can infer the season, the length of day and night and the climate changes at that time according to changes in the thickness of the laminae and other features of stromatolites.

That's amazing! Scientists are really

great to be able to infer climate changes one billion years ago based on fossil stromatolites. How can they do so?

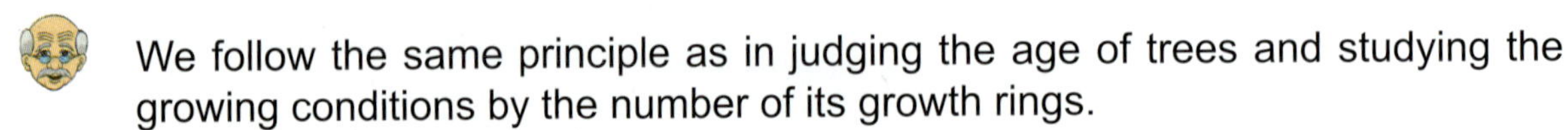

We follow the same principle as in judging the age of trees and studying the growing conditions by the number of its growth rings.

That's simple. I once counted the number of tree rings.

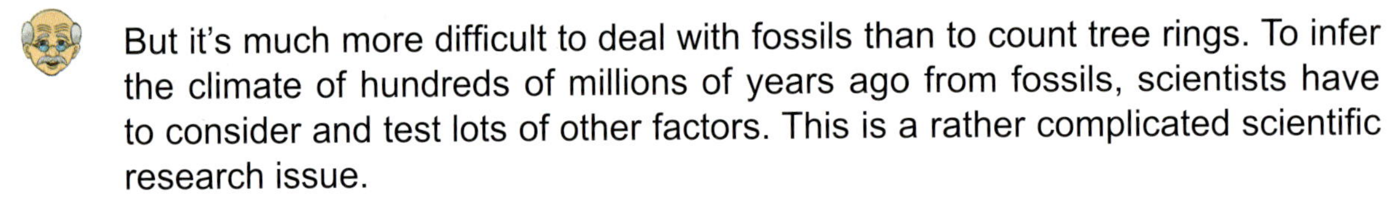

But it's much more difficult to deal with fossils than to count tree rings. To infer the climate of hundreds of millions of years ago from fossils, scientists have to consider and test lots of other factors. This is a rather complicated scientific research issue.

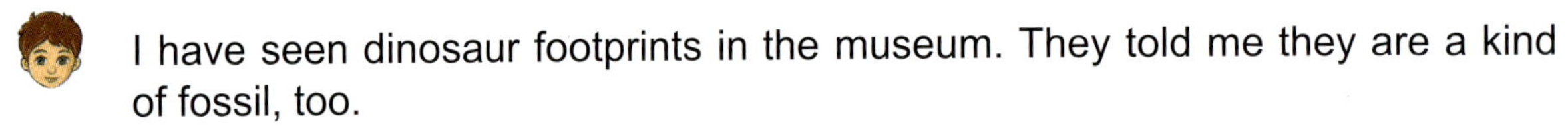

I have seen dinosaur footprints in the museum. They told me they are a kind of fossil, too.

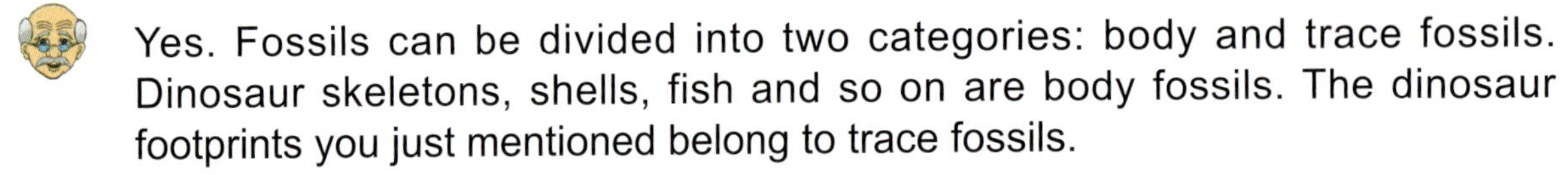

Yes. Fossils can be divided into two categories: body and trace fossils. Dinosaur skeletons, shells, fish and so on are body fossils. The dinosaur footprints you just mentioned belong to trace fossils.

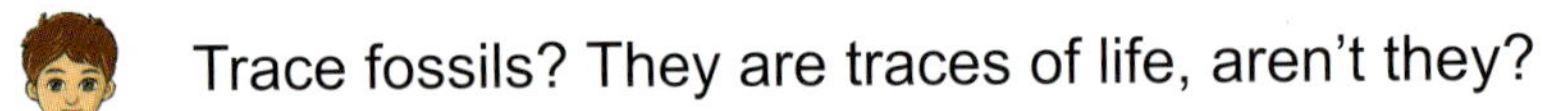

Trace fossils? They are traces of life, aren't they?

Yes. Traces of life include various traces of organisms such as their footprints, crawling marks, fecal matter, holes they dug and drilled and so on.

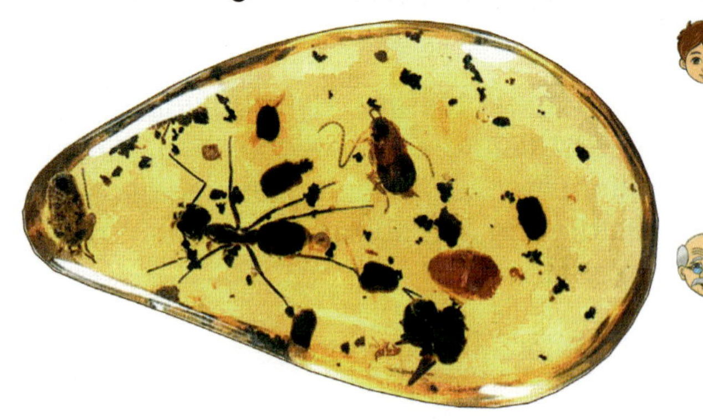

Oh, do you mean those things have become fossils? Grandpa, I've seen ants in amber. Are the ants also fossils? How did they form?

yes. Insects preserved in amber are also fossils. Do you know how amber was formed? Amber is a fossil resin. With insects trapped

inside as a result of resin flow, the plant resin has undergone oxidation and consolidation and was buried for millions of year until it gradually became amber. So amber is regarded as a special organic fossil.

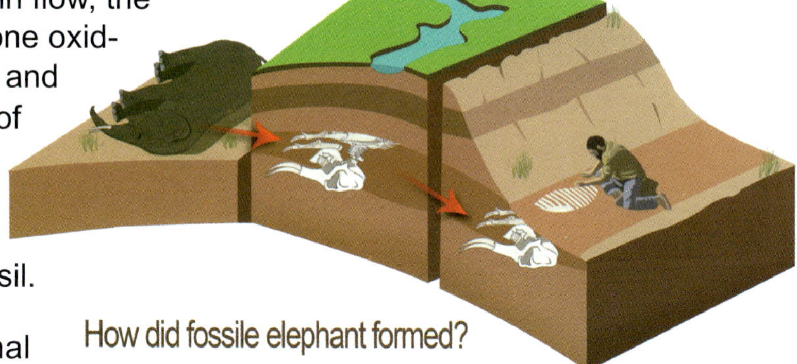

How did fossile elephant formed?

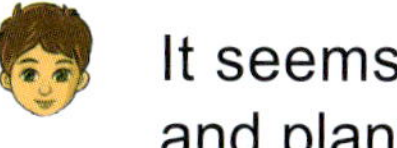

It seems that any animal and plant can form fossils, right?

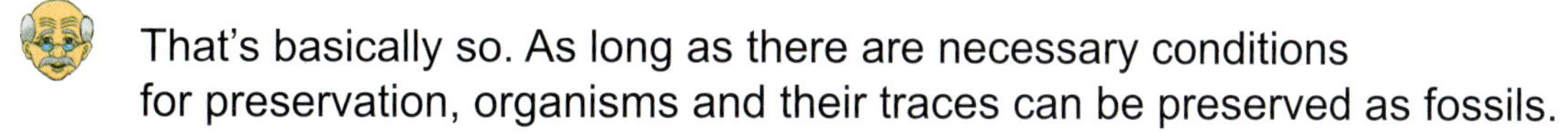

That's basically so. As long as there are necessary conditions for preservation, organisms and their traces can be preserved as fossils.

But what are the necessary conditions then?

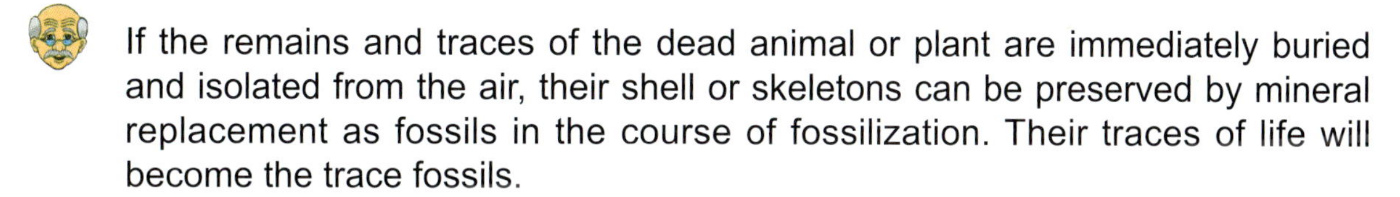

If the remains and traces of the dead animal or plant are immediately buried and isolated from the air, their shell or skeletons can be preserved by mineral replacement as fossils in the course of fossilization. Their traces of life will become the trace fossils.

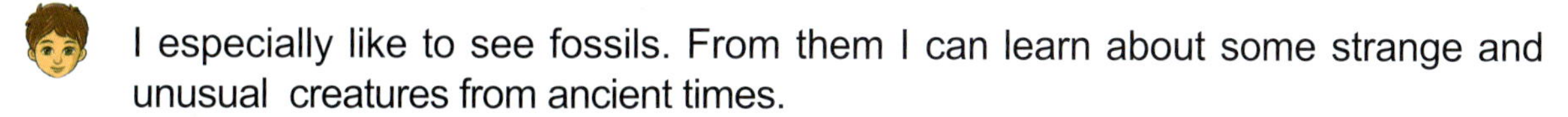

I especially like to see fossils. From them I can learn about some strange and unusual creatures from ancient times.

Fossils can also tell us stories in geological history. By studying fossils, scientists can not only know the shape, structure and category of organisms in the remote past but also speculate the process of their origin, evolution and development. It is also possible to restore the ecological environment of the earth at all stages in the long history of geology.

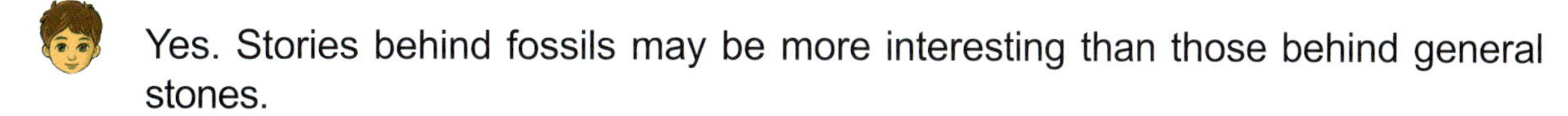

Yes. Stories behind fossils may be more interesting than those behind general stones.

That's true. However, it is not so easy to make inferences from fossils. The basic principle in this work is also uniformitaranism, which is always an important fundamental for inference.

Yes, I think so. The laws of nature are not so easy to change as the specific natural conditions.

That's right. It has proved an effective way for geologists to apply the natural laws we understand to the analysis of the environment or geological events on the earth in remote ages.

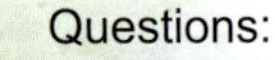

Questions:

1.Explain the magical phenomenon "odd outlet" in Guanmenshan.

2.Can you describe the great changes in the Shennongjia area after you have visited the Reconstructed Logging Field?

3.Why does the Shennongjia area have such diverse biological communities?

4.What are stromatolites? What kind of role have they played in the geological development of an area?

5.What do you know about fossils? What is a body (the solid) fossil and a trace fossil?

21

The senior professor and the little boy were talking as they were walking along the pedestrian plank path while the dog was excited, running and jumping. On a lawn, it found a beautiful yellow butterfly. It crept quietly to get close to it. This cunning butterfly, pretending not to notice the coming of the dog, was still unhurried from one flower to another. When the dog raised its forepaw, ready to catch it, it suddenly got out of the way, and rounded a stone tablet out of sight. Disappointed, the dog was barking over the stone.

Its bowwow led the professor and the boy to the front of the stone tablet.

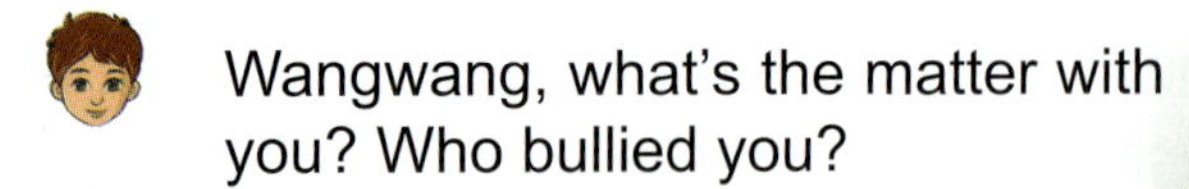
Wangwang, what's the matter with you? Who bullied you?

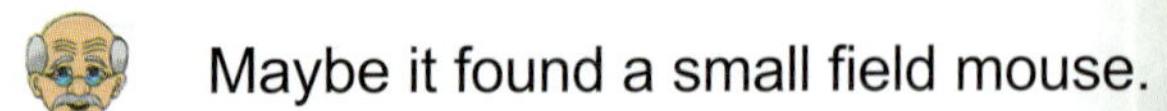
Maybe it found a small field mouse.

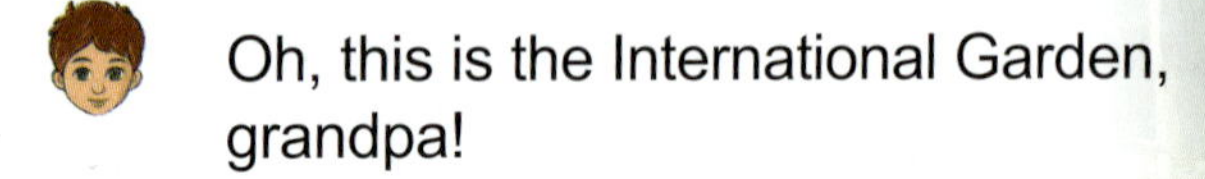
Oh, this is the International Garden, grandpa!

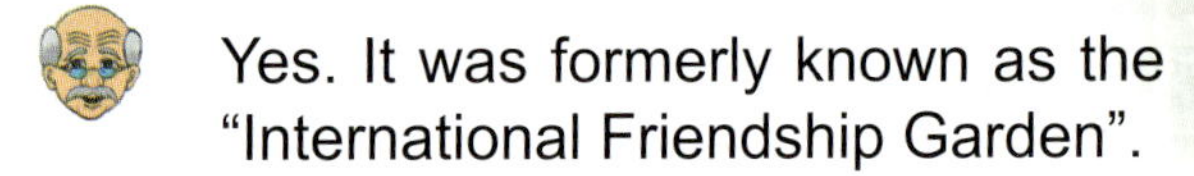
Yes. It was formerly known as the "International Friendship Garden".

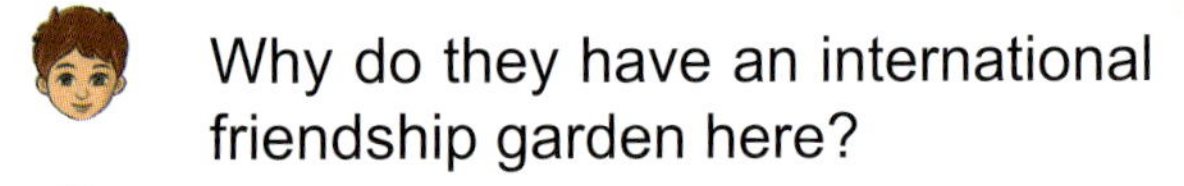
Why do they have an international friendship garden here?

It is to commemorate the 2007 Third World Botanical Garden Conference held in Wuhan. Shennongjia was then one of the venues. During the conference, there were 31 Chinese and foreign experts and scholars from 10 countries who came here for a survey and planted trees here as a memorial.

This does make a very good memorial!

The International Garden, located at an altitude of 1290m, covers an area of nearly 20,000m^2. It is divided into 6 sections, including an Asian section, North American, South American, European, African, and Oceania sections. It can be described as "a place where world plants collect and where rare plants are countless".

Wow, it seems that Shennongjia has long embraced the world!

In the Garden, over 200 rare species of trees in China are also planted, such as yew, dove tree, ginkgo, berchemiella wilsonii.

It is a grand garden of plants!

Yes. Every garden and every plantation base in the Guanmenshan Subarea has its own story and history of development.

It's an outdoor classroom of plants indeed!

Not only are the plants here abundant and luxuriant, but also there is a snake museum, an animal breeding research center for giant salamander, Chrysolophus pictus, spotted deer, etc. and bee gardens.

Definitely this is an ideal place as a centre for scientific research, investigation, ecological tourism, popular science teaching, life cultivation and relaxation.

In Guanmenshan, science popularization activities and park facilities cater for families with both aged and young people. A family can ramble about for pleasure the whole day, enjoying the beauty of nature. The elderly can relax from sightseeing whereas the young can have access to nature and learn a lot about plants and animals which can't be learned in the classroom.

Yes indeed. I saw quite a lot of signs along the way. On the signs scenic spots are shown with their interpretations. It's a good way, helping people learn a lot.

I think so. These interpretation signs are intended to help people understand the various phenomena of nature. For example, the place where we are standing now is a very important geological point.

"Geological point"? What do you mean?

A geological point is the place where an important geological event occurred.

But I can’t see any differences between here and elsewhere?

When you look at this interpretation sign, you’ll understand what it is. This is the place from which the “Shicaohe Formation of the Shennongjia Group” was named.

Let me have a look. The Shicaohe Formation dates back about 1.3 billion years. Wow, it's so old! What is “group” and what is “formation”?

“Group” and “formation” are professional terms of geology. As most of the rock strata we see are hundreds of kilometers thick, people must differentiate them in order to facilitate their study and reveal the environment and process of their formation. “Group” and “formation” are the units used by geologists to divide the rock strata.

湖北神农架国家地质公园
Shennongjia Group Shicaohe Formation
Shicaohe Formation(1300 Ma) refers to the lithostratigraphic unit of Upper Shennongjia Group with the thickness of 1655.83m, which gets its name from Guanmenshan Shicaohe(Shicao River). The major rock types are purple silty dolomite, siliceous banded dolomite, siliceous siltstone, and abundant layered-periclinal-columnar stromatolites.

How is it divided? Is it divided into parts of the same thickness?

No. Geologists base their division on the composition and lithology of the rock, from which they can deduce the process of the rock formation and the environment. According to their place in the hierarchy four levels are classified and generalized for this purpose. They are “group”, “formation”, “member” and “bed” respectively.

Wow! So complicated!

Yes. However, after the complex formation is classified into different stratigraphic units,

things will be simpler. “Group” is the largest unit, often reflecting the formation of a set of rocks throughout the geological stage and process. For example, the “Shennongjia Group” we are now concerned about is a record of marine sedimentary history which lasts hundreds of millions of years.

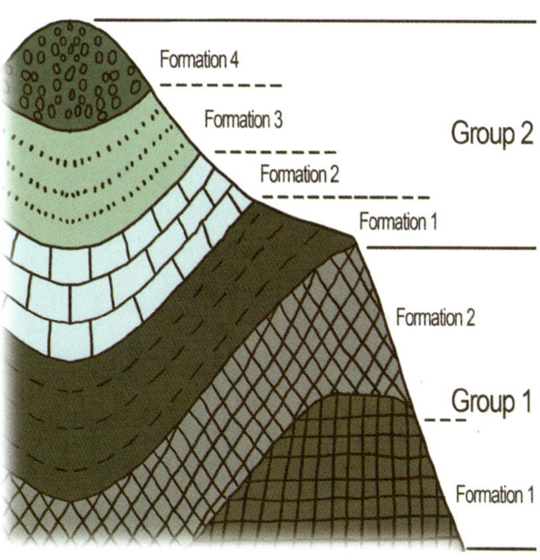

Subdivision of “Group” and “Formation”

Oh! It's very interesting!

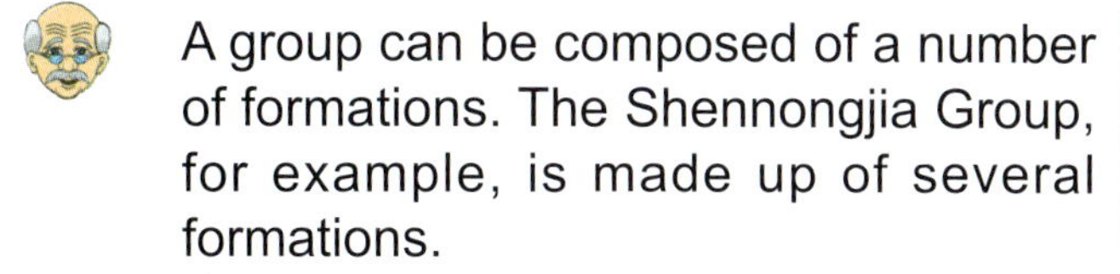
A group can be composed of a number of formations. The Shennongjia Group, for example, is made up of several formations.

Then, how is “formation”classified?

“Formation” is the unit below “group”. It often reflects a particular sedimentary environment, or shows a change in the sedimentary cycle. For example, a formation composed of black shale is likely to represent a deep-sea anoxic sedimentary environment. A formation with interactive bed of sand and shale tells the regularity of changes in the source of the sediment. Possibly, it is the season that causes the difference in water kinetic energy. For example, what is carried by water during the rainy seasons is sand because of the great water kinetic energy while silty sand or mud, which is the source of shale, is moved in the dry seasons because of the weaker water kinetic energy.

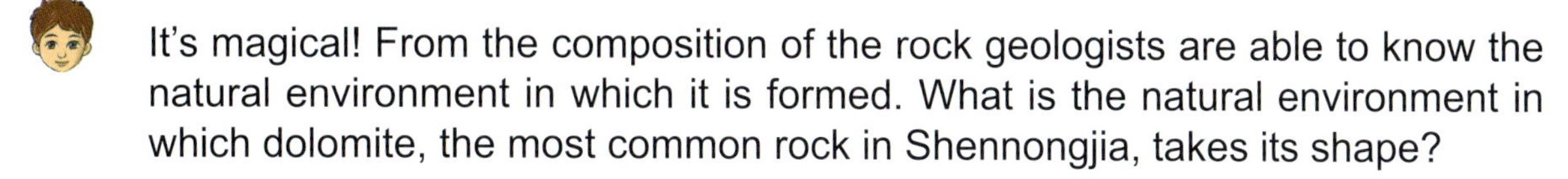
It’s magical! From the composition of the rock geologists are able to know the natural environment in which it is formed. What is the natural environment in which dolomite, the most common rock in Shennongjia, takes its shape?

Dolomite is shaped through the chemical reaction that occurs in the ocean.

The sea water is salty, isn't it? That is because the sea water dissolves a large variety of salt compounds. Calcium carbonate and magnesium carbonate are the most common components in the ocean.

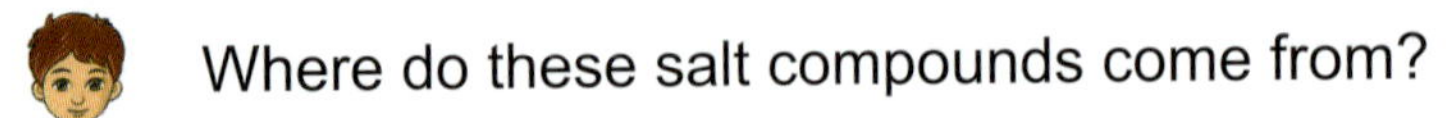

Where do these salt compounds come from?

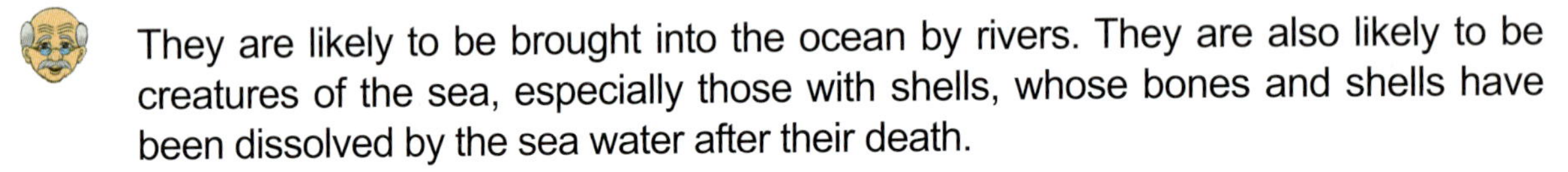

They are likely to be brought into the ocean by rivers. They are also likely to be creatures of the sea, especially those with shells, whose bones and shells have been dissolved by the sea water after their death.

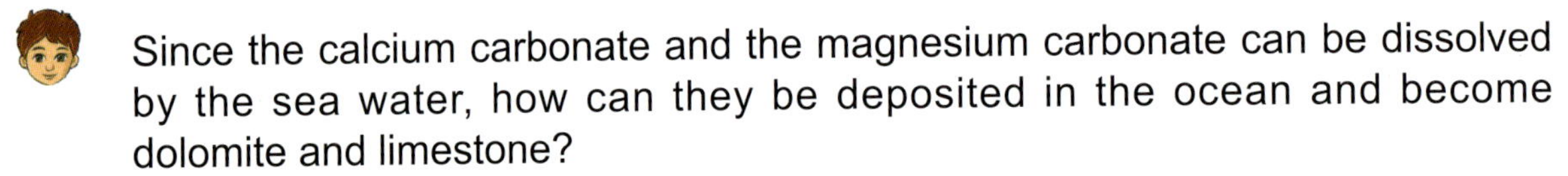

Since the calcium carbonate and the magnesium carbonate can be dissolved by the sea water, how can they be deposited in the ocean and become dolomite and limestone?

That's the wonder of nature. Many chemical reactions are reversible, and they change with the change of conditions which control the chemical reaction. For example, when the temperature rises, calcium carbonate is more easily dissolved. But when the temperature falls and the pressure increases, it falls out of solution. So under specific conditions of temperature and pressure, the calcium carbonate and the magnesium carbonate dissolved in the ocean will deposit and become dolomite or limestone.

Nature is so wonderful. Can we infer the change of the natural environment from the composition and lithology of the rock?

Yes. This is the problem that geology and sedimentology try to solve. It is also the purpose of our stratigraphic division which further aims at finding useful minerals.

Oh! Stratigraphic division can help find minerals? What is there in the sea?

The most important purpose of geology is to find useful mineral resources. A lot of minerals are formed in the sea. The famous "Anshan-type iron deposit" in China is closely related to sedimentation during its geological history.

Is it? How is it related to sedimentation?

Geologists label "Anshan-type iron deposit" as a "metamorphic volcanic sedimentary deposit". Its mineralization covers the following basic steps: first, giant thick volcanic sediments containing iron are deposited in the sea; then, in the diagenetic process and after the diagenetic evolution, they have to undergo a multi-phased metamorphism which enriches the iron component and results in useful mineral deposits.

Ah. I think the iron component is caused by volcanic deposits, right?

Most of it is. But there is also the possibility that it is produced by later metamorphism. Sedimentation can give rise to many different mineral deposits. That means mineral-forming materials on the ground, when they are carried into water by wind, river, glaciers, or living creatures and then settle and accumulate there, can form deposits. Deposits, formed in this way, are called "sedimentary deposits". Coal, oil and natural gas are the most typical. The fact that coal is rich in North China attributes much to the climate and the environment then and there: large tracts of forest provided the material basis for the formation of coal. In addition, iron, manganese, aluminum, phosphorus and other elements can also form mineral deposits by sedimentation.

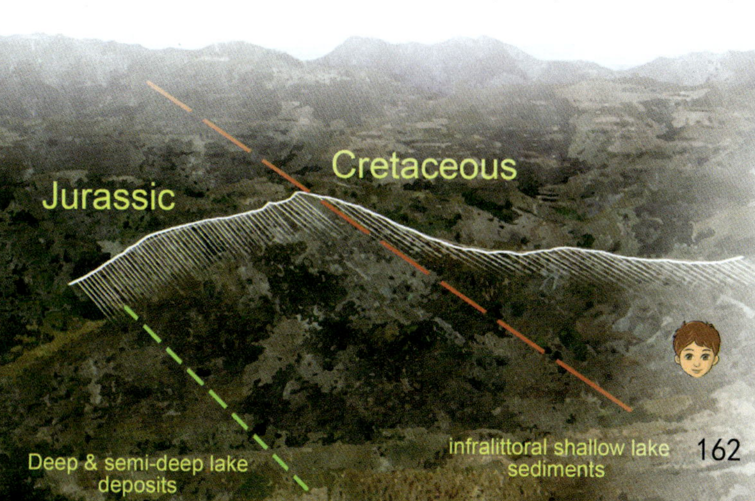

Geology is so great. I am determined to be a geologist in the future, travelling all

over the mountains and rivers and prospecting for the country.

Prospecting without a target will not get you far even though you travel all over the mountains and plains. The first thing to do is to make an analysis of the basic geological conditions of a region to see whether there is any deposit there or not and what kind of deposit there may be. So, fundamental geological research is the most important step. The study of stratigraphy is the most basic geological prospecting method.

Why is the study of stratigraphy the most fundamental one?

Because stratigraphic research can tell you the geological development history of the region: whether it used to be ocean or land in the past, whether in deep ocean or neritic shelf, whether there was any volcanic activity, what kind of tectonic movement it experienced and so on.

How does stratigraphy get this information?

Oh, that is a professional problem, and it's very complicated. We say that the geologic history of a region is reconstructed mainly according to its stratigraphic sequence. Understanding this sequence is actually based on research on the sedimentary environment of the region, analysis of its environmental evolution, a comprehensive analysis of the geologic history of each stage and its characteristics. Therefore, the results of the stratigraphic study must be a summary of that region's geological history.

Questions:

1.Why do science popularization activities and park facilities in the Guanmenshan Subarea cater for families with both aged and young people?

2.What can be learned from the interpretation of the "geological point" shown on the sign in the Geological Park?

3.Why do geologists classify strata as "group", "formation"?

4.Please refer to relevant information and talk about the basic principles of the formation of "sedimentary deposit".

5.Try to learn about the important sedimentary deposits in our country.

22

The senior professor talked on and on enthusiastically about the scientific principles of stratigraphy. Xiaoming listened with attention and put up some new questions at intervals. Wangwang ran back and forth, apparently showing no interest in what the old professor talked about. Its interest was the butterfly in the grass down the plank path.

A very important issue in a stratigraphic study is to judge whether the strata are continuous.

How do you judge that?

Geologists make their judgment from the contact relationship between rock strata. There are three types of contact relationship between rock strata, namely, "conformable contact", "parallel unconformity" and "angular unconformity".

I suppose "conformable contact" is a consequence of continuous deposition, right?

You are right. By concordant contact, we mean there is no obvious gap between strata. It is reflected by continuous deposition.

How about "parallel unconformity" and "angular unconformity"?

First of all, the "unconformity" means that there is a break in deposition or a sedimentary discontinuity between the strata. "Angular unconformity" occurs when the upper and the lower strata are not consistent. They intersect at an angle.

Oh, I understand that, after the formation of the previous strata, there must have been great changes in the earth's crust which make them no longer horizontal.

Yes. You have learned how to make inferences now. Very good! "Angular unconformity" illustrates that the ages of the two sets of strata are not continuous. There is a time interval between the two. There

Angular unconformity

may be even some tectonic fold movements, and some will leave a significant "erosion surface".

What is the "erosion surface"?

An "erosion surface" results when a rock stratum is exposed at the surface. It will become oxidized and broken by weathering and erosion, leaving a layer of gravel or sand. This is an "erosion surface" or "ancient weathering crust".

"Ancient weathering crust"? The name is funny. Upon hearing it, you could understand that weathering had occurred here in the past.

Yes. If a stable layer of gravel emerges in a large area, we should take particular notice of it, because it may be a remaining trace of "ancient weathering crust".

That shows the whole area went through weathering and erosion in the past.

Right. After the exposed rock surface is broken after oxidation, some gravel will be left. But when sediments pile up once again, often a gravel stratum will be formed. Geologists call it "a basal conglomerate".

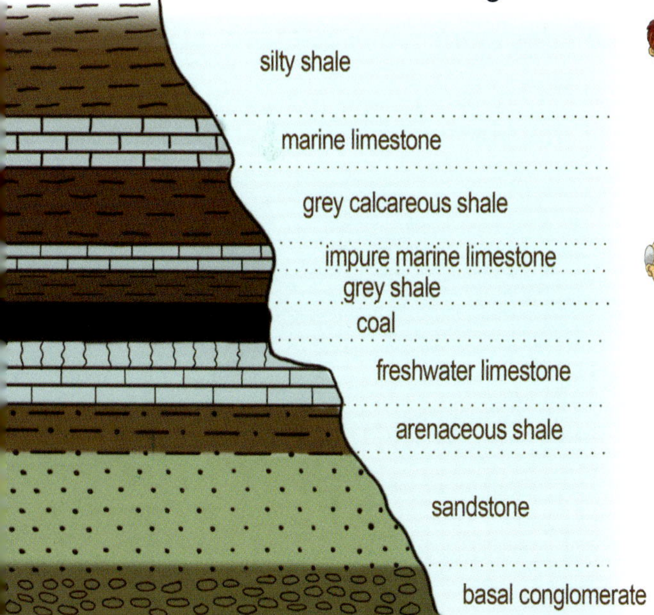

"Basal conglomerate" is a term easy to understand. The gravel may be covered by later sediments. Because it is at the bottom, it is called "a basal conglomerate". Is that right?

You are smart, Xiaomng. We can take the "basal conglomerate" as a "marker". It marks the beginning of another sedimentary cycle. In the Shennongjia area there is such a gravel layer called "a basal conglomerate", which the local

people call “a gem conglomerate”.

“Gem conglomerate”? That should be worth a lot, right?

Not necessarily. But the “gem conglomerate” is truly attractive because the gravel consists of agate, a very beautiful stone with different colors.

What is agate? I’ve often heard of it. Is it a kind of precious stone?

The main component of agate is silicon dioxide, which forms crystals. But silicon dioxide in crystal is crystalline whereas that in agate is cryptocrystalline. In other words, the mineral particles in agate are so small as to be nearly invisible. Agate with beautiful colors or patterns has high aesthetic value and is popular with collectors. The famous “Yuhua stone” in Nanjing is a good example.

I have seen that stone, very nice. Nature is really wonderful, creating so many pleasing miracles for us! Grandpa, what is a “parallel unconformity”?

“Parallel unconformity”, also known as “false conformity”, refers to the occurrence of the upper and lower strata with substantially the same orientation, but with a time gap in their formation, that is, there is a longer-term hiatus between the formations of these two sets of strata.

Then how do we know whether there is a hiatus or not?

We can judge according to changes of the rock properties in the upper and lower strata, but the more important evidence should be whether the fossils contained in the two strata are continuous in time.

Yes. The fossil is the most qualified to speak.

Yes. Xiaoming, I want to talk to you about the conglomerate. Do you know where gravel usually appears?

In rivers, lakes, and seas!

Right. Gravel generally stays on the water's edge. When the water deepens, it will be covered by sand and mud, right? In oceans, when the water gradually deepens, gravel will be covered first by sand. When the water continues to deepen, the water's kinetic energy will become weaker and weaker and sediments will gradually turn into silt and mud. If the water still continues to deepen, it is possible that there will be chemical disposition and limestone or dolomite sediments will develop.

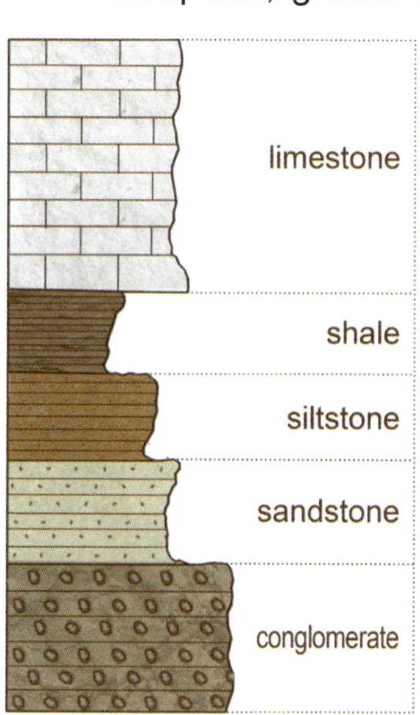

Transgression deposit sequence

Oh! It seems that, as the water gets deeper, sediment particles will become smaller, right?

Yes. The deeper the water, the weaker the water's kinetic energy, and the smaller the sediment particles which the water can carry. We refer to this sedimentary sequence, i.e. with the deepening of the water, sediment particles will gradually become finer and finer, as a "transgressive sedimentary sequence".

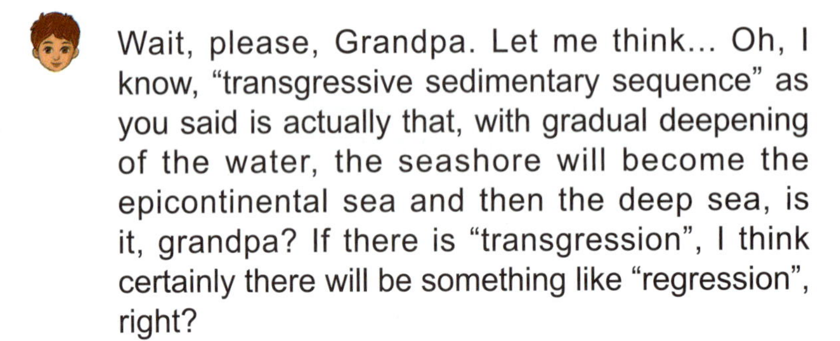

Wait, please, Grandpa. Let me think... Oh, I know, “transgressive sedimentary sequence” as you said is actually that, with gradual deepening of the water, the seashore will become the epicontinental sea and then the deep sea, is it, grandpa? If there is “transgression”, I think certainly there will be something like “regression”, right?

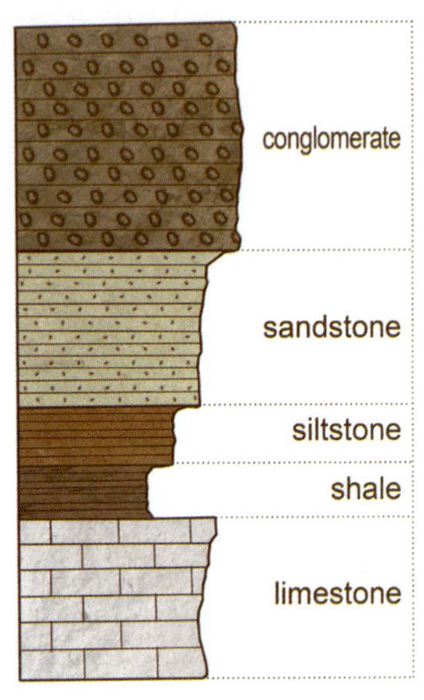

Regression deposit sequence

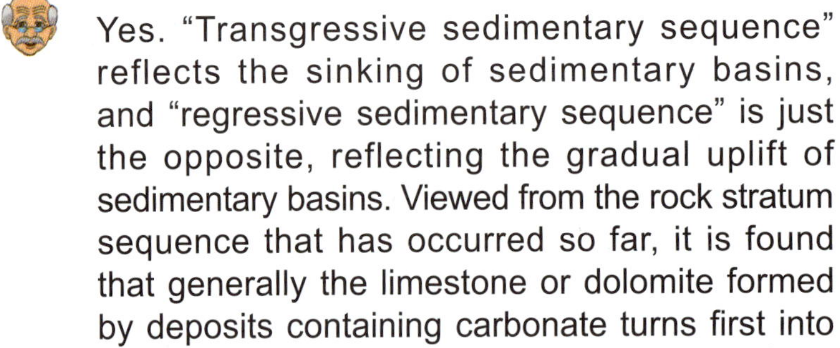

Yes. “Transgressive sedimentary sequence” reflects the sinking of sedimentary basins, and “regressive sedimentary sequence” is just the opposite, reflecting the gradual uplift of sedimentary basins. Viewed from the rock stratum sequence that has occurred so far, it is found that generally the limestone or dolomite formed by deposits containing carbonate turns first into mudstone, siltstone, or sandstone, and then into coastal conglomerate.

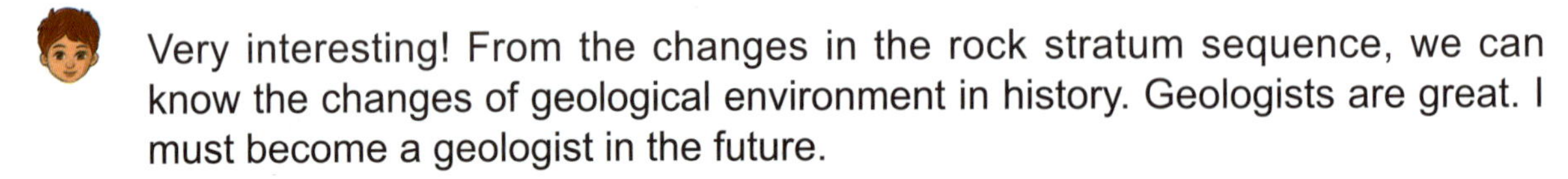

Very interesting! From the changes in the rock stratum sequence, we can know the changes of geological environment in history. Geologists are great. I must become a geologist in the future.

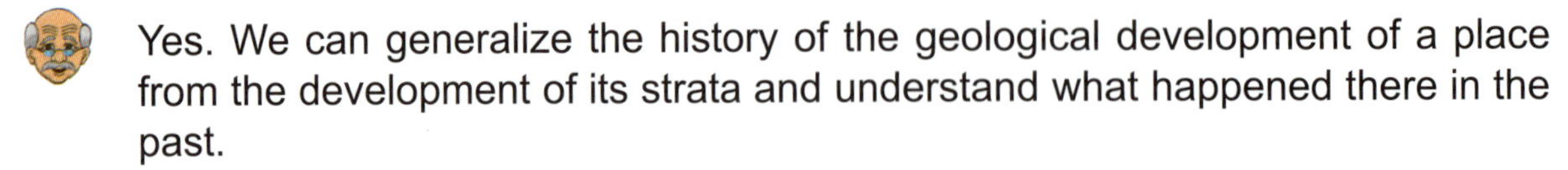

Yes. We can generalize the history of the geological development of a place from the development of its strata and understand what happened there in the past.

Geology is quite an interesting science!

Think of it further, Xiaoming. If we follow the chronological order to get hold of the strata of all parts of the world and then make a horizontal comparative analysis of them and some complementary study between them, won’t we

know all the events happening on the earth in different geological periods of history?

Yeah! This is the best way to study the history of the earth.

In fact, the International Commission on Stratigraphy and the Geological Sciences Federation of the United Nations Educational, Scientific and Cultural Organization (UNESCO) started this research long ago. They have already worked out the "International Stratigraphic Table" and the "International Geological Chronology". Though still in the process of improving, the two documents have provided a basis for the analysis of geological events worldwide. Think of fossils and glaciers in continental drift. Are they not based on them?

Yes. Great!

In order to determine the exact boundary of each geological time in unmapped rock strata,

Eon	Era	Period	Epoch	Age (Ma)	Biological Evolution
Phanerozoic	Cenozoic	Quaternary	Holocene	modern	Age of Man, Modern Plants
			Pleistocene	0.01	
		Tertiary	Pliocene	2.4	Mammal, Angiosperm
			Miocene	5.3	
			Oligocene	23	
			Eocene	36.5	
			Paleocene	53	
	Mesozoic	Cretaceous	Late	65	Reptile, Gymnosperm
			Middle		
			Early		
		Jurassic	Late	135	
			Middle		
			Early		
		Triassic	Late	205	
			Middle		
			Early		
	Paleozoic	Permian	Late	250	Amphibian, Pteridophyte
			Middle		
			Early		
		Carboniferous	Late	290	
			Middle		
			Early		
		Devonian	Late	355	
			Middle		
			Early		
		Silurian	Late	410	Fish, Pteridophyte
			Middle		
			Early		
		Ordovician	Late	438	
			Middle		
			Early		
		Cambrian	Late	510	Invertebrate
			Middle		
			Early		
Proterozoic		Sinian		570	Ancient Homonemeae
				800	
Archaean				2500	
				4000	

geologists have advanced the concept of “golden spike”. It is set as the reference mark in identification and recognition of the stratigraphic boundary between two eras around the globe by the International Commission on Stratigraphy and the Geological Sciences Federation, which have driven worldwide for a sole criterion or model boundary to define and differentiate rock strata formed in different eras and to get it marked in the specific place and specific rock strata sequence.

Grandpa, what does this mean? I don’t know much.

Let me give you an example. The Paleozoic Permian and the Mesozoic Triassic are two consecutive geological eras. Through detailed investigation and study of the field profile, it is found that the boundary of these two geological eras is indicated by a point appearing in this cross-section. This point is the most accurate cut-off point for the end of the Permian and the start of the Triassic in the world, representing the stratigraphic boundary between the two ages. You can set a “golden spike” here. Well, the “golden spike” which demarcates Permian and Triassic and gains international recognition is set in Meishan, Changxing County, Zhejiang Province, China.

Eon	Era	Period	Epoch	Age	Age (Ma)	GSSP
					199.6±0.6	
Phanerozoic	Mesozoic	Triassic	Late Triassic	Rhaetian		
					203.6±1.5	
				Norian		
					216.5±2.0	
				Carnian		
					228.0±2.0	
			Middle Triassic	Ladinian		
					237.0±2.0	
				Anisian		
					245.0±1.5	
			Early Triassic	Olenekian		
					249.7±0.7	
				Induan		
					251.0±0.4	✓
	Paleozoic	Permian	Lopingian	Changhsingian		
					253.8±0.7	
				Wuchiapingian		
					260.4±0.7	✓
			Guadalupian	Capitanian		
					265.8±0.7	✓
				Wordian		
					268.0±0.7	✓
				Roadian		
					270.6±0.7	✓
			Cisuralian	Kungurian		
					275.6±0.7	
				Artinskian		
					284.4±0.7	
				Sakmarian		
					294.6±0.8	
				Asselian		
					299.0±0.8	✓

Wow, that’s amazing! How many “golden spikes” in total in the world?

About 110, according to the Global Stratigraphic Chronology. Nearly 60 of them are confirmed at present. 10 of them

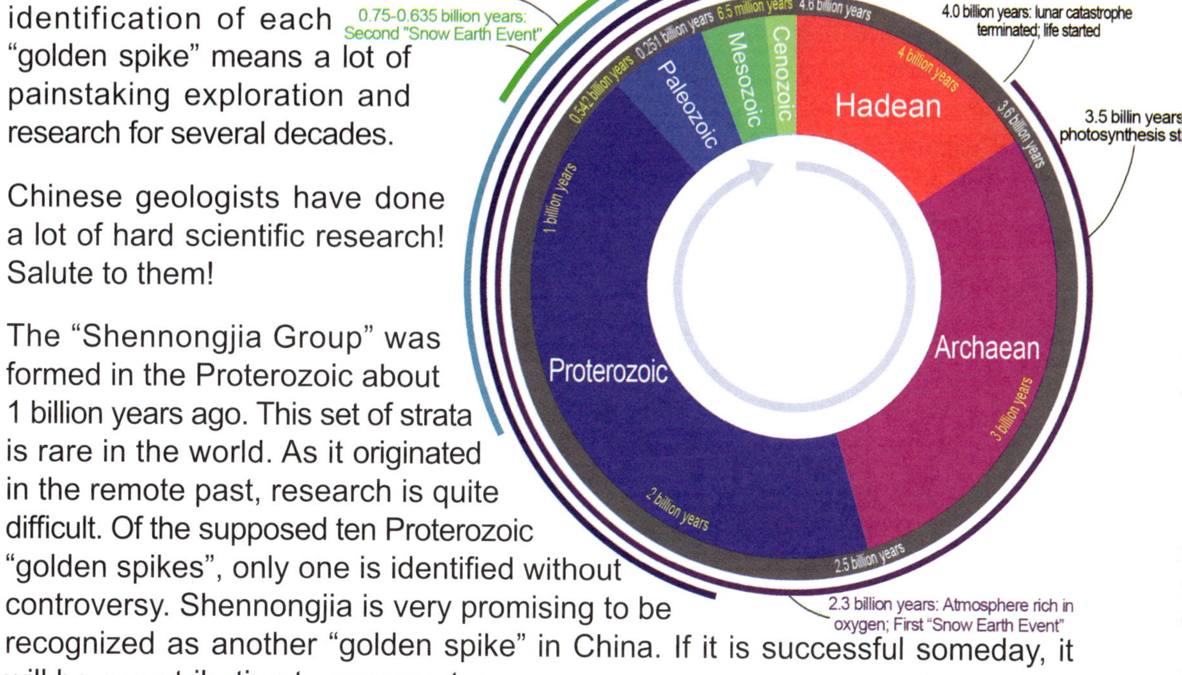

are found in China. The identification of each "golden spike" means a lot of painstaking exploration and research for several decades.

Chinese geologists have done a lot of hard scientific research! Salute to them!

The "Shennongjia Group" was formed in the Proterozoic about 1 billion years ago. This set of strata is rare in the world. As it originated in the remote past, research is quite difficult. Of the supposed ten Proterozoic "golden spikes", only one is identified without controversy. Shennongjia is very promising to be recognized as another "golden spike" in China. If it is successful someday, it will be a contribution to our country.

Really, Grandpa? I'll try my best to become a geologist, and set more "golden spikes" in China!

Okay, Xiaoming. You are ambitious. Study hard. There is a lot to learn if you want to be an outstanding geologist.

Questions:

1.What is the stratigraphic "conformable contact", "parallel unconformity" and "angular unconformity"?

2.What is a "basal conglomerate"? What is the significance of identifying "basal conglomerates"?

3.What do "transgressive sedimentary sequence" and "regressive sedimentary sequence" mean respectively? What kind of environment do they reflect?

4.Why should the "International Stratigraphic Table" and "International Geological Chronology" be established?

5. What do you know about "golden spikes"?

23

Built at the foot of the mountain and beside a stream, the pedestrian plank path took the senior professor and Xiaoming to the entrance gate of the Guanmenshan Subarea. Before them stood the famous statue named "Maternal Love", which startled Xiaoming greatly. It is a life-like statue depicting an ape-like mother and her son hugging each other. The mother appears somewhat rough, squatting on the ground with arms around her son, and passionately kissing him. The little wild man looks simple and lovely, showing his dependence on and love for the mother. The universal values and human emotions of "love" find expression in the statue so much that anyone will be moved by it.

Xiaoming gazed at the statue, and then walked out of the gate at a leisurely pace. Wangwang was a little uneasy, following the boy quietly. It glanced back repeatedly at the two wild creatures, timidly guarding against them.

Grandpa, this statue is perfect! It is in great harmony with the environment here.

Yes. The entrance to this subarea is uniquely designed, full of strong humanistic spirit, embodying noble universal maternal love. Meanwhile, it also shows people's respect for nature, and the idea of harmony with nature. The square outside the gate is called "Maternal Love Square".

"Maternal Love Square", a good name. Regardless of humans or animals, maternal love is one of the most common precious feelings.

It is of special significance for the Guanmenshan Subarea to set geology and geography, biology and ecology, scientific research and science popularization, and folk shows as a whole, and the "Maternal Love Square" as the starting point of the tour. By so doing, people are guided not only to cherish mutual love but also to cultivate their love for the biological world and nature. If everyone can love nature and its environment as a mother loves her children, our green planet will become even more beautiful.

Grandpa, what you said is great! Hey, what are the words carved on the two big stone tablets?

Carved on the smaller one are words "Hubei Shennongjia National Geopark". It was erected in September 2005 when the park officially opened.

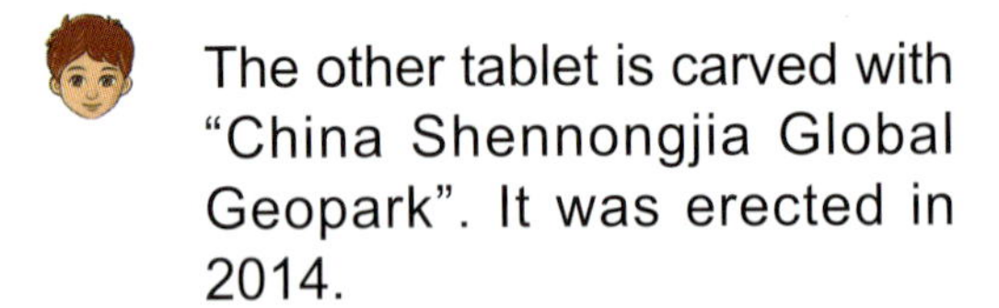

The other tablet is carved with "China Shennongjia Global Geopark". It was erected in 2014.

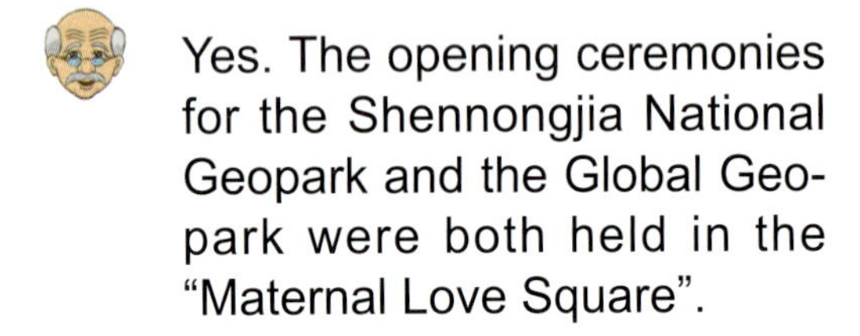

Yes. The opening ceremonies for the Shennongjia National Geopark and the Global Geopark were both held in the "Maternal Love Square".

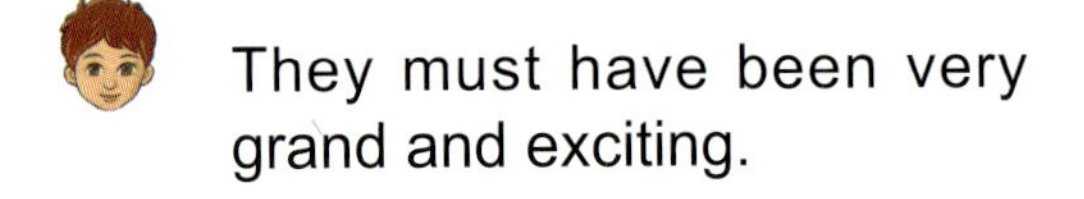

They must have been very grand and exciting.

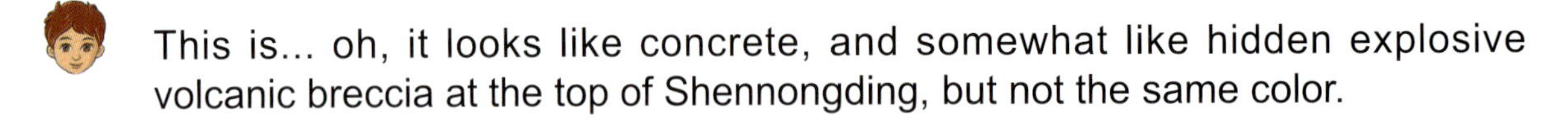

Yes. Xiaoming, come and look! Do you know the rock around the square?

This is... oh, it looks like concrete, and somewhat like hidden explosive volcanic breccia at the top of Shennongding, but not the same color.

It is indeed a kind of breccia, but it has nothing to do with the volcanic action. On the contrary, its formation is closely related to snow and ice.

Related to snow and ice?!

Yes, it's ice-related. It is called "tillite".

Wow, Grandpa, nature is so marvelous. Volcanoes can bring about breccia, and snow and ice can do it also.

This stratum of “tillite” was named “Nantuo Formation”. Look, the rock is mainly composed of grey green sandy conglomerate. Gravel of this kind is complex in composition, with angular pieces, or clasts, varying in size and shape. Roundness and sorting are very poor.

What is roundness? What is sorting?

Roundness and sorting are common terms used to describe fragments in sedimentary rocks. Roundness refers to angular rock fragments becoming more round. For example, stone fragments and mineral particles, in their movement in the flowing water, have to experience repeated long-term erosion, rolling and collision and their edges and corners are getting rounded off. Generally speaking, the longer they are moved, the better their roundness is. Of course, roundness has a lot to do with the hardness of the stone fragments.

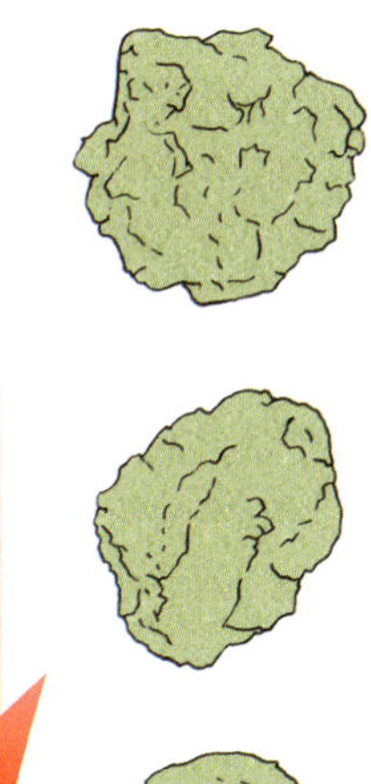

Diagram of Rock Roundness

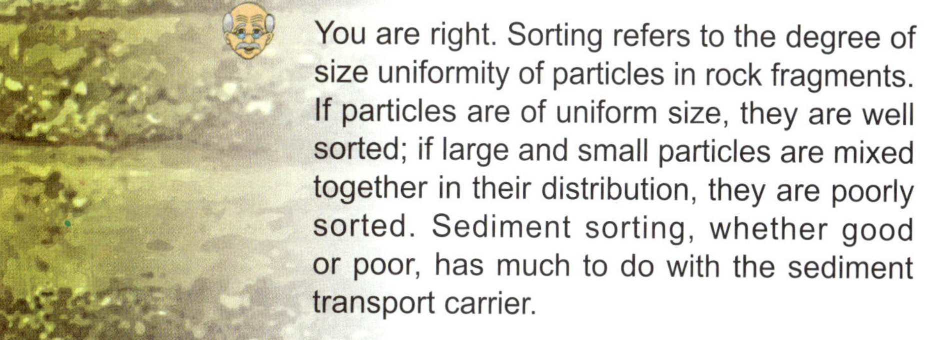

After their long-term, long-distance journey of rolling and colliding, stone fragments will certainly become round and smooth, like pebbles we see in the stream.

You are right. Sorting refers to the degree of size uniformity of particles in rock fragments. If particles are of uniform size, they are well sorted; if large and small particles are mixed together in their distribution, they are poorly sorted. Sediment sorting, whether good or poor, has much to do with the sediment transport carrier.

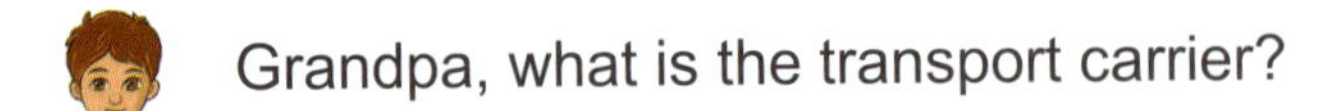

Grandpa, what is the transport carrier?

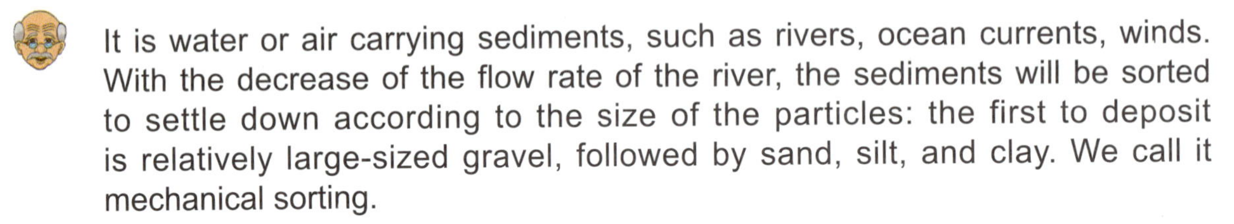

It is water or air carrying sediments, such as rivers, ocean currents, winds. With the decrease of the flow rate of the river, the sediments will be sorted to settle down according to the size of the particles: the first to deposit is relatively large-sized gravel, followed by sand, silt, and clay. We call it mechanical sorting.

No wonder we see a lot of huge stones in the valley, and often sand and mud in the river. The reason for it turns out to be that, with the change of the water flow, sediments are deposited in different places according to their volume or weight.

Yes! Xiaoming, look at the tillite in the "Nantuo Formation". What characteristics do they have?

Gravel of different sizes is mixed up together, out of order, like a pile of rubble.

So they are obviously not brought about by the river, but by the glacier.

Glacier?

Yes. A glacier is a transport carrier. The glacier, in the process of migration, will take a large amount of sand, stones, etc. with it. When the glacier melts, the mud, sand and stones, all of a sudden, begin to deposit in heaps.

Oh, that's the way it was! But glaciers exist in the Antarctic and the Arctic, or in such high mountains as the Himalaya Range, right? Here in Shennongjia were

there any glaciers in the past?

Simple List of Geological Time

Geologic Time Chart		Age (Ma)
Cenozoic		66.0
Mesozoic		298.9
Paleozoic	Late Paleozoic	419.2
	Early Paleozoic	541.0
Proterozoic	Neoproterozoic	1000
	Mesoproterozoic	1600
	Paleoproterozoic	2500
Archaean		4600

That'll be a long story. The tillite in Shennongjia formed in the Neoproterozoic, about seven or eight hundred million years ago. Very, very old!

What is the Neoproterozoic?

The Neoproterozoic is an era in the Geological Chronology. By means of absolute dating and fossil analysis, geologists have divided the last 4.6 billion years of the earth's history into many chronological units. The Neoproterozoic is one of them. It lasted approximately from 1 billion to 540 million years.

Wow. One billion years ago! It's old enough! But I do not understand, Grandpa, why should we divide the 4.6 billion years of the earth history?

There have been many major geological events in the long history of the earth, and these events are not the same in different parts of the globe. When we study the history of the earth now, we must use the same time scale to discuss them. Therefore, we must have a unified division of geological time, hence, the "International Geological Chronology" as mentioned before. All of us should follow the same time divisions and record the local geological events.

Oh, I see. In the same period, things that happen on the earth must be different in different places. For example, when the North and South Poles are

a world of ice and snow, the equatorial region is probably under the scorching sun, or attacked by heavy rain, or flash floods.

Yes, right. In the same period, here on the earth there may be an outbreak of flash floods which may have washed hillsides away, but in other parts of the globe deposition may be taking place and thick sandstone is being formed.

Yes. Every place has a different history of development.

Yes. If every place keeps a record of its local geological history on the basis of their specific strata and the unified "International Geological Chronology", we will be able to get them together and draw conclusions about the global geological development.

<table>
<tr><th>Era</th><th colspan="3">Group & Formation</th></tr>
<tr><td rowspan="4">Low Paleozoic</td><td colspan="3">Silurian</td></tr>
<tr><td colspan="3">Ordovician</td></tr>
<tr><td colspan="2" rowspan="2">Cambrian</td><td>Loushanguan Form.
——Niutitang Form</td></tr>
<tr><td rowspan="2">Dengyin Form.</td></tr>
<tr><td rowspan="7">Neoproterozoic</td><td rowspan="2">Sinian System</td><td>Upper Sinian</td></tr>
<tr><td>Lower Sinian</td><td>Doushantuo Form.</td></tr>
<tr><td rowspan="4">Nanhua System</td><td rowspan="3">Upper Nanhua</td><td>Nantuo Form.</td></tr>
<tr><td>Datangpo Form.</td></tr>
<tr><td>Gucheng Form.</td></tr>
<tr><td>Lower Nanhua</td><td>Liantuo Form.</td></tr>
<tr><td colspan="2">Qingbaikou System</td><td>Liangfengya Form.</td></tr>
<tr><td rowspan="4">Mesoproterozoic</td><td rowspan="2">Jixianian System</td><td>Upper Jixianian</td><td rowspan="4">Shennongjia Group</td></tr>
<tr><td>Lower Jixianian</td></tr>
<tr><td rowspan="2">Changcheng System</td><td>Upper Changcheng</td></tr>
<tr><td>Lower Changcheng</td></tr>
</table>

Wow, that'll be very complicated, but it'll be great fun! The local records will certainly vary immensely: here deposition happening, there erosion taking place; here transgression, there regression; here heavy rains, there volcanic eruptions.

Yes, it'll be really complicated. However, in the development of the earth's history, there have been some significant geological events which are common in nature.

I don't think it's possible! Grandpa, the earth's so big, how can there be a common occurrence?

Let me give you an example before our eyes. The rock stratum of the "Nantuo

Formation" we are facing now was formed in very cold conditions. At that time, or about 700 to 800 million years ago, Shennongjia was not the only place to suffer severe cold, and many parts of the world were covered with snow and ice.

Is that so? How do you know about it?

As early as in 1964, Professor Harland (Harland B.W.) of Cambridge University found that, during the period 800 million years to 550 million years ago, glacial deposits emerged in every continent of the world. In 2000, Yale University geology professor Evans (Evans DAD, et al) through the study of glacial sedimentary stratigraphy, geochronology and paleomagnetism, noted that within the period of 800 million years to 550 million years ago, the mixed conglomerate from continental glacial deposition was mostly distributed within the north and south latitude 10°. Xiaoming, think of it. What does this mean?

The equator of the earth is 0°, within the north and south latitude 10°? Oh, my God! That means the glacial sediments occurred near the equator. How could that be?

But such is the case!In the period of 800 to 550 million years ago, or in the Neoproterozoic Age, the earth's land was essentially a single piece, called "Pangea". Scientists have speculated that the earth was once covered with snow and ice everywhere at that time, and the equatorial region was no exception. The entire planet was a giant snow ball. The ice sheet was 1 kilometer thick, the earth's temperature dropped to –50°C or so, and lasted for about 100 million years. This is the famous "Snowball Earth Event" in the

history of the earth.

Oh, that's terrible! Fortunately, we were not born then.

Hahahaha! Humans only appeared about 600 million years later after that event. Mankind has only a history of about 3 million years.

Grandpa, then what happened? How did the earth evolve as it is now?

Later, due to the volcanic eruption, large quantities of carbon dioxide gas were released. After more than 10 million years of accumulation, the carbon dioxide finally became strong enough to cause the "greenhouse effect", which quickly melted the "Snowball Earth" and the global ocean temperature rose to more than 50°C.

Ah ah! More than 50°C! Becoming so hot all of a sudden, sure people will have a heat stroke! It's hard to stand when the temperature in summer gets up to 37-38°C!

Yes. Later "Pangea" broke up into many blocks of land. Marine and terrestrial organisms multiplied and the earth began to turn green. Finally, the earth was gradually formed as it is now, suitable for humans and all kinds of living things to live in.

Not easy! Grandpa, I think our planet seems to be a self-regulating, living thing. It will try to adjust the environment to the best.

Yes, a good metaphor you used. Our earth does have the ability to self-regulate. However, when mankind achieved industrialization, we began to make endless demands on nature, such as ore mining and deforestation. Meanwhile, we have constructed a large number of factories which continuously emit carbon dioxide and a variety of toxic gases into the atmosphere. As a result, the water and the land are polluted and the environment is damaged. Today, the earth can hardly sustain and is about to collapse.

Then what should we do? Grandpa, we must help the earth restore a good ecological environment.

Yes. The earth is currently facing a very grim situation. Each year nearly 6 million hm^2($1hm^2=0.01km^2$) of land become desert, and 20 million hectares of forest disappear. One living species becomes extinct per hour on average!

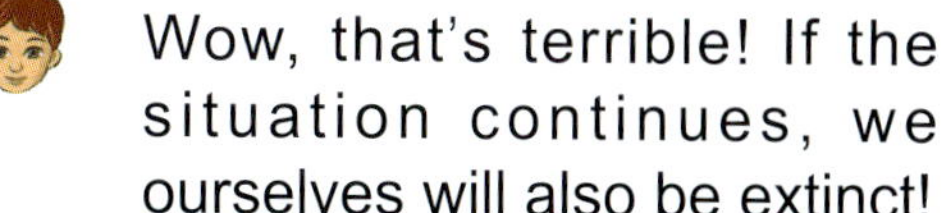

Wow, that's terrible! If the situation continues, we ourselves will also be extinct!

So, each of us should undertake responsibility and obligation to protect the environment and the earth's resources so that the earth will be kept evergreen as Shennongjia is. Well, the sun is setting. Let's go to the "Tiansheng Bridge" for a walk.

Questions:

1.What is the symbolic meaning of the "Maternal Love Square" in front of the gate of the Guangmenshan Subarea?

2.Please tell something about the tillite in the "Nantuo Group". Why are its "roundness" and "sorting" very poor?

3.What is the "Snowball Earth Event" 800 to 550 million years ago?

4.What are the challenges the earth faces in the present ecological environment?

5.How should we help the earth restore a good ecological environment? Please list the environmental protection measures that can be implemented in daily life.

24

The senior professor was explaining to Xiaoming when he found himself in the middle of the "Maternal Love Square". He sat on the flying carpet with the boy. The sun was going down, leaving the mountain peaks around afire with orange and green light. The flying carpet took off smoothly against the sunset glow, heading for the Tiansheng Bridge Subarea surrounded by mountains.

Viewed from above, a flat asphalt road zigzagged through the green mountains like a tape winding on the mountainside. Thick white clouds moved over the top of the mountain like a waterfall running into the valley. After the flying carpet passed through the clouds, what came in sight was a huge gourd-shaped cave.

This is a nice penetrating karst cave. Seen from afar, it looks like a bridge lying on the mountain. Flying close to it, the scenery is more beautiful. In the rock cliff next to the cave, a spring is gushing from the rock seam, forming a waterfall as if a water curtain hangs on the cliff. Inside the cave, the roaring water is rushing down the stream, coming out of the cave with a waterfall, as high as 10 m, shining dazzling white, and then dashing to a string of deep pools below. A plank path is built upwards along the cliff inside the cave; after many twists and turns, the path extends itself through the cave to the tranquil green valley beyond the cave. The flying carpet the old professor and the little boy took slowed down toward the entrance to the cave. The raging torrent running through the cave blends with the waterfall on the steep cliff on the other side, with a thunderous roar.

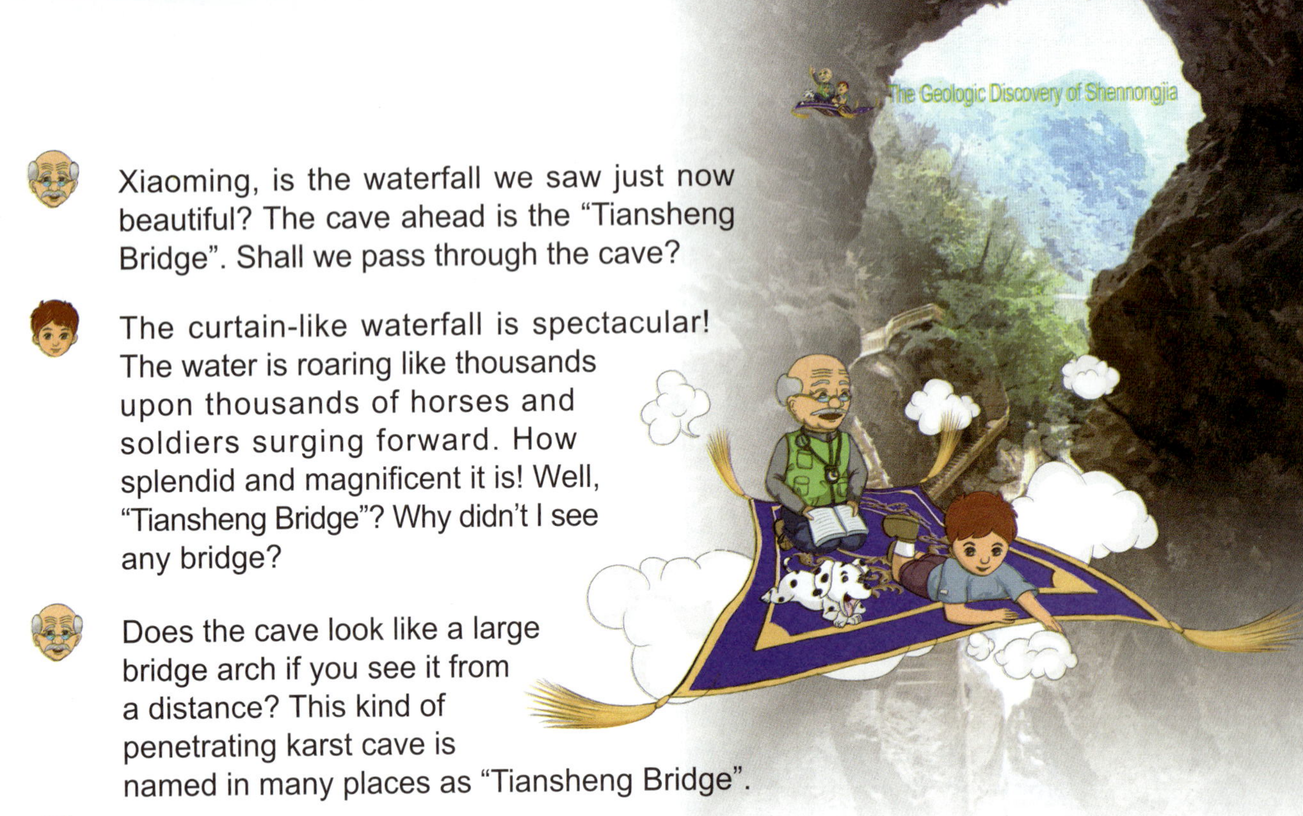

Xiaoming, is the waterfall we saw just now beautiful? The cave ahead is the "Tiansheng Bridge". Shall we pass through the cave?

The curtain-like waterfall is spectacular! The water is roaring like thousands upon thousands of horses and soldiers surging forward. How splendid and magnificent it is! Well, "Tiansheng Bridge"? Why didn't I see any bridge?

Does the cave look like a large bridge arch if you see it from a distance? This kind of penetrating karst cave is named in many places as "Tiansheng Bridge".

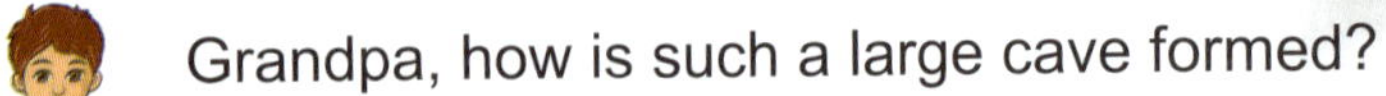

Grandpa, how is such a large cave formed?

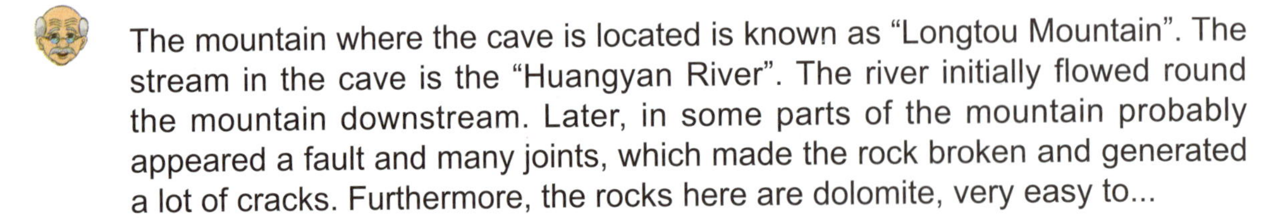

The mountain where the cave is located is known as "Longtou Mountain". The stream in the cave is the "Huangyan River". The river initially flowed round the mountain downstream. Later, in some parts of the mountain probably appeared a fault and many joints, which made the rock broken and generated a lot of cracks. Furthermore, the rocks here are dolomite, very easy to...

Oh, I see! The rocks here are very easily corroded. These broken rocks were constantly lashed and corroded by the flowing water for so long that a cave was formed.

Yes. In the beginning, the cracks in the rock might have been small. The water

kept on infiltrating into them, and little by little they became larger, more and more water flowing down along the cracks. Then water diversion began to occur.

Oh, I see. The curtain-like waterfall we saw before we entered the cave must be a diversion of the Huangyan River, isn't it?

Yes. The waterfall is basically flowing out of the karst hole formed along the bedding planes of the rock strata. With further development of the karstification and the crumbling process, the Huangyan River crushed all rocks in its way bit by bit for an underground cavern system, leaving a large cave. Finally, the Huangyan River was completely diverted and the old river which used to bypass Longtou Mountain discontinued. Some part of the water from the "Huangyan River" forms a curtain-like waterfall, and the rest runs through the huge cave, making its way constantly downstream.

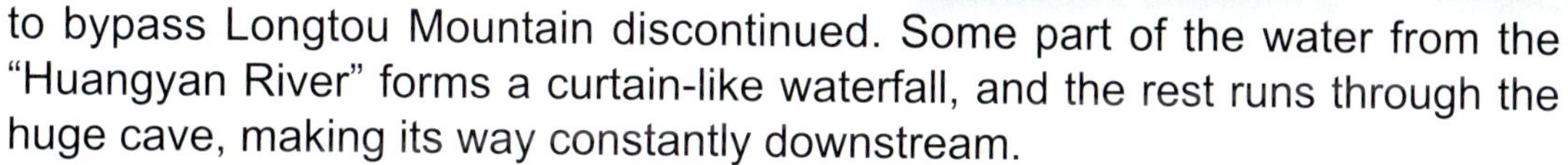

The "Huangyan River" is really great, building a natural bridge for Shennongjia!

The flowing water in nature is a great craftsman. The formation of most modern landscape is connected to it. In regions that contain carbonate rocks in particular, penetrating karst caves similar to the Tiansheng Bridge are very common.

I feel that the karst landform with a penetrating cave like the Tiansheng Bridge is particularly impressive, extremely beautiful.

Yes, that's what we call a geological heritage. Such features not only reveal a very interesting geological history, but also constitute a long-lasting tourism resource in geology parks, attracting more people to come to study and appreciate nature's creation. They will be inspired to love nature and consciously protect the environment .

Grandpa, as you said, flowing water in nature is a great craftsman. This "craftsmen" is so persistent, and our true role model.

Good! This is the revelation made by nature to human beings. Those who are able to appreciate nature can often get a lot of inspiration from the natural landscape and become better informed.

Yes. It is also said in books that the history of mankind is a process of exploring and learning from it.

Yes, that's right. Xiaoming, look, there is a well under the shady trees. This is a geological relic that I'd especially like to introduce to you.

Is it? I saw it. The interpretation sign beside it writes "Tianquan". Why is it called "Tianquan"? What is special about it?

This is a miraculous well. Regardless of seasons, this well will have a steady stream of water flowing out, never becoming dry.

Is this so, grandpa? How can it be?

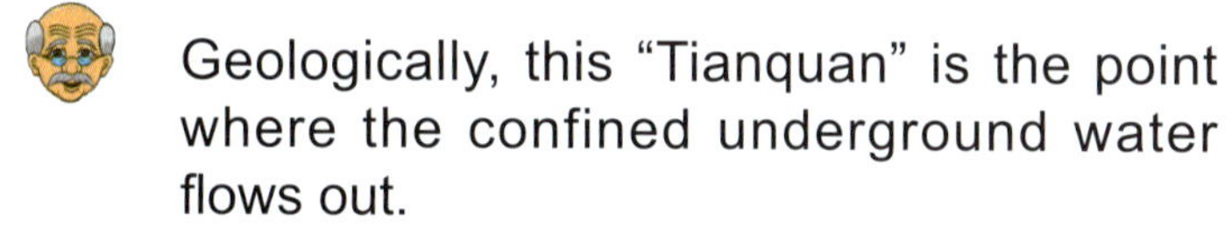
Geologically, this "Tianquan" is the point where the confined underground water flows out.

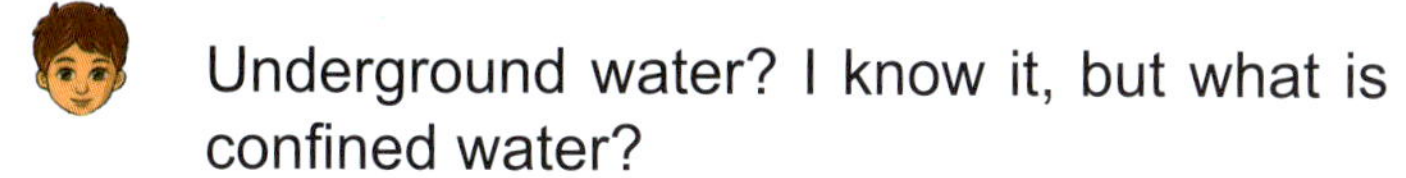
Underground water? I know it, but what is confined water?

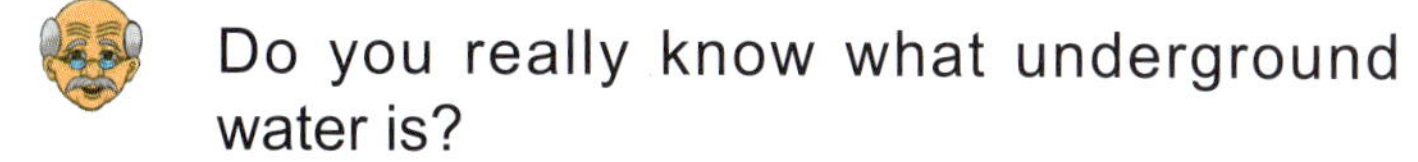
Do you really know what underground water is?

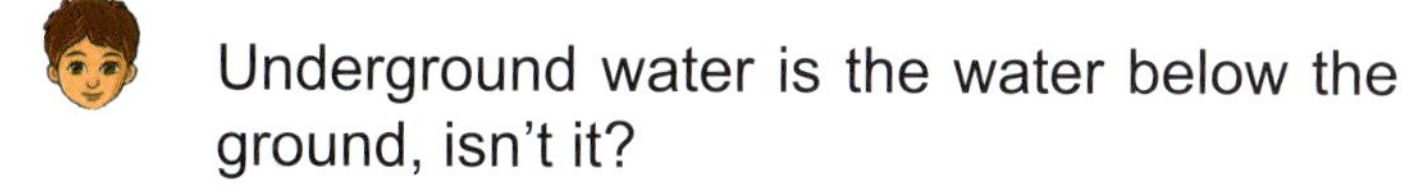
Underground water is the water below the ground, isn't it?

Yes, but not exactly. According to the burial conditions, underground water is divided in hydrogeology into three categories, namely, upper lagging water, phreatic water and confined water.

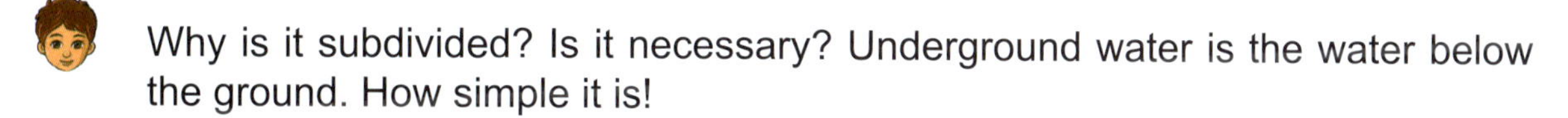
Why is it subdivided? Is it necessary? Underground water is the water below the ground. How simple it is!

Scientific research is particular about precision. Every scientific term has a definite academic significance. "Underground water is the water below the ground" is the general understanding of people. However, Xiaoming, don't you want to be a scientist?

Yes. So let's get it right. Would you please tell me what the upper lagging water is?

The upper lagging water is the rain water or melted snow soaked in the surface soil or kept in rock fractures. Due to the fact that some part of a stratum is impermeable, such as a layer of fine clay or a layer of dense rock

under the ground, the water will stay in shallow rock cracks or in the surface soil. It will gradually seep from the low-lying ground, getting into ditches, or gradually evaporate from the soil, or infiltrate downwards to join the phreatic water.

What kind of underground water is phreatic water?

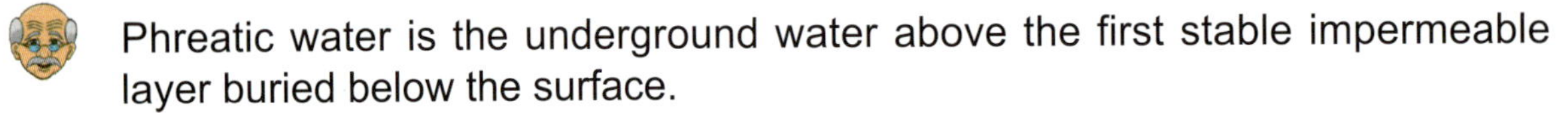

Phreatic water is the underground water above the first stable impermeable layer buried below the surface.

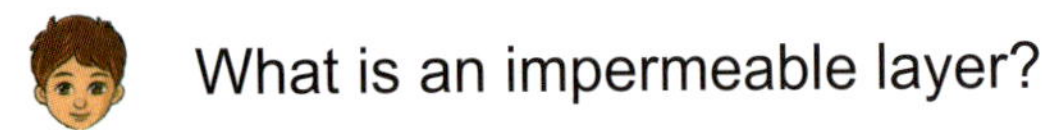

What is an impermeable layer?

An impermeable layer is a layer of rock which is water resisting. Many rocks look very hard, but they have a lot of small pores inside. Sandstone and conglomerate which are not well-cemented are examples of that kind. The pores are filled with water which leaks along these pores. However, a lot of other rock layers are basically impermeable. Most of magmatic rock and mudstone are good examples. It is difficult for water to pass through them, so they are called the impermeable layer.

Oh I see. It's the kind of stone whose particles are densely arranged inside.

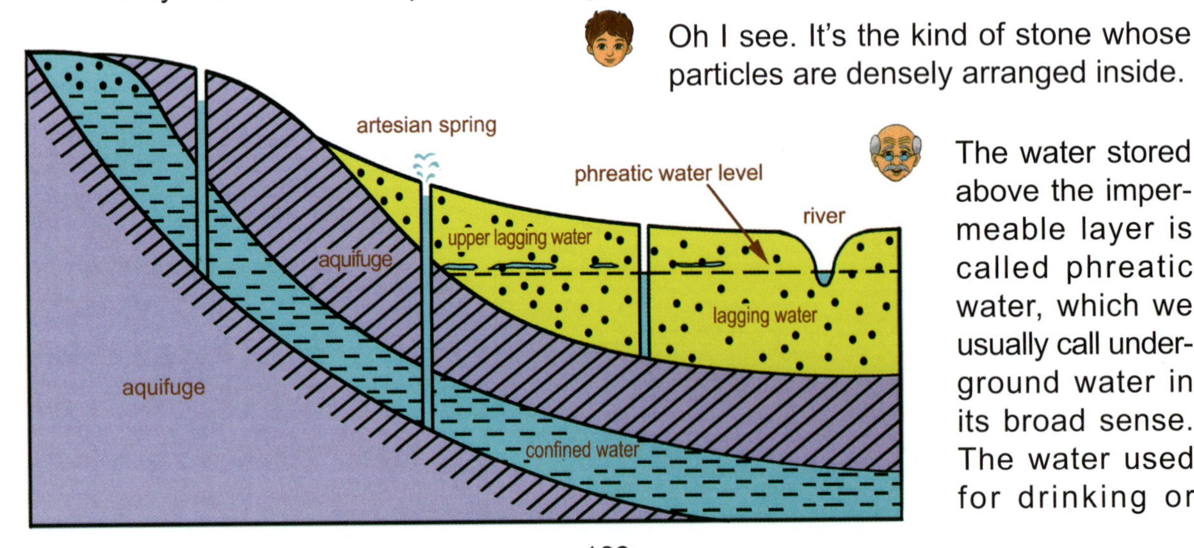

The water stored above the impermeable layer is called phreatic water, which we usually call underground water in its broad sense. The water used for drinking or

irrigation is mostly phreatic water. The surface of the phreatic water is called the ground water level, which is generally known as ground water table. Depending on the terrain and the amount of rainfall, the depth of the ground water level is different in different places, and will frequently change.

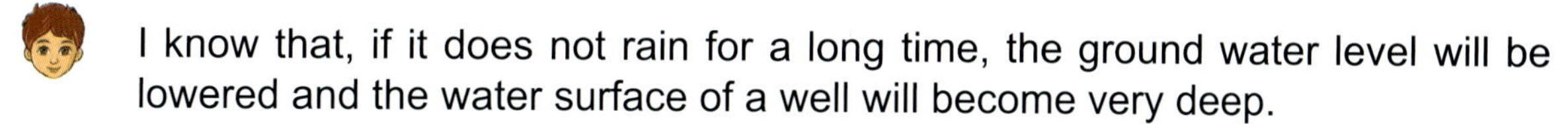

I know that, if it does not rain for a long time, the ground water level will be lowered and the water surface of a well will become very deep.

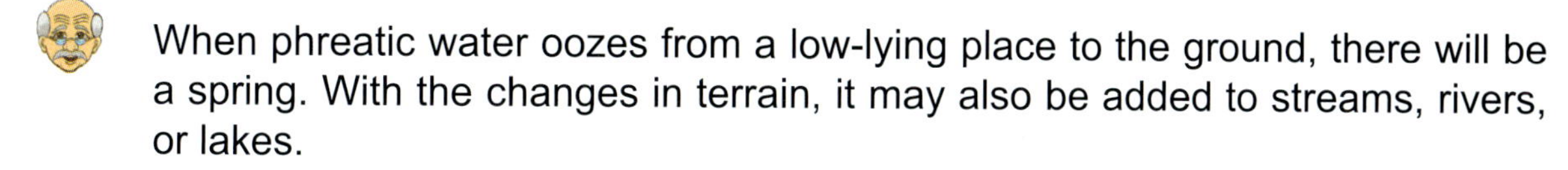

When phreatic water oozes from a low-lying place to the ground, there will be a spring. With the changes in terrain, it may also be added to streams, rivers, or lakes.

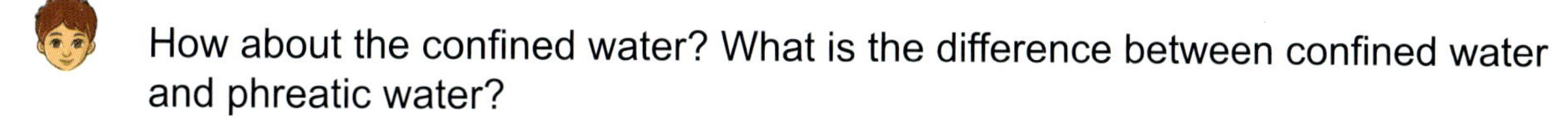

How about the confined water? What is the difference between confined water and phreatic water?

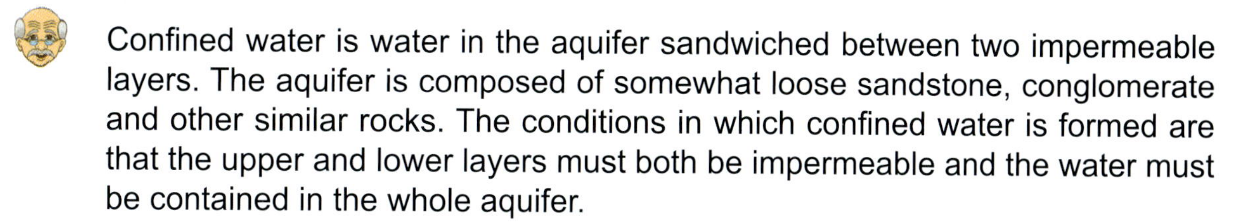

Confined water is water in the aquifer sandwiched between two impermeable layers. The aquifer is composed of somewhat loose sandstone, conglomerate and other similar rocks. The conditions in which confined water is formed are that the upper and lower layers must both be impermeable and the water must be contained in the whole aquifer.

Why is it called "confined water"? What pressure does it bear?

The underground water, sandwiched between the two impermeable layers and controlled by the terrain, is often confined to the underground. It is subject to hydrostatic pressure, just like the water enclosed in the "U" shaped tube, so it is called "confined water". As the water table is connected to the imperme-

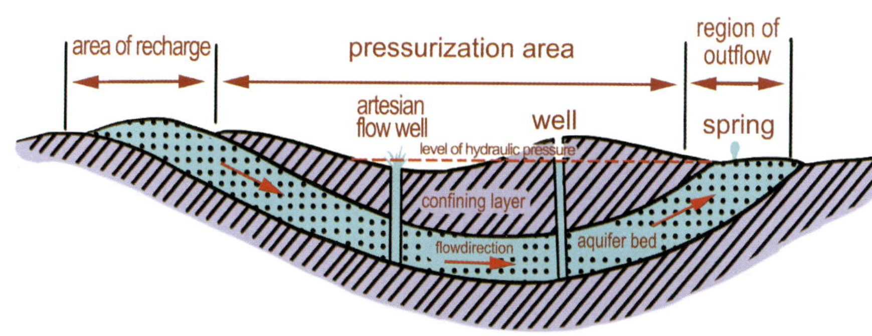

able layer, the water supply area is much smaller than the confined water distribution area.

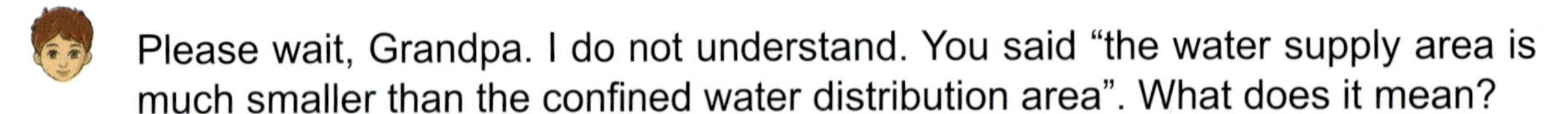

Please wait, Grandpa. I do not understand. You said "the water supply area is much smaller than the confined water distribution area". What does it mean?

The answer is very simple. Let's make a comparison. Here is a multi-layered chocolate cookie. The upper and the lower layers are cookies which are like the two impermeable layers. The layer in between is a chocolate chip which is like the aquifer.

Well, it's really like a multi-layered chocolate cookie!

If you want to add some "confined water" to the aquifer, you can only do it from the edge of the cookie, right?

Yes, yes... Ah, I understand, the confined water can only be added from the broken edge of the impermeable layer!

You have a good understanding. Owing to the limitation of the watertight roof, the confined water, unlike phreatic water which is often involved in the groundwater hydrological cycle, does not have much interaction with and relation to the atmosphere and the surface hydrosphere. Therefore, confined water is not vulnerable to contamination. Most of the confined water is infiltration water, so it is often relatively clean and fresh water, well suited to drinking.

Do you mean the water from Tianquan is very clean, and we can drink it directly?

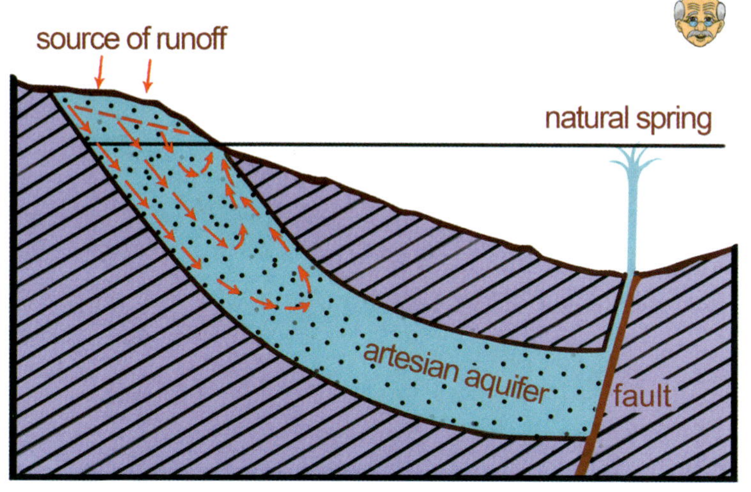

Yeah. Shennongjia has long been sparsely populated with little pollution. The water in many valley streams is safe to drink, not just Tianquan water.

Shennongjia is really a good place!

The formation of the confined water is closely related to the geological structure. The most favorable geological structure for its formation is a syncline and monocline that constitute artesian basin and artesian slope respectively.

Why is it called "artesian" basin? Will the water come out by itself?

Because the confined water in the aquifer bears hydrostatic pressure, it will flow out by itself, just as the water in a fully filled U-shaped tube will come out immediately by itself at one end of the tube under pressure when the other end of it is elevated, right?

Yes, yes.

Then a natural fountain can emerge under appropriate conditions of terrain and geological structure. For example, a fault from the earth's surface to the aquifer can possibly constitute a channel for the confined water which will be made to bear the hydrostatic pressure. When the water gushes onto the surface with changes of the geological structure, a natural spring will come into being. The Baotu Spring in Jinan is a good example.

Oh, I see. The Tianquan in Shennongjia is formed this way, isn't it?

Yes. There are many large "artesian basins" in the world, such as Paris Artesian Basin, China Sichuan Artesian Basin, Great Australian Basin. They provide high quality water for humans.

Are they much larger than the Shennongjia's Tianquan?

Hahahaha! Xiaoming, do you still remember the "geological thinking"? When I introduced it to you, I mentioned that geological problems should be taken into consideration in the light of time length. Now, it seems one more factor must be added in the "geological thinking", that is, the scale of the object of study. For example, an anticline can span several tens of kilometers and a large fault can extend across several provinces.

Wow, so big scale?!

It's not too big! The Great Australian Basin I referred to just now is even bigger. It is located in the middle of the Australian mainland, between the Great Dividing Range in the east and the desert plateau in the west. It covers an area of more than 1.75 million km^2, accounting for nearly 1/4 of the total area of Australia.

Is there really such a big "confined water" basin? It is simply an underground sea!

It can be said to be an underground sea. This artesian basin is situated to the east of the Great Dividing Range.

Influenced by the southeast wind from the Pacific Ocean, there is abundant rainwater for the basin.

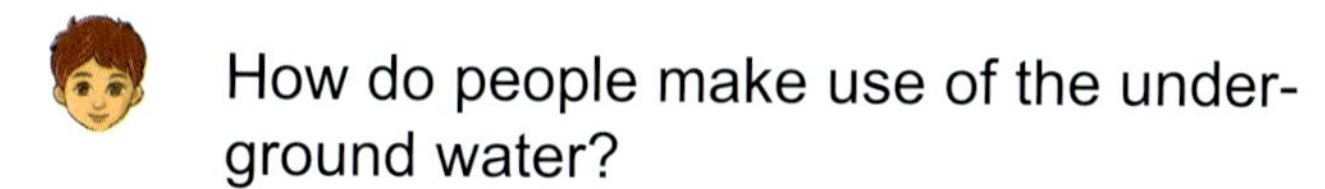

How do people make use of the underground water?

They make holes in the ground through the impermeable layer by drilling. This is so-called confined well drilling. The water in the confined well will go upwards along the borehole under hydrostatic pressure and reach the height of the local confined water level. When the ground is lower than the confined water level, the confined water will pour out to the surface and form an artesian well. People may use canals or plastic pipes to carry the water to local farmland, pastures or residential areas. The daily average water yield in the Great Australian Basin amounts up to 1.3 billion L.

Great! Nature has built a great underground reservoir for people!

Yes, of course. Nature has created many beautiful things for human beings. But we must carefully plan their use and cherish the resources so as to ensure the sustainable development of human society.

Yes. Water is a precious natural resource, and it is the basic material to support human existence.

But, Xiaoming, you know, China is a country with very scarce water resources.

It can't be true. You can see water everywhere here!

Yes, there is water everywhere here, but many places in our country are waterless and even drinking water is very difficult to get.

That must be the case in the desert, right? Of course there is no water there.

Not only in the desert. According to the global statistics, China is a country with an extreme lack of water resources, ranking 121st in the world. It is one of the 13 countries in the world where water resources per capita are the poorest. The water resource per capita in China is only 2200m^3, only 1/4 of the world's water resources per capita.

Ah! That is to say, when each person in the world can drink 4 glasses of water on average, we Chinese can only drink 1 glass. Is it true? Is the problem really that serious? Every day when we turn on the tap, the water will stream out of it, and it never occurs to us that China is a country with a very poor water resource.

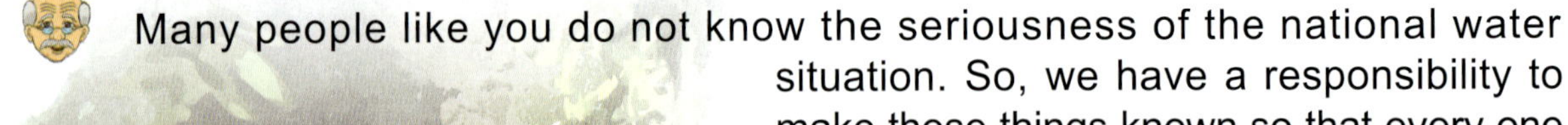

Many people like you do not know the seriousness of the national water situation. So, we have a responsibility to make these things known so that every one of us should understand the crisis, try their best to protect the water resources and save water at any time and in any place.

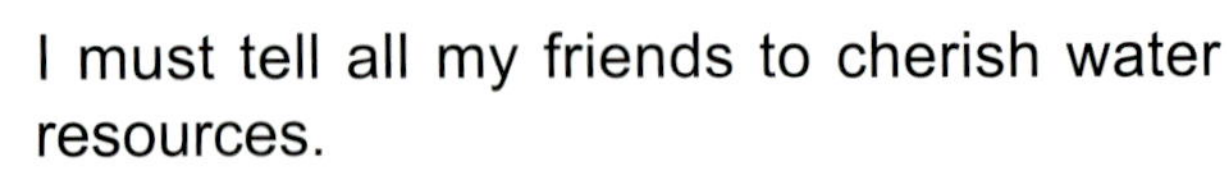

I must tell all my friends to cherish water resources.

Good. Xiaoming, it's getting dark. Let's go back.

Go back? Haven't we stayed here just for a while? Is there no other place worth seeing in the Tiansheng Bridge Subarea?

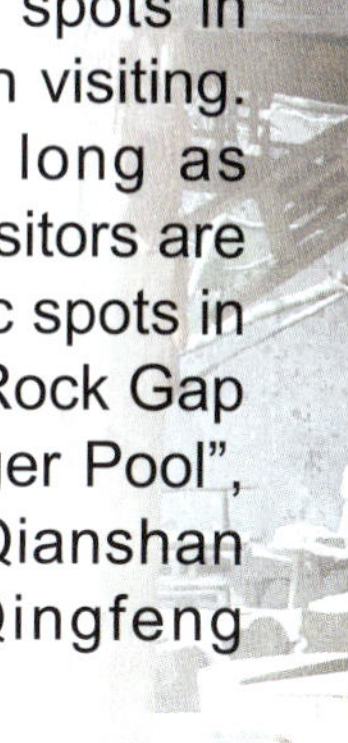

There are too many scenic spots in this subarea which are worth visiting. A circular plank path, as long as about 3 km, is built so that visitors are led to those wonderful scenic spots in it, such as “Yishui Bridge”, “Rock Gap Waterfall”, “Eagle Pool”, “Tiger Pool”, “Moss and Vine Creek”, “Qianshan Bridge”, “Laojun Inn”, “Qingfeng Bridge”, and “Mingyue Pool”.

Wow, so many scenic spots!

Walking along the plank path, tourists can also stop to see the remodeled “Ba tribe” of ancient human life, the local folk arts and crafts, and various workshops where the local daily food was produced in the past, such as “Sugar Workshop”, “Wine Workshop”, “Oil Extraction Workshop”, “Bean Curd Workshop”, “Noodle Workshop”. These workshops recreate the ancient folk methods of production, showing the industriousness and the wisdom of the people who lived here at that time.

Let’s go and have a look, grandpa!

There is not enough time. You see, it is getting dark. Or let’s have a quick view of everything around. Next time we come, we’ll take our time to see them carefully, OK?

All right... Hey, grandpa, look, what’s that round building?

Ah, that is a stage for the performance called “Tangxi”.

What is “Tangxi”?

"Tangxi" is the cultural recreational activities performed by the local people in the central room of their house. The Shennongjia people love to sing folk songs that are concerned about their work and life, including farming in the field, collecting Chinese medicinal herbs or hunting in the mountains, sacrificial ceremony and festival celebration, weddings and funerals, expression of love and even military campaigns.

I went to listen to the Shennongjia Bang drum performance. Very nice, very heroic!

A stage is deliberately set up in the subarea to bring out the ancient folk customs in mountainous areas.

Wow, terrific! Next time I come here, I'll surely watch "Tangxi" carefully and have a good tour of the place.

OK! Well, the sun has already set and it's time we went back to Wuhan.

OK, let's get going. Wangwang, we are going home now. Goodbye, "Tiansheng Bridge"! Goodbye, Shennongjia!

Questions:

1.Can you analyze how the Tiansheng Bridge, a penetrating karst cave is formed?

2.What are the important factors that control the formation of karst caves?

3.Please explain what confined water is, using the Tiansheng Bridge as an example.

4.What is your understanding of "underground water" after reading this book?

5.How should we cherish and protect our water resources? Support your argument with examples from our daily life.

Epilogue

As the carpet flew over Longtoushan, Xiaoming took a last wistful look at the forested mountains. In the twilight, he bade farewell to Shennongjia. “Goodbye, Shennongjia! I'm coming to see you again with my friends.” Like a big stone thrown into a calm pond, the overlapping echoes were surging throughout the mountain fields. In the red sunset, the carpet was flying farther and farther like a boat travelling in green waves until it became a small black spot on the horizon. On his way back from Shennongjia, Xiaoming was planning his next visit to this worldwide famous geopark.